AF901430

White, Kim
Twin Screw Extrusion

James L. White
Eung Kyu Kim

Twin Screw Extrusion

Technology and Principles

2nd Edition

HANSER

Hanser Publishers, Munich Hanser Publications, Cincinnati

The Authors:

Prof. Dr. James L. White (†), Institute of Polymer Engineering, University of Akron, Ohio, USA
Ph.D Eung Kyu Kim, The Dow Chemical Company, Midland, MI 48667, USA

Distributed in the USA and in Canada by
Hanser Publications
6915 Valley Avenue, Cincinnati, Ohio 45244-3029, USA
Fax: (513) 527-8801
Phone: (513) 527-8896 or 1-800-950-8977
www.hanserpublications.com

Distributed in all other countries by
Carl Hanser Verlag
Postfach 86 04 20, 81631 München, Germany
Fax: +49 (89) 98 48 09
www.hanser.de

The use of general descriptive names, trademarks, etc., in this publication, even if the former are not especially identified, is not to be taken as a sign that such names, as understood by the Trade Marks and Merchandise Marks Act, may accordingly be used freely by anyone.
While the advice and information in this book are believed to be true and accurate at the date of going to press, neither the authors nor the editors nor the publisher can accept any legal responsibility for any errors or omissions that may be made. The publisher makes no warranty, express or implied, with respect to the material contained herein.

Library of Congress Cataloging-in-Publication Data

White, James Lindsay, 1938-
Extrusion : technology and principles. -- 2nd ed. / James L. White, Eung Kyu Kim.
p. cm.
ISBN-13: 978-1-56990-471-8 (hardcover)
ISBN-10: 1-56990-471-5 (hardcover)
1. Plastics--Extrusion. I. Kim, Eung Kyu. II. Title.
TP1175.E9W48 2010
668.4'13--dc22
2010015244

Bibliografische Information Der Deutschen Bibliothek
Die Deutsche Bibliothek verzeichnet diese Publikation in der Deutschen Nationalbibliografie; detaillierte bibliografische Daten sind im Internet über <http://dnb.d-nb.de> abrufbar.

ISBN 978-3-446-42272-8

All rights reserved. No part of this book may be reproduced or transmitted in any form or by any means, electronic or mechanical, including photocopying or by any information storage and retrieval system, without permission in writing from the publisher.

© Carl Hanser Verlag, Munich 2010
Production Management: Steffen Jörg
Coverconcept: Marc Müller-Bremer, www.rebranding.de, München
Coverdesign: Stephan Rönigk
Typeset: Manuela Treindl, Laaber
Printed and bound by Druckhaus "Thomas Müntzer", Bad Langensalza, Germany
Printed in Germany

Professor James Lindsay White 1938–2009

James Lindsay White was born to Robert Lindsay and Margaret (Young) White in Brooklyn on January 3, 1938. His father's family originated from Scotland in the 1880's from a long line of coal miners and later machine shop owners. These roots seeded his lifelong interest in the history of mining and industrial technology. He attended Brooklyn Technical High School and later studied Chemical Engineering at the Brooklyn Polytechnic Institute. He then joined the Department of Chemical Engineering, at the University of Delaware, in the research group of Prof.

A. B. Metzner and finished his MS degree in 1962 and PhD degree in 1965. They jointly developed the now famous White-Metzner rheological model which is still widely used for polymer processing simulation.

Prof. White's early career started in industry working for the Uniroyal Company as a research engineer and then a group leader from 1963–1967 where he acquired his lifelong interest in rubber science and engineering to which he has made numerous contributions. He was also introduced to his first wife Yoko by his supervisor Dr. Noboru Tokita at Uniroyal.

In 1967 Prof. White joined the University of Tennessee as an associate professor and quickly rose through the academic ranks to become a professor, ultimately becoming the originator and professor in charge of the Polymer Science and Engineering MS and PhD degree programs. His interest took a turn into experimental and theoretical rheology of filled polymers and process induced polymer structuring and he rapidly dove into the new areas of polymer melt/solution, fiber spinning, film blowing and biaxial stretching processes, as well as injection molding. During this time, his contributions expanded into liquid crystalline polymers where he pioneered unique processes to manufacture biaxially oriented lyotropic liquid crystalline films and developed fundamental understanding of solidification from the solution processing. While at the University of Tennessee, he founded the *Journal of Polymer Engineering* of which he was Editor-in-Chief until 1984.

In 1983, Prof. White moved to The University of Akron where he founded the Institute and Department of Polymer Engineering, serving as both Center Director and Department Chair. At Akron, he turned his attention to rubber processing and compounding, experimentally studying and simulating flow in internal mixers and pin barrel extruders as well as twin screw extruders without and with chemical reactions taking place in the latter process. These activities culminated with the first commercially successful Akro-Twin software to simulate flow in twin screw extrusion.

It was during this time that Prof. White originated a new professional society now known as the Polymer Processing Society. From its beginning in 1985, this society was organized to be international in character. A number of key individuals worked with Dr. White to found the

PPS, including: Lloyd Goettler, Musa Kamal, Tom Lindt, Jim Stevenson, Leszek Utracki, and Jim Throne. With this core group and under the leadership of Professor White, many others joined together and the inaugural meeting of the PPS was held in 1985 in Akron. Shortly after formation of the society, the journal *International Polymer Processing* was initiated and Prof. White successfully served as its Editor-in-Chief from 1986 until 2004.

Prof. White published over 500 papers in international journals. He has also published eight books on the subjects of rheology, twin screw extrusion, rubber processing, polyolefins, polymer compounds and thermoplastic elastomers.

Prof. White was a pioneer in establishing Polymer Engineering as an academic discipline. Various polymer societies, institutions and industries around the globe have recognized his many research contributions in both rubber and plastics. Prof. White was a man of encyclopedic knowledge and erudition and is the most cited researcher at the University of Akron.

Prof. White received numerous awards over his career for his accomplishments, including: Bingham Medal of the Society of Rheology; Yuko-Sho Award, Society of Rheology, Japan 1984; Society of Plastics Engineers, Education Award 1987; Research Award, Society of Plastics Engineers, 1992; Heinz Herrmann Twin Screw Extrusion award of the Extrusion Division of Society of Plastics Engineers; and in 2009, the Charles Goodyear Medal from the Rubber Division of the American Chemical Society. Most recently, Prof. White served The University of Akron as the Harold A. Morton Professor of Polymer Engineering.

Prof. White had broad range of interests and encyclopedic knowledge of history, from the development of machinery in various industries through to social issues in many parts of the world. He leaves his wife Alganesh and a world-wide community of friends and colleagues in the polymer engineering community.

On behalf of the Editors, Editorial Board
and Hanser Publishers

Andrew Hrymak,
Editor-in-Chief, International Polymer Processing

Wolfgang Beisler,
Chief Managing Director, Hanser

Preface to Second Edition

This volume is the second edition of the book "Twin Screw Extrusion: Technology and Principles" published in 1990 by the senior author. The earlier book was the first effort to write a comprehensive volume on twin screw extrusion technology and scientific understanding. Since that time there have been many changes and advances in twin screw extrusion technology, the industries associated with it, and our understanding and ability to model the process and predict the performance of these machines. Most striking has been the continuous rise and domination of the industry by intermeshing co-rotating machines.

In this edition, we bring the various chapters of the first edition up to date. We expand the treatment of co-rotating machines and move its placement from Chapters 10 to 12 to Chapters 4 to 6. We recognize other developments by separating the original Chapters 4 to 6 in the first edition on tangential counter-rotating twin screw machines into the new Chapters 10 to 12 on Welding Engineers machines and new Chapters 13 to 15 on continuous mixers. Separate chapters on Reactive Extrusion (new Chapter 17) and on Devolatilization and Dewatering (new Chapter 18) have been added. In all, the book has grown from the 13 chapters of the first edition to 19 chapters in the second edition.

All of the chapters have been updated including discussions of computer aided design in co-rotating twin screw extruders.

There are other books on twin screw extrusion. The early volumes by Janssen ("Twin Screw Extrusion") and Martelli ("Twin Screw Extrusion: A Basic Understanding") are really not comprehensive in their scope and are outdated. The more recent Manas-Zloczower and Tadmor's 1994 and 2009 "Mixing and Compounding of Polymers: Theory and Practice" is a very broad treatment by many authors with differing viewpoints and makes very interesting reading. Todd's 1998 "Plastic Compounding: Equipment and Processing" is a good technological treatment from machinery manufacturers but lacks consideration of flow fundamentals. Kohlgrueber's 2007 "Co-rotating Twin Screw Extruders" considers only co-rotating machines and is light in its treatment of flow fundamentals.

There are many general books on extrusion which have individual chapters on twin screw extrusion.

June 2010

J. L. White
E. K. Kim

Contents

1 Overview of Multi-Screw Extruders

1.1 Introduction

Screw extruders represent the major processing machine of the thermoplastics industry, and play an important role in many related industries. Certainly, the major continuous processing machine in the polymer industry has always been the single screw extruder. However, we have observed through the years a growing industrial role for multiple screw machines, especially twin screw extruders. These machines have generally been treated together as a single category and not carefully distinguished.

Twin screw extruders have various practical advantages. Intermeshing counter-rotating twin screw extruders operate as close as to a positive displacement pump allowing compared to single screw extruders high output rates, short residence times and outputs independent of die pressure. Twin screw extruders, especially non-intermeshing machines have high interfacial areas for more effective devolatilization. Twin screw extruders with modular design are very versatile for use in compounding, reactive extrusion, and other applications.

It is our purpose in this book to carefully (i) describe these twin screw extruders and distinguish between them, (ii) indicate how they were developed and their applications, (iii) describe the flow mechanisms which operate in them and present models of their behavior, (iv) present experimental studies of their performance, and (v) describe competitive machines such as the Buss Kokneter and continuous mixers.

Our concern in the present chapter is to give an overview and classification of the different types of multiple screw extruders. First, in Section 1.2 we will discuss screw geometry. We then turn to classification of multi-screw machines. In Section 1.3, we consider the relative direction of screw rotation and number of screws in the machine. The manner of contact of the screws is considered in Section 1.4. The question of segmentation and modularity of construction is discussed in Section 1.5.

1.2 Screw Geometry

Before we can discuss twin and multi-screw extruders, we must consider the geometry of individual screws. The geometry of an extruder screw is summarized in Figure 1.1a. The screw fits inside of a cylindrical barrel with internal diameter D_B. The distance between the root of the screw and the internal surface of the barrel is H. It should be noted that the value of H varies with position between the flights. The screw root may be curved and the flight need not be perpendicular to the root, though it is to the screw axis. The radial clearance between the crest of the screw flight, and the inner barrel surface is δ_f.

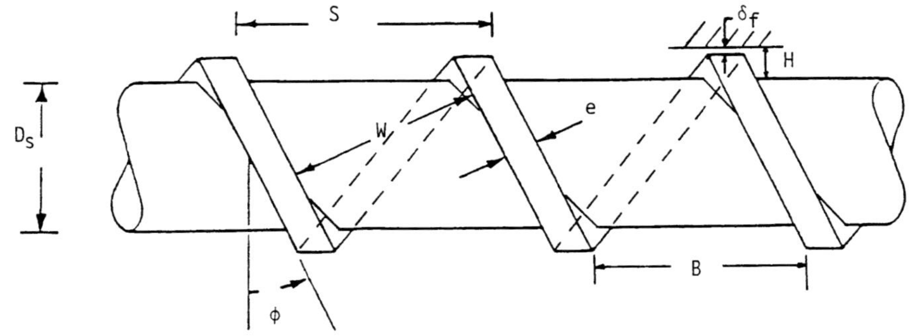

Figure 1.1 a: Single-flighted extruder screw

The axial distance of one full turn of the screw (screw lead or pitch) is S. The axial distance between flights is B, and e is the perpendicular width of the flights. The perpendicular distance between the flights along the helical path of the screw is W. The angle of the helix is ϕ. The distance W may vary (increase) with increasing altitude of the flight. This is associated with a decreasing flight thickness.

The helix angle ϕ will vary with radius because the pitch S is constant and ϕ is defined so as to relate S to the circumference through:

$$S = 2\,\pi\,r\,\tan\phi(r) \tag{1.1}$$

Thus, at the screw root, we have:

$$S = \pi\,D_s\,\tan\phi_s \tag{1.2a}$$

and at the barrel:

$$S = \pi\,D_b\,\tan\phi_b \tag{1.2b}$$

Angle ϕ decreases as one proceeds from the screw root to the barrel.

The quantities defined above are not necessarily constant along a screw. Quantities such as W, e, and ϕ vary not only then with the screw radius from the screw root to the barrel but along the length of the screw. The pitch can vary along the length of the screw.

Many geometric relationships exist between the quantities defined in the above paragraphs. The pitch S is related to B and e through:

$$S = B + \frac{e}{\cos\phi} \tag{1.3}$$

The channel width W is related to B through:

$$W = B\cos\phi \tag{1.4}$$

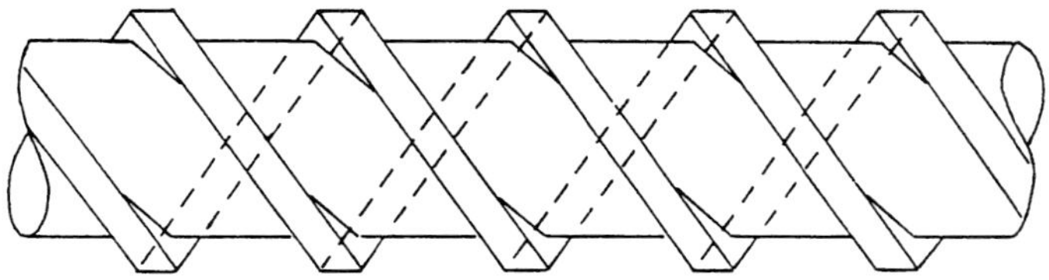

Figure 1.1 b: Double-flighted extruder screw

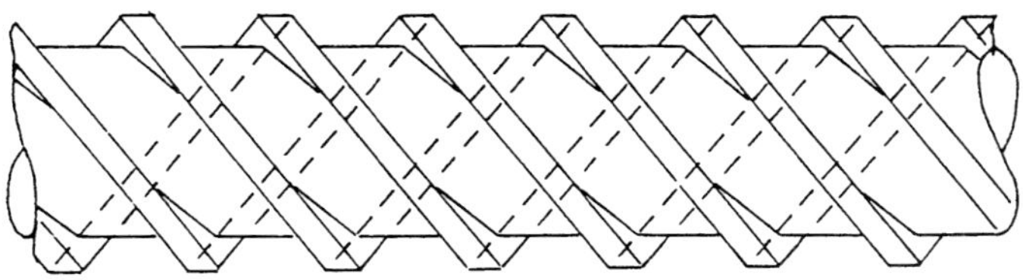

Figure 1.1 c: Triple-flighted extruder screws

and to pitch S through:

$$W = \left(S - \frac{e}{\cos\phi} \right) \cos\phi = S\cos\phi - e \tag{1.5}$$

The relationships given in the above paragraph are for a single-flighted screw as shown in Figure 1.1a. More generally, extruder screws are multi-flighted, i.e., fluid added to the inlet will simultaneously travel along two or more parallel flights. A double-flighted screw is shown in Figure 1.1b and a triple-flighted screw in Figure 1.1c. If a screw has p parallel flights, we have:

$$|S| = p\left(B + \frac{e}{\cos\phi} \right) \tag{1.6}$$

$$\text{and } W = \frac{|S|}{p}\cos\phi - e \tag{1.7}$$

Screw elements may be forward or backward pumping (often described as right-handed or left-handed when screw rotation is in a clockwise direction). This is shown in Figure 1.2. Backward pumping screws have negative helix angles ϕ and pitch S. It is for this reason that absolute values are used in Eqs. (1.6) and (1.7).

Single screw extruders and twin screw extruders also contain elements that are not true screws. In twin screw extruders, these are often assemblies of incomplete discs, gear, or rotor designs intended for kneading and mixing. However, the incomplete discs may be arranged in such a way as to approximate forward or backward pumping screws. They then act as a leaky screw in practice.

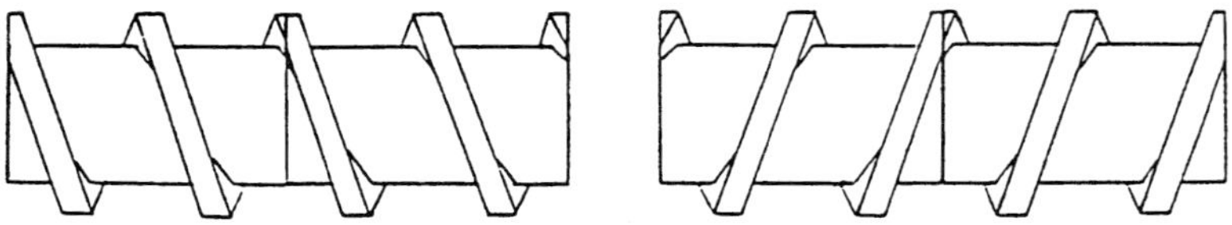

Figure 1.2: Forward pumping (right-handed) and backward (left-handed) screws

1.3 Classification According to Direction of Rotation and Number of Screws

We now turn to twin screw extruders. The most obvious classification in such a machine is whether the two screws rotate in the same or in opposite directions. This is shown in Figure 1.3. Both co-rotating (in German: *gleichläufig*) and counter-rotating (in German: *gegenläufig*) are of great industrial importance and are manufactured throughout the world.

Let us consider multiple screw extruders. Descriptions of extrusion machines with not only 2, but 3, 4, 5, 6, 7, 8, or more screws are contained in the patent literature. The screws may be arranged in various geometric manners. Figure 1.4 presents drawings from patents by Roberto Colombo [1–3], one of the pioneers of this technology (see Chapter 4), which contains designs for multiple screw extruders. The arrangements of the screws may vary. The 3- and 5-screw machines are set up in a linear manner. The 6- and 8-screw machines are set up in a circumferential manner. The 4-screw machine has a control screw and 3 spurs. These screw arrangements are in no sense unique. Other arrangements are discussed in the patent literature. V-shaped 3-screw arrangements are described for example by various authors [4–5]. Figure 1.5 shows such an arrangement given by Meskat and Erdmenger [5].

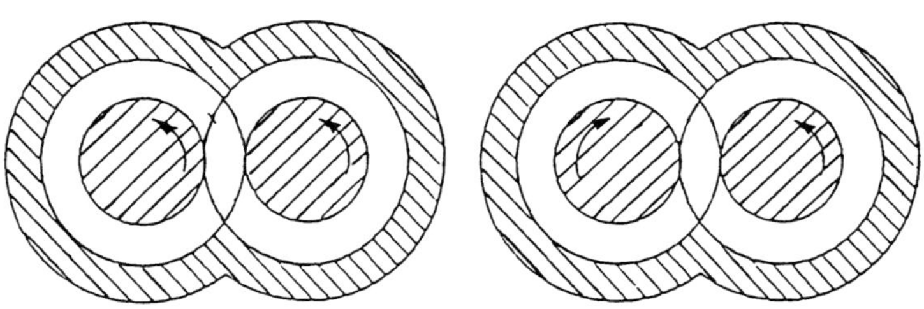

Figure 1.3: Co-rotating and counter-rotating twin screw extruders

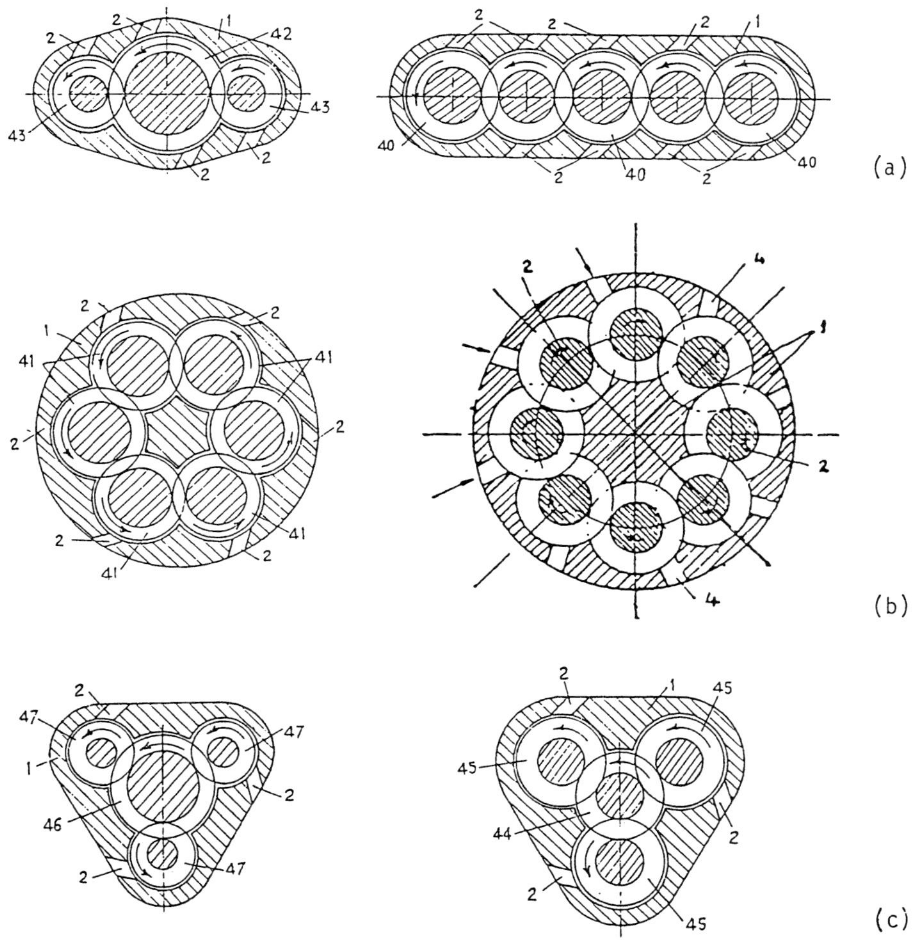

Figure 1.4: Multiple co-rotating screw extruders after Colombo [1–3] indicating
(a) 3- and 5-screw linear arrangements,
(b) 6- and 8-screw circumferential arrangements, and
(c) 4-screw arrangement with contact spurs

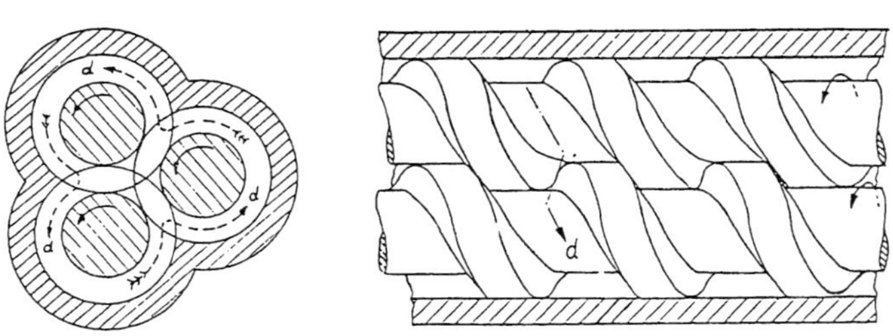

Figure 1.5: V-shaped co-rotating 3-screw arrangement after Meskat and Erdmenger [5]

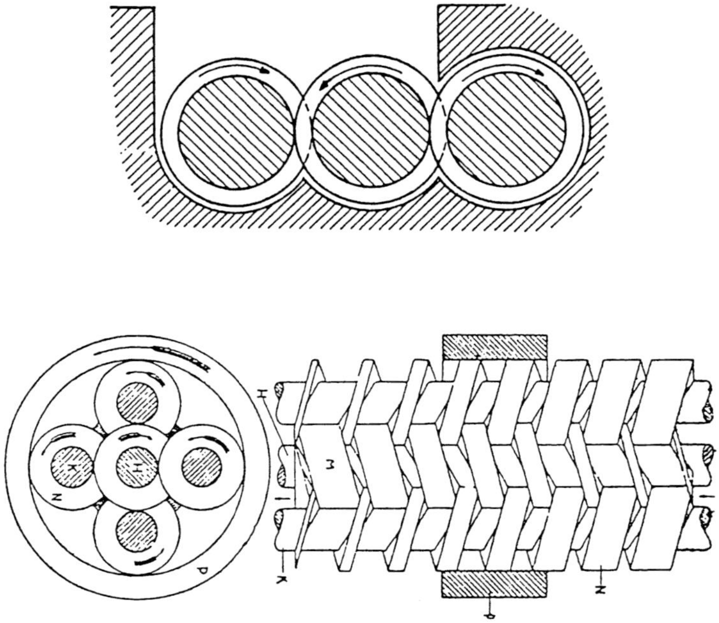

(a)

(b)

Figure 1.6: (a) Counter-rotating 3 screw extruder after Burghauser and Erb [6],
(b) Counter-rotating 5 screw extruder after Anonymous [7]

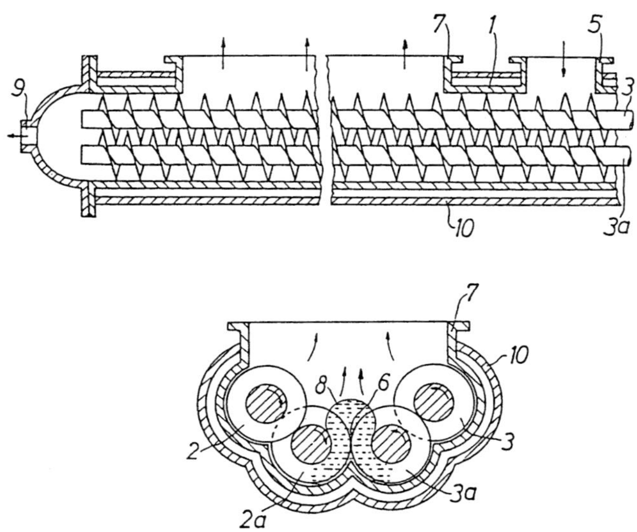

Figure 1.7: 4-screw extruder with 2 co-rotating pairs and 1 counter-rotating pair after Erdmenger and Oetke [9]

With multiple screw extruders, one may have co-rotation, counter-rotation, or mixed situations. The screws of Colombo's [1–3] 3-, 4-, 5-, 6-, and 8-screw extruders all rotate in the same direction. Meskat and Erdmenger's [5] design is also co-rotating. However, multiple screw extruders may be counter-rotating as well. The first 3-screw extruder is described in Pfleiderer's 1881 patent [4]. It was counter-rotating. Figure 1.6 shows 3-, 4-, and 5-screw counter-rotating machines taken from the patent literature [4, 6–8]. It is also possible to have both co- and counter-rotating screws in the same multiple screw machine. This is seen in Erdmenger and Oetke's [9] 4-screw machine shown in Figure 1.7.

1.4 Classification According to Contact of Screws

Another classification of multiple screw extruders is whether the screws are separated, tangential, or intermeshing. The intermeshing designs can further be divided into partially intermeshing and fully intermeshing. This is indicated in Figure 1.8 for twin screw extruders. This leads to eight different categories when one considers co- and counter-rotating screws. Generally, in this book we will divide our considerations between non-intermeshing (separated or tangential) and intermeshing twin screw machines.

Intermeshing co-rotating [10] and counter-rotating [11] twin screw machines are not new. They are discussed by various inventors of the 19th century.

A fully intermeshing screw, whether co-rotating or counter-rotating, can have a self-wiping quality. This involves special root profiles, i.e., the variation of H along the screw axis. This effect has been realized since the earliest patents for such machines.

Self-wiping co-rotating twin screw extruders were first described in detail by Wunsche [12] in 1901. They were subsequently discussed by Easton [13, 14]. Easton [14] and later Meskat and Erdmenger [5] argued that their closely matching co-rotating twin screw profiles have a special full self-cleaning character.

With intermeshing screws, one must always be wary as to whether they may actually be intermeshing the way you wish. The use of models, a good geometric imagination, and experience is needed. D. B. Todd [15] has pointed out to the author that the design of Figure 1.5 will work with single and triple but not with double-flighted screws.

It is possible in multiple screw machines to have some of the screw flights intermeshing and some non-intermeshing. An example of this is seen in the four screw machine of Erdmenger and Oetke [9] in Figure 1.7. This involves two pairs of intermeshing co-rotating screws which interact in a tangential counter-rotating manner.

1.5 Classification According to Modularity of Design

Multiple screw extruders may contain uniformly machined screws or segmented/modular screws. The latter machines contain not only forward pumping right-handed screw elements, but also special mixing elements which can exert pumping action. Left-handed screw elements which pump backwards are also found. Today, manufacturers of twin screw extruders intended for applications other than pumping usually produce segmented or modular designs.

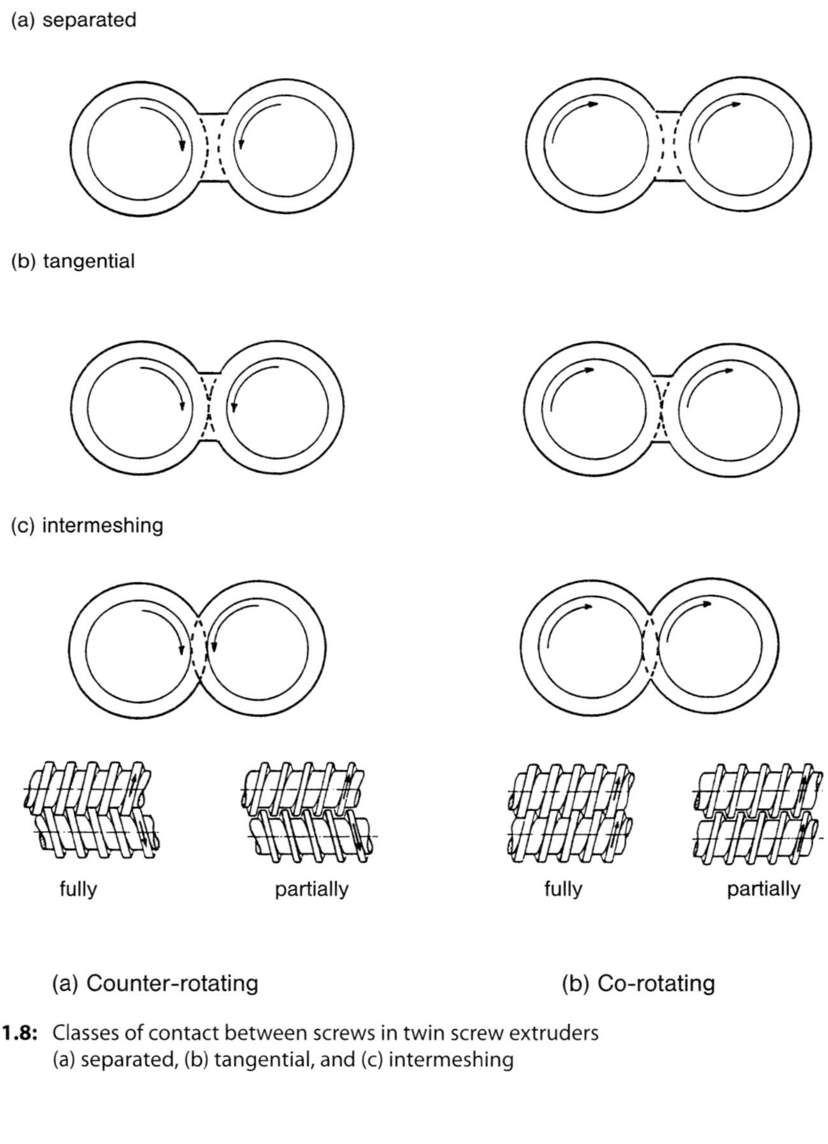

Figure 1.8: Classes of contact between screws in twin screw extruders
(a) separated, (b) tangential, and (c) intermeshing

The existence of modular elements along the length of the screw shafts dates to the earliest patent technology in this area. An 1881 patent of Pfleiderer [4] has both screw pumping and mixing sections. Pfleiderer's machine and many of the machines to the present consist of machined shafts. An alternate technology is to fabricate separate elements and key or otherwise fit the screw elements together on a shaft. These modules may at any time be separated and re-arranged along the length of the shaft. This principle is first seen in patents in the 1950s by Fuller [16] and by Meskat and Pawlowski [17], who combine forward and backward pumping screw elements for tangential counter-rotating and intermeshing co-rotating machines, respectively.

References

1. R. Colombo, Italian Patent 370,578 (1939).
2. R. Colombo, U.S. Patent 2,563,396 (1951).
3. R. Colombo, Canadian Patent 517,911 (1955).
4. P. Pfleiderer, German Patent 18,797 (1881).
5. W. Meskat and R. Erdmenger, German Patent 862,668 (1953).
6. F. Burghauser and K. Erb, German Patent 690,990 (1940).
7. Anonymous (IG Farbenindustrie-Hoechst), Italian Patent 373,183 (1939).
8. M. B. Sennet, U.S. Patent 2,581,451 (1952).
9. R. Erdmenger and W. Oetke, German Auslegungschrift 1,111,154 (1961).
10. F. Coignet, U.S. Patent 98,035 (1869).
11. S. L. Wiegand, U.S. Patent 155,602 (1874).
12. A. Wunsche, German Patent 131,392 (1901).
13. R. W. Easton, British Patent 109,663 (1912).
14. R. W. Easton, U.S. Patent 1,468,379 (1923).
15. D. B. Todd, *Personal Communication* (1989).
16. L. J. Fuller, U.S. Patent 2,615,199 (1952).
17. W. Meskat and J. Pawlowski, German Patent 949,162 (1956).

2 Background: Polymer Properties and Flow Principles

2.1 Introduction

In this book we are concerned with the processing of materials in twin screw extrusion machines. In order to understand the operation of the machines, it is not only useful but necessary to have some background regarding the properties of the substances to be processed, the characteristics of flow, and the occurring phenomena. Screw machines generally show better pumping characteristics with viscous liquids. They have become most important in their applications to polymer melts and their compounds. It is the purpose of this chapter to provide this background.

We begin with a section on the types of polymers and their transition behavior (Section 2.2). The polymerization process and its kinetics are described in Section 2.3. We then describe the basic principles of mass conservation and mechanics (Section 2.4). The rheological behavior of polymer melts and solutions is summarized together with 3-dimensional constitutive representations (Sections 2.5 and 2.6).

We will then move on to other considerations of flow of viscous materials of importance to extrusion machines. This includes discussions of energy balances and viscous dissipation heating (Section 2.7). We subsequently present a critical discussion of hydrodynamic lubrication theory which is the primary method used to analyze flow of viscous liquids in machinery with moving parts including screw extruders (Section 2.8).

2.2 Nature of Polymers

Polymers are compounds, usually but not necessarily organic, of very high molecular weight, generally 20,000 to 200,000 or more. They are usually linear chains with one or more repeating units. The most common polymers (homopolymers) possess only a single unit which repeats itself. In this category are materials, such as polyethylene, polypropylene, polyvinyl chloride, and polystyrene which have the repeating unit:

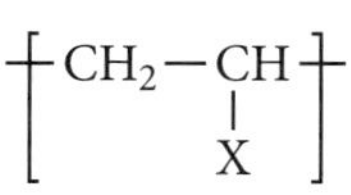

(Structure 2.1)

where X is H, CH_3, Cl, and a phenyl ring, respectively. Another important group of homopolymers are polydiene elastomers, which have the repeat unit:

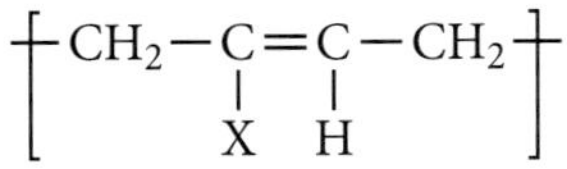

(Structure 2.2)

For polybutadienes, X is H, for polyisoprene, it is CH_3 and for polychloroprene it is Cl.

Homopolymers, produced by condensation polymerization, should also be considered among the important families. These include polyamides, polyesters, polycarbonates, and polyetherketones. The most important of the polyamides have the structural repeat unit:

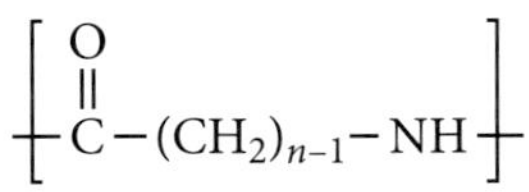

(Structure 2.3a)

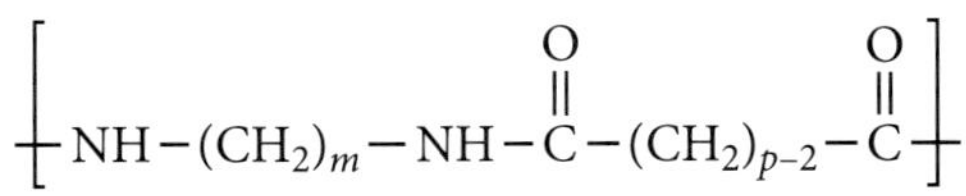

(Structure 2.3b)

These represent polyamide-n and polyamide-mp. These polymers are frequently referred to as nylons, the original trademark of DuPont. Polyamide-6, polyamide-11, and polyamide-12 are commercially available versions of polyamide-n. The most important polyamide-mp is polyamide-66, but polyamide-46 and polyamide-1212 are also available.

The most important polyesters are of the form:

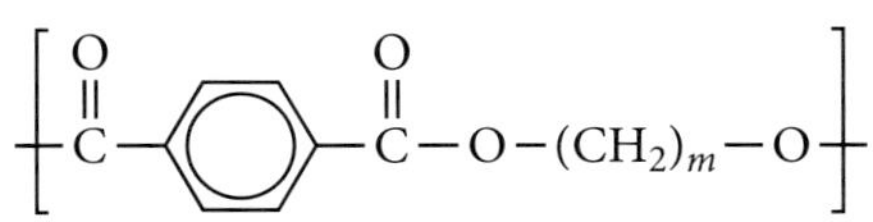

(Structure 2.4)

In polyethylene terephthalate (PET), *m* is 2, while in polybutylene terephthalate, *m* is 4. Other polyesters include polycaprolactone, which has the structural unit:

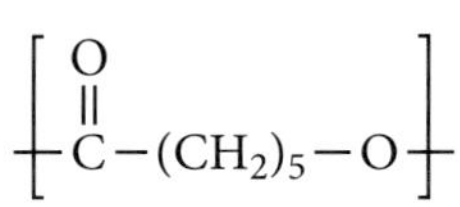

(Structure 2.5)

Commercial polycarbonate has the structural unit:

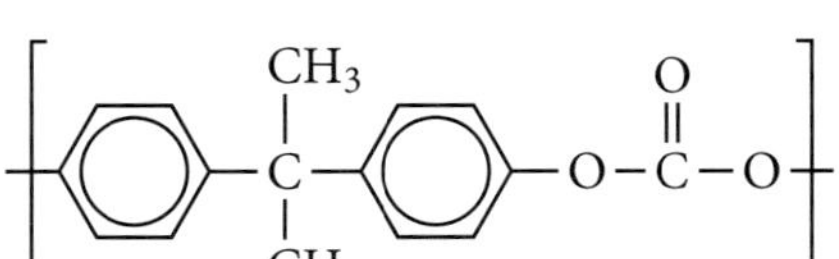

(Structure 2.6)

Polyetherketones are the newest family of engineering thermoplastics. The most important of these materials, polyetheretherketone (PEEK), has the structural unit:

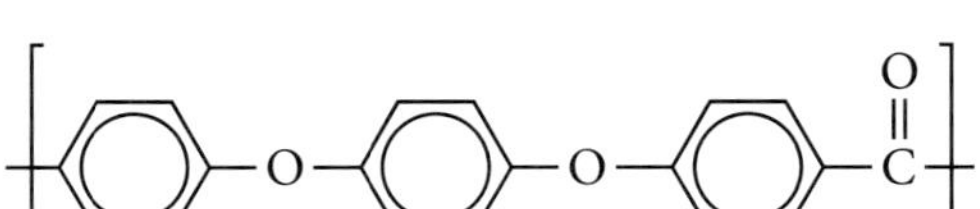

(Structure 2.7)

In addition to homopolymers, we must consider copolymers where there are two different structural units in the backbone. These may be arranged variously in random, block, or alternating manners as shown below:

Random

A B A A B A B B A B A A A B A B

Simple A B Block

A A A A A A A A A B B B B B B B B B B B B

A B A Block

A A A A A A B B B B B B B B B A A A A A A

Alternating

A B A B A B A B A B A B A B A B A B A B A B

Other varieties are possible. One may also codify structures in terms of contents of AA, AB, BB diads or AAA, ABA, BAB, BBB triads, etc. There are many important commercial copolymer systems, including butadiene-styrene, styrene-acrylonitrile, ethylene-propylene, butadiene-acrylonitrile, and vinyl chloride-vinylidene chloride copolymers.

Twin screw extruders are also widely used in the food industry where the materials processed are also polymeric and include carbohydrates and proteins. These are condensation polymers, protein being polyamide (or better, polyamino acid) copolymers with many different structural units.

The most important characteristic of polymers we need to specify is their molecular weight. Generally, molecular weight is determined in terms of different averages, specifically:

$$M_\mathrm{n} = \frac{\sum_i N_i\, M_i}{\sum_i N_i} \tag{2.1a}$$

$$M_\mathrm{w} = \frac{\sum_i W_i\, M_i}{\sum_i W_i} = \frac{\sum_i N_i\, M_i^2}{\sum_i N_i\, M_i} \tag{2.1b}$$

$$M_\mathrm{z} = \frac{\sum_i N_i\, M_i^3}{\sum_i N_i\, M_i^2} \tag{2.1c}$$

$$M_{\mathrm{z}+1} = \frac{\sum_i N_i\, M_i^4}{\sum_i N_i\, M_i^3} \tag{2.1d}$$

where N_i represents the number of molecules of molecular weight M_i. The above molecular weight average are known, respectively, as the number average molecular weight, M_n, the weight average molecular weight, M_w, the z-average molecular weight, M_z, and the z+1-average molecular weight, M_{z+1}. The number average molecular weight is based on averaging equally among all different polymer molecules. The weight average molecular weight averages according to the masses W_i of individual molecular weights present.

If only one species of a particular molecular weight is present, all of its molecular weight averages M_n, M_w, M_z, and M_{z+1} are equal. If there is a distribution of molecular weights we then have:

$$M_{z+1} > M_z > M_w > M_n \tag{2.2}$$

Typically molecular weights of commercial polymers are in the range from 10,000 to 200,000, depending on the polymer. Breadths of molecular weight distribution vary considerably, generally from 1.5 to 10. This depends on the method of polymerization. However, molecular weight distributions are often engineered for the application for which the polymer is designed. As an example, narrow molecular weight distribution linear polyolefins process in a more stable manner during melt spinning than broader distribution polyolefins [5–9]. Technologies involving reactive extrusion have been developed by Kowalski and his coworkers [10, 11] of Exxon and others which degraded broad molecular weight polyolefins to narrower molecular weight distributions to achieve improved processability. Similarly, broadening molecular weight distribution seems to improve processability in tubular film extrusion [8, 9, 12]. The presence of high molecular weight fractions certainly upgrades mechanical properties.

We now turn to the thermodynamic characteristics of polymers. Generally, pure low molecular weight compounds exist in three different states, a crystalline state possessing a high level of 3-dimensional structural order at low temperatures, a molten state at intermediate temperatures, and a gaseous state at elevated temperatures. One distinguishes a melting temperature, T_m, which represents the transition from a crystalline state to a molten state, and a vaporization temperature, T_v, which represents the transition of the molten state to the gaseous state. Water exists in these three states which we know as ice (crystal), water (melt), and steam (gas), which have transition temperatures, T_m of 0 °C (273.2 K) and T_v of 100 °C (373.2 K).

The situation in polymers is much more complex. First, unless polymeric materials have uniform structures, they will not crystallize at low temperatures. Rather, the molten or liquid state supercools or vitrifies to form a glass, a relatively rigid substance which does not have 3-dimensional order of a crystal. It may often be considered a highly viscous liquid. This character is, of course, also commonly found in inorganic substances such as silicates. A transition temperature, T_g, also called the glass transition temperature, is associated with this vitrification process.

As stated above, polymers with regular structures crystallize while those with irregular structures vitrify. Thus, in general random copolymers form glasses rather than crystallizing. Homopolymers may or may not crystallize. Consider the vinyl polymers of Structure (2.1), which we repeat here:

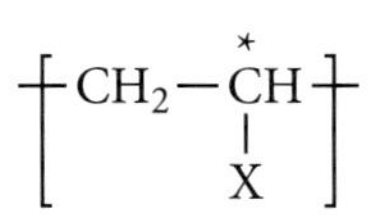

(Structure 2.8)

The asterisked carbon atom is asymmetric and has two possible mirror image configurations. These polymers only have regular structures when the asterisked asymmetric carbon atom of Structure (2.8) has the same configuration along the chain. Vinyl polymers having the same configuration of their asymmetric carbon atoms along their backbone are said to be *isotactic*. Polymers, in which the configuration of the asymmetric carbon atom alternates, are called *syndiotactic*. If there is no order to the asymmetric configurations, the polymer is said to be *atactic*. The isotactic and syndiotactic forms of vinyl polymers are generally crystalline. The atactic forms are amorphous.

Among important commercial vinyl homopolymers, polyethylene, polypropylene (which is isotactic) and polybutene-1 (which is isotactic) are crystalline. Polystyrene (which is atactic) is amorphous. 1,2-polybutadiene has been available in a syndiotactic form, which is crystalline, and an atactic form, which is amorphous. Polyvinyl chloride has enough syndiotactic content to produce a small level of crystallinity. However, it is frequently considered a glassy polymer.

Consider now other classes of polymers. Polyamides are generally crystalline. Polymers with phenyl rings in their backbones crystallize very slowly compared to aliphatic vinyl polymers and polyamides. Polybutylene terephthalate crystallizes more rapidly than polyethylene terephthalate, both of which are considered crystalline. Similarly polyetheretherketone, Structure (2.7), also crystallizes very slowly. Polycarbonate, Structure (2.6), crystallizes with much more difficulty and is generally considered amorphous.

Transition temperatures for important polymers are summarized in Table 2.1.

Table 2.1: Major Transitions for Important Polymers

Polymers	T_m (°C)	T_g (°C)
Polyethylene (linear high density)	135	~ −100
Polyethylene (low density)	110	~ −100
Polypropylene (isotactic)	160	~ −15
Polystyrene	–	95
Polyvinylchloride	240	70
Polyamide 6	215	40
Polyamide 66	260	40
Polyethylene terephthalate	260	70
Polycarbonate	–	140

2.3 Polymerization

Twin screw extruders have been used as polymerization reactors. Some background on polymerization should be provided, so we can usefully discuss this area. In this section, we will discuss the mechanisms of polymerization [1–4]. Polymers may be formed by addition or by condensation polymerization mechanism. We will describe each of these in turn.

The vinyl polymers of Structure (2.1) are polymerized by addition polymerization through an initiator R according to:

$$R^{*} + CH_2{=}\underset{\displaystyle X}{\underset{|}{CH}} \xrightarrow{k_p} RCH_2{-}\underset{\displaystyle X}{\underset{|}{CH^{*}}} \qquad \text{(Structure 2.9a)}$$

$$RCH_2{-}\underset{\displaystyle X}{\underset{|}{CH^{*}}} + CH_2{=}\underset{\displaystyle X}{\underset{|}{CH}} \xrightarrow{k_p} RCH_2{-}\underset{\displaystyle X}{\underset{|}{CH}}{-}CH_2{-}\underset{\displaystyle X}{\underset{|}{CH^{*}}} \qquad \text{(Structure 2.9b)}$$

$$R\left[CH_2{-}\underset{\displaystyle X}{\underset{|}{CH}}\right]_n\left[CH_2{-}\underset{\displaystyle X}{\underset{|}{CH}}\right]^{*} + CH_2{=}\underset{\displaystyle X}{\underset{|}{CH}} \xrightarrow{k_p} R\left[CH_2{-}\underset{\displaystyle X}{\underset{|}{CH}}\right]_{n+1}\left[CH_2{-}\underset{\displaystyle X}{\underset{|}{CH}}\right]^{*}$$

(Structure 2.9c)

Diene monomers such as butadiene may be similarly polymerized. The rate of propagation r_P and monomer depletion r_M are:

$$r_P = r_M = k_P\, C_M \sum_n C_{R\,M_n^{*}} \qquad (2.3)$$

where C_M is the monomer concentration and $C_{R\,M_n^{*}}$ is the concentration of active species of degree of polymerization 'n'.

There are various addition polymerization mechanisms, which have been characterized as free radical, anionic, and cationic. In free radical polymerization, the free radical initiator, e.g., a peroxide, degrades as:

$$I \xrightarrow{k_i} R'\cdot + R''\cdot \qquad \text{(Structure 2.10)}$$

$$r_I = 2\, r_R = 2\, k_i\, C_I \qquad (2.4)$$

Here C_I is the concentration of initiator.

Radicals are destroyed according to the mechanism:

$$R'\, M_n \cdot + R''\, M_m \cdot \xrightarrow{k_t} R'\, M_{n+m}\, R'' \qquad \text{(Structure 2.11)}$$

or

$$R'\, M_n + R''\, M_m$$

The propagation reaction is very rapid compared to the initiation. This leads to the development of a steady state between initiation and termination. We then have the kinetic expression:

$$2\, k_i\, C_I = k_t \left(\sum C_{R\,M_r^{\bullet}}\right)^2 \qquad (2.5)$$

Here $C_{R\,M_r^{\bullet}}$ is the concentration of the $R\, M_r^{\bullet}$ species.

The rate of polymerization or monomer depletion is from Eq. (2.3) equal to:

$$r_P = r_M = k_P \sqrt{\frac{2\,k_i}{k_t}}\; C_I^{1/2}\; C_M \tag{2.6}$$

In anionic polymerization, we have a compound, such as an alkyl lithium (RLi), which reacts quickly with monomer relative to the rate of subsequent addition. This is of the form:

$$\mathrm{RLi} + \mathrm{CH_2{=}\underset{\displaystyle X}{\underset{|}{C}H}} \xrightarrow{k_i} \mathrm{RCH_2{-}\underset{\displaystyle X}{\underset{|}{C}H^{-}}Li^{+}} \qquad \text{(Structure 2.12)}$$

with rate:

$$r_I = k_i\; C_{RLi}\; C_M \tag{2.7}$$

The rate of polymerization is much slower than the rate of initiation. It is the rate determining step. From Eq. (2.3):

$$r_P = r_M = k_P\; C_{RLi}\; C_M \tag{2.8}$$

There is often an association reaction of alkyl lithiums, i.e.:

$$(\mathrm{RLi})_n \rightleftharpoons n\mathrm{RLi} \qquad \text{(Structure 2.13)}$$

The rate of polymerization is determined from the concentration of RLi, which in turn is determined by the equilibrium with $(RLi)_n$.

In condensation polymerization, we are involved with reactions of the form:

$$\begin{aligned} \mathrm{AXB + AXB} &\xrightarrow{k_P} \mathrm{AXCXB} \\ \mathrm{AXCXB + AXB} &\xrightarrow{k_P} \mathrm{AXCXCXB} \\ \mathrm{AXCXCXB + AXCXB} &\xrightarrow{k_P} \mathrm{AXCXCXCXCXB} \end{aligned} \qquad \text{(Structure 2.14)}$$

etc., where end groups A and B react to form C or of the form:

$$\begin{aligned} \mathrm{AX_1A + BX_2B} &\xrightarrow{k_P} \mathrm{AX_1CX_2B} \\ \mathrm{AX_1A + AX_1CX_2B} &\xrightarrow{k_P} \mathrm{AX_1CX_2CX_1B} \end{aligned} \qquad \text{(Structure 2.15)}$$

Specifically A and B are reactive units, such as carboxylic acid, amines, and alcohols and C is an ester, amide or other linkage. Examples are the synthesis of polyesters from dicarboxylic acids and glycols and polyamide from dicarboxylic acids and diamines:

$$\mathrm{HO\,X_1\,OH + HOOC\,X_2\,COOH} \xrightarrow{k_p} \mathrm{HO}\left[\mathrm{X_1{-}O{-}\overset{\displaystyle O}{\overset{\|}{C}}{-}X_2{-}\overset{\displaystyle O}{\overset{\|}{C}}{-}O}\right]_n \mathrm{X_1{-}OH + H_2O}$$

(Structure 2.16)

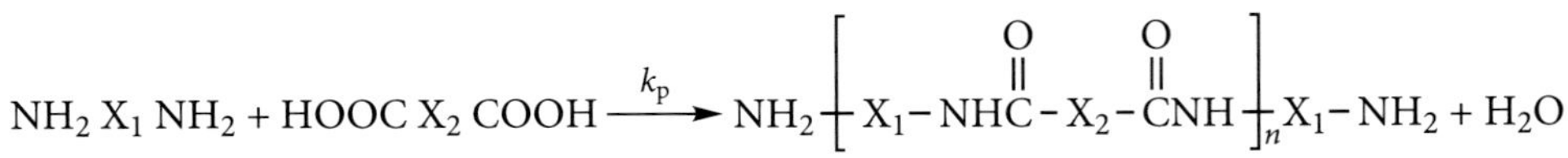

(Structure 2.17)

The synthesis of polyamides from amino acids and lactams or polyesters from lactones can be carried out using condensation or ionic polymerization mechanisms [13]. In reactive extrusion, it is generally the more rapid anionic mechanism which is used:

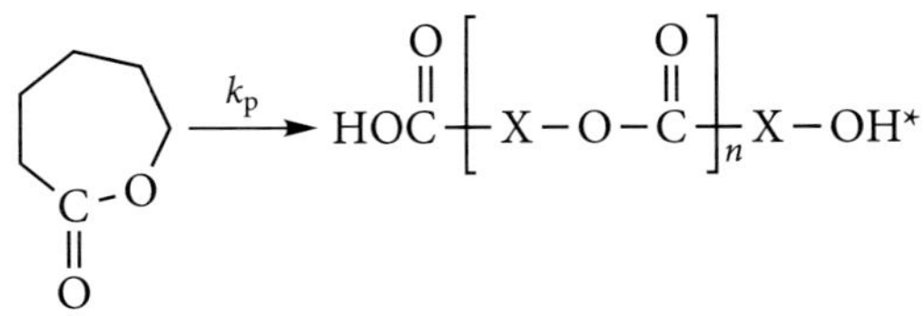

(Structure 2.18)

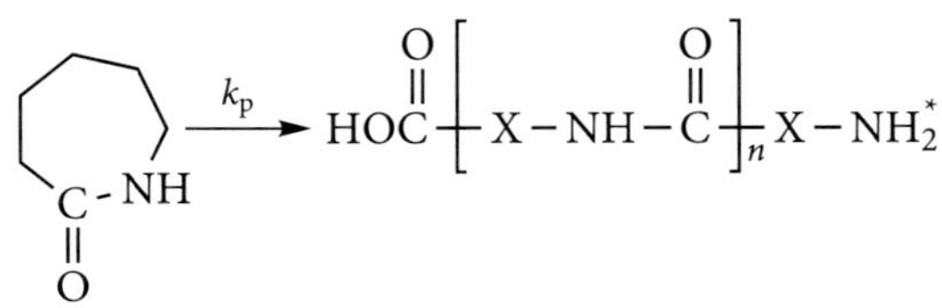

(Structure 2.19)

Polyurethanes are formed from diisocyanates and glycols:

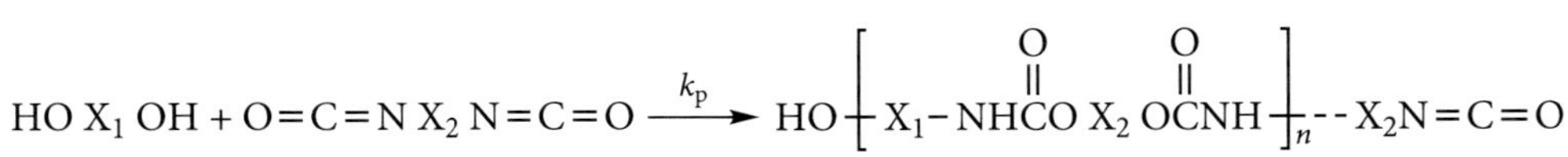

(Structure 2.20)

2.4 Mass Conservation and Mechanics

The simulation of flow in extrusion is based on the application of the principles of conservation of mass and balance of forces [14–17]. It is the purpose of this chapter to review these principles.

The conservation of mass in fluid mechanics may be thought of in terms of a material balance on a small cubic volume of dimensions Δx_1, Δx_2, Δx_3 located in a Cartesian space. If ρ is the density and velocity v has the components:

$$\underset{\sim}{v} = v_1\, \underline{e}_1 + v_2\, \underline{e}_2 + v_3\, \underline{e}_3 \tag{2.9}$$

the conservation of mass has the form:

$$\frac{\partial \rho}{\partial t} + \frac{\partial}{\partial x_1} \rho\, v_1 + \frac{\partial}{\partial x_2} \rho\, v_2 + \frac{\partial}{\partial x_3} \rho\, v_3 = 0 \tag{2.10}$$

This may be rewritten in a vector form as:

$$\frac{\partial \rho}{\partial t} + \nabla \cdot \rho \, \underset{\sim}{v} = 0 \tag{2.11}$$

where ∇ is the 'del' operator defined as:

$$\nabla = \underset{\sim}{e}_1 \frac{\partial}{\partial x_1} + \underset{\sim}{e}_2 \frac{\partial}{\partial x_2} + \underset{\sim}{e}_3 \frac{\partial}{\partial x_3} \tag{2.12}$$

In cylindrical coordinates, the continuity equation has the form:

$$\frac{\partial \rho}{\partial t} + \frac{\partial}{\partial z}(\rho \, v_z) + \frac{1}{r}\frac{\partial}{\partial r}(r \, \rho \, v_r) + \frac{1}{r}\frac{\partial}{\partial \theta}(\rho \, v_\theta) = 0 \tag{2.13}$$

Usually, in the processing of fluids including polymer melts, we may consider these materials as being incompressible, i.e., the density ρ is constant. This simplifies Eqs. (2.10), (2.11), and (2.13), respectively, to:

$$\nabla \cdot \underset{\sim}{v} = 0 \tag{2.14a}$$

$$\frac{\partial v_1}{\partial x_1} + \frac{\partial v_2}{\partial x_2} + \frac{\partial v_3}{\partial x_3} = 0 \tag{2.14b}$$

$$\frac{\partial v_z}{\partial z} + \frac{1}{r}\frac{\partial}{\partial r}(r \, v_r) + \frac{1}{r}\frac{\partial v_\theta}{\partial \theta} = 0 \tag{2.14c}$$

The principles of mechanics are based on a balance of forces with rates of change of linear momentum, and of torques with angular momentum. The balance of force with variations in linear momentum, with which we are most familiar, is for a particle.

Here:

$$\frac{\mathrm{d}}{\mathrm{d}t}(m \, \underset{\sim}{v}) = \underset{\sim}{F} \tag{2.15}$$

This form is commonly associated with Newton. In a deforming material, the external force $\underset{\sim}{F}$ is considered to consist of two parts, contact forces and body forces due to force fields. These are written in terms of an integral over the surface of the body and a volume integral:

$$\underset{\sim}{F} = \oint \underset{\sim}{t} \, \mathrm{d}a + \int \rho \, \underset{\sim}{f} \, \mathrm{d}V \tag{2.16}$$

where $\underset{\sim}{t}$ is a stress vector and $\underset{\sim}{f}$ is a body force per unit mass. The stress vector may be expressed in terms of a second order tensor quantity, the stress tensor, $\underset{\sim}{\sigma}$, through:

$$\underset{\sim}{t} = \underset{\sim}{\sigma} \cdot \underset{\sim}{n} = \sum_i \underset{\sim}{e}_i \left(\sum_j \sigma_{ij} \, n_j \right) \tag{2.17}$$

We may identify two types of stress components in Eq. (2.17). These are (i) normal stresses σ_{ii}, where the stresses act normal to the surface on which they are applied, and (ii) shearing stress σ_{ij} $(i \neq j)$, which act parallel to the surface to which they are applied. Both normal and shearing stresses occur in deforming and flowing polymer melts.

Using Eq. (2.17), we are able to reduce Eq. (2.16) to an expression of the form of Newton's Law of motion, but applicable to a point in a deforming continuum.

This has the form in vector notation:

$$\rho\left[\frac{\partial \underset{\sim}{v}}{\partial t}+(\underset{\sim}{v}\cdot\nabla)\,\underset{\sim}{v}\right]=\nabla\cdot\underset{\sim}{\sigma}+\rho\,\underset{\sim}{f} \tag{2.18}$$

Equation (2.18) is known as Cauchy's Law of Motion. Here, ∇ is the "del" operator of Eq. (2.12). In Cartesian components, Eq. (2.18) is expressed as:

$$\begin{aligned}
\rho\left[\frac{\partial v_1}{\partial t}+v_1\frac{\partial v_1}{\partial x_1}+v_2\frac{\partial v_1}{\partial x_2}+v_3\frac{\partial v_1}{\partial x_3}\right]&=\frac{\partial\sigma_{11}}{\partial x_1}+\frac{\partial\sigma_{12}}{\partial x_2}+\frac{\partial\sigma_{13}}{\partial x_3}+\rho\,\underset{\sim}{f_1}\\
\rho\left[\frac{\partial v_2}{\partial t}+v_1\frac{\partial v_2}{\partial x_1}+v_2\frac{\partial v_2}{\partial x_2}+v_3\frac{\partial v_2}{\partial x_3}\right]&=\frac{\partial\sigma_{21}}{\partial x_1}+\frac{\partial\sigma_{22}}{\partial x_2}+\frac{\partial\sigma_{23}}{\partial x_3}+\rho\,\underset{\sim}{f_2}\\
\rho\left[\frac{\partial v_3}{\partial t}+v_1\frac{\partial v_3}{\partial x_1}+v_2\frac{\partial v_3}{\partial x_2}+v_3\frac{\partial v_3}{\partial x_3}\right]&=\frac{\partial\sigma_{31}}{\partial x_1}+\frac{\partial\sigma_{32}}{\partial x_2}+\frac{\partial\sigma_{33}}{\partial x_3}+\rho\,\underset{\sim}{f_3}
\end{aligned} \tag{2.19}$$

In cylindrical components, this is:

$$\begin{aligned}
\rho\left[\frac{\partial v_z}{\partial t}+v_z\frac{\partial v_z}{\partial z}+v_r\frac{\partial v_z}{\partial r}+\frac{v_\theta}{r}\frac{\partial v_z}{\partial\theta}\right]&=\frac{\partial\sigma_{zz}}{\partial z}+\frac{1}{r}\frac{\partial}{\partial r}(r\,\sigma_{zr})+\frac{1}{r}\frac{\partial\sigma_{r\theta}}{\partial\theta}+\rho f_z\\
\rho\left[\frac{\partial v_r}{\partial t}+v_z\frac{\partial v_r}{\partial z}+v_r\frac{\partial v_r}{\partial r}+\frac{v_\theta}{r}\frac{\partial v_r}{\partial\theta}-\frac{v_\theta^2}{r}\right]&=\frac{\partial\sigma_{rz}}{\partial z}+\frac{1}{r}\frac{\partial}{\partial r}(r\,\sigma_{rr})+\frac{1}{r}\frac{\partial\sigma_{r\theta}}{\partial\theta}-\frac{\sigma_{\theta\theta}}{r}+\rho f_r\\
\rho\left[\frac{\partial v_\theta}{\partial t}+v_z\frac{\partial v_\theta}{\partial z}+v_r\frac{\partial v_\theta}{\partial r}+\frac{v_\theta}{r}\frac{\partial v_\theta}{\partial\theta}-\frac{v_r\,v_\theta}{r}\right]&=\frac{\partial\sigma_{\theta z}}{\partial z}+\frac{1}{r^2}\frac{\partial}{\partial\theta}(r^2\,\sigma_{\theta r})+\frac{1}{r}\frac{\partial\sigma_{\theta z}}{\partial\theta}+\rho f_\theta
\end{aligned} \tag{2.20}$$

Conservation of torques and angular momentum yields a restriction on the stress tensor. It necessitates the stress tensor to be symmetric. Thus:

$$\sigma_{ij}=\sigma_{ji} \tag{2.21}$$

2.5 Flow of Polymer Fluids

The processing, including compounding, of polymer systems is generally done in the molten state, particularly in shear flow. It is thus necessary to understand the flow (or better *rheological*) characteristics of polymer melts [16–18]. We begin our discussion with simple shear flow defined as:

$$\underset{\sim}{v} = \dot{\gamma}\, x_2\, \underset{\sim}{e}_1 + 0\, \underset{\sim}{e}_2 + 0\, \underset{\sim}{e}_3 \tag{2.22}$$

where '1' is the direction of flow, '2' is the direction of shear, and '3' is the neutral direction. The stress field σ_{ij} developed in response to this flow for a Newtonian fluid is:

$$\begin{aligned} \sigma_{12} &= \eta\, \dot{\gamma} \\ \sigma_{11} &= \sigma_{22} = \sigma_{33} = -p \\ \sigma_{32} &= \sigma_{13} = 0 \end{aligned} \tag{2.23}$$

where η is the viscosity and p the pressure. For a polymer melt, the stress field is of form:

$$\begin{aligned} \sigma_{12} &= \eta\, \dot{\gamma} \\ \sigma_{11} &= -p + \beta_1\, \dot{\gamma}^2 \\ \sigma_{22} &= -p + \beta_2\, \dot{\gamma}^2 \\ \sigma_{33} &= -p \\ \sigma_{23} &= \sigma_{13} = 0 \end{aligned} \tag{2.24}$$

It is seen that normal stresses as well as shearing stresses arise. Normal stresses are well known in polymer solutions and melts. They give rise to 'rod climbing' effects in closed streamline flows that typically occur in flow in a beaker or between coaxial cylinders (Figure 2.1). Indeed, it was observations of this type which first led Weissenberg [19] to hypothesize the occurrence of normal stresses in shear flows of these materials.

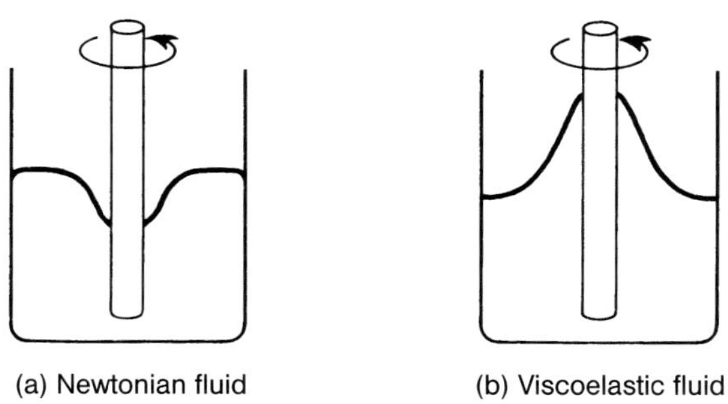

Figure 2.1: Weissenberg normal stress effect

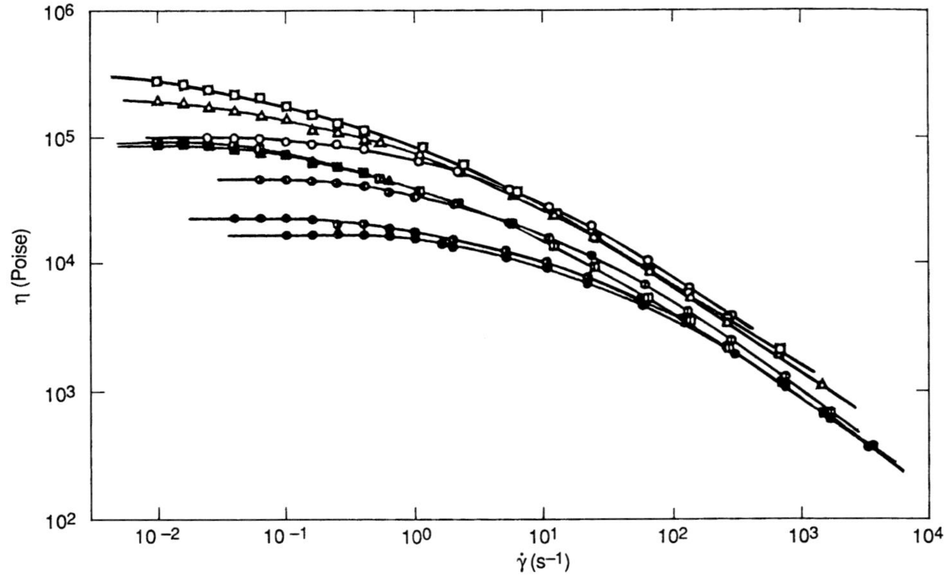

Figure 2.2: Shear viscosity η as a function of shear rate for polymer melts. The data is for a series of polypropylenes of varying molecular weight and distribution [22]

The shear viscosity in polymer melts and solutions has been investigated for more than a half century. Certain general conclusions may be drawn. The magnitudes of the shear viscosity are quite large, 10,000 Pascal-seconds (Pa-s) (100,000 poise) is not uncommon. On these scales, water is 0.001 Pa-s (0.01 poise). At low shear rates, polymer melts and solutions exhibit a constant or 'zero shear' viscosity, η_0. At higher shear rates, the shear viscosity decreases. For polymer solutions, a high shear rate Newtonian asymptote η_∞ is observed. The viscosity-shear rate behavior of polymer melts and solutions is summarized in Figure 2.2.

A wide range of empirical viscosity-shear rate relationships have been proposed through the years. The most popular of these is the power law, which represents a high shear rate asymptote in polymer melts and an intermediate shear rate characteristic for polymer solutions. This has the form:

$$\begin{aligned} \sigma_{12} &= K\,\dot{\gamma}^{n} \\ \eta &= K\,\dot{\gamma}^{n-1} \end{aligned} \tag{2.25}$$

Many empirical equations have been fashioned which combine two or more of the regions [20]. Typical are forms such as:

$$\eta = \frac{\eta_0}{1 + (\eta_0 / K)\,\dot{\gamma}^{1-n}} \tag{2.26}$$

which might be applied to polymer melts or the low and intermediate shear rate behavior of polymer solutions. Another form which has been used for polymer solutions is:

$$\eta = \eta_\infty + \frac{\eta_0 - \eta_\infty}{1 + A\,\dot{\gamma}^2} \tag{2.27}$$

This exhibits the low shear rate asymptote η_0 and the high shear rate asymptote η_∞.

The viscosity-shear rate behavior of polymer melts and solutions is strongly affected by the structure of the polymers. Most experimental studies have involved flexible chain polymers. At low molecular weights, these polymers exhibit Newtonian behavior and the viscosity increases linearly with molecular weight. Above a critical molecular weight which corresponds to roughly 350 chain carbon atoms, the shear viscosity becomes non-Newtonian and η_0 increases according to:

$$\eta_0 = K\, M_w^{3.5} \tag{2.28}$$

where the weight-average molecular weight M_w is to be noted. The dependence of η on shear rate has been found by Vinogradov and Malkin [18, 21] to be representable in a temperature independent form:

$$\frac{\eta}{\eta_0} = F\,(\eta_0\,\dot{\gamma}) \tag{2.29}$$

Various researchers [22–25] find that the functional form of Eq. (2.29) depends on the breadth of the molecular weight distribution, with η/η_0 decreasing more rapidly with $\eta_0\,\dot{\gamma}$ as the distribution broadens (Figure 2.3).

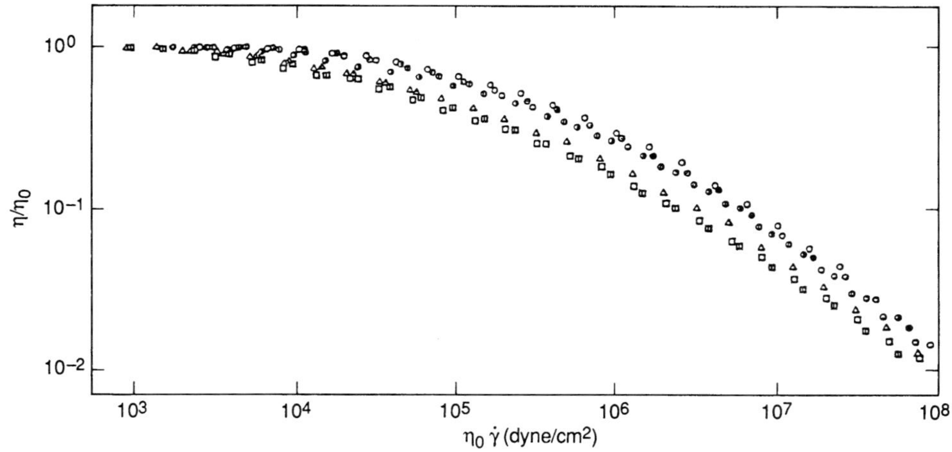

Figure 2.3: Effect of molecular weight distribution on plots of η/η_0 versus $\eta_0\,\dot{\gamma}$ for polypropylenes of varying molecular weight distribution [22]

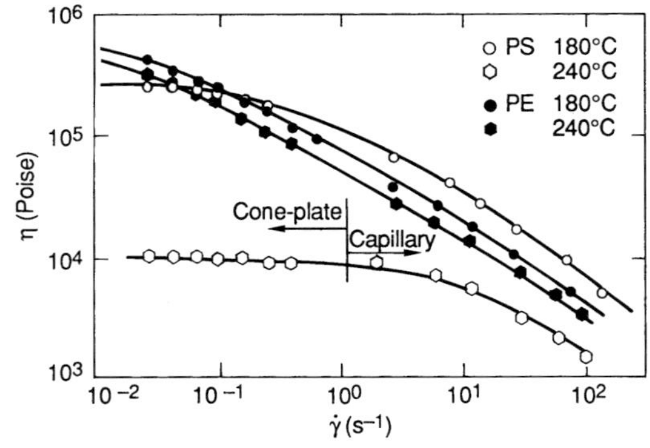

Figure 2.4: Influence of temperature on melt viscosity for polyethylene (PE) and polystyrene (PS) melts

Generally, shear viscosity decreases with increasing temperature (Figure 2.4). The temperature dependence of the zero shear viscosity η_0 has been linked to the glass transition temperature through the Williams-Landel-Ferry equation [26]:

$$\eta_0(T) \cong \eta_0\,(T_g)\,e^{-[17.44\,(T-T_g)]/[51.6+(T-T_g)]} \tag{2.30}$$

The temperature dependence is the greatest when T is near T_g. Figure 2.4 illustrates the strong temperature dependence of the viscosity of polystyrene ($T_g \sim 100$ °C) and the weak dependence of polyethylene ($T_g \sim -100$ °C).

Other empirical relationships have been used for the temperature dependence. This includes simple exponentials and activation energy forms, i.e.:

$$\eta = A\,e^{-B\,(T-T_0)} \tag{2.31a}$$

$$\eta = C\,e^{-E/RT} \tag{2.31b}$$

These have also often been used in combined formulations with the power law non-Newtonian shear viscosity equation:

$$\eta = A'\,\dot{\gamma}^{n-1}\,e^{-B\,(T-T_0)} \tag{2.32a}$$

$$\eta = C'\,\dot{\gamma}^{n-1}\,e^{-E/RT} \tag{2.32b}$$

The shear viscosity of polymers is reduced by the addition of solvent. For flexible chain polymers, the following empirical equation has been found to be valid for concentrated solutions [27, 28] (see Figure 2.5).

$$\eta_0 = K'\,M_w^{3.5}\,c^5 \tag{2.33}$$

The behavior of solutions of rigid macromolecules is different. Solutions of polypeptides [29, 30] and p-linked aromatic polyamides [30, 31] exhibit shear viscosities which at first increase with concentration, then exhibit maxima at a critical concentration, followed by a viscosity

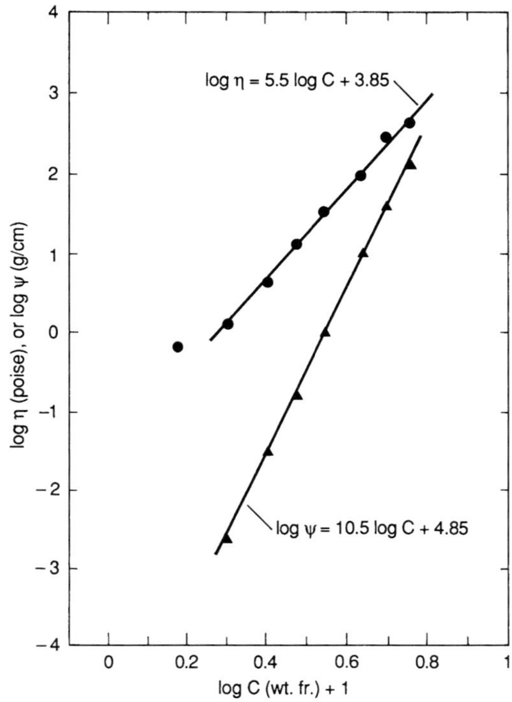

Figure 2.5: Viscosity and principal normal stress difference coefficient $\Psi_1 = N_1 / \dot{\gamma}^2$ of flexible chain polymer solutions as a function of concentration based on polystyrene-styrene solutions [28]

decrease at higher concentrations. This is associated with the formation of liquid crystalline phases. At higher concentrations, the shear viscosity again increases.

Polymer solutions and melts exhibit normal stresses as well as shearing stresses in steady shear flow. There have been many measurements of these quantities. Generally, one determines experimentally the principal N_1 and second N_2 normal stress differences defined by:

$$N_1 = \sigma_{11} - \sigma_{22} \tag{2.34a}$$

$$N_2 = \sigma_{22} - \sigma_{33} \tag{2.34b}$$

There have been many more measurements of N_1 as opposed to N_2. Certain general conclusions have risen from those studies. Specifically [32]:

$$N_1 > 0 \qquad N_2 < 0 \tag{2.35a}$$

$$N_2 \sim -(0.1 - 0.3)\, N_1 \tag{2.35b}$$

Generally, N_1 is found to be strongly influenced by molecular weight distribution, being increased by the high molecular weight tail. Oda et al. [33] have shown that the relationship between N_1 and σ_{12} is independent of temperature for polystyrene melts. It may be expressed in a power law form:

$$N_1 = A\,\sigma_{12}^{a} \tag{2.36}$$

The coefficient 'A' and exponent 'a' depend upon the breadth of the molecular weight distribution. With increasing breadth of distribution, N_1 is larger. Generally 'A' increases and 'a' decreases. This was later confirmed for polyolefins [22–25].

Normal stresses are sometimes expressed in terms of normal stress coefficients Ψ_1 and Ψ_2 defined as:

$$N_1 = \Psi_1\,\dot{\gamma}^2 \qquad N_2 = \Psi_2\,\dot{\gamma}^2 \tag{2.37}$$

where Ψ_1 equals $\beta_1 - \beta_2$ of Eq. (2.24) and Ψ_2 is β_2. For polymer melts, Ψ_1 has been found to be correlated with molecular weight and its distribution through [22–25]:

$$\Psi_1 \sim M_w^{3.5}\,M_z^{3.5} \tag{2.38}$$

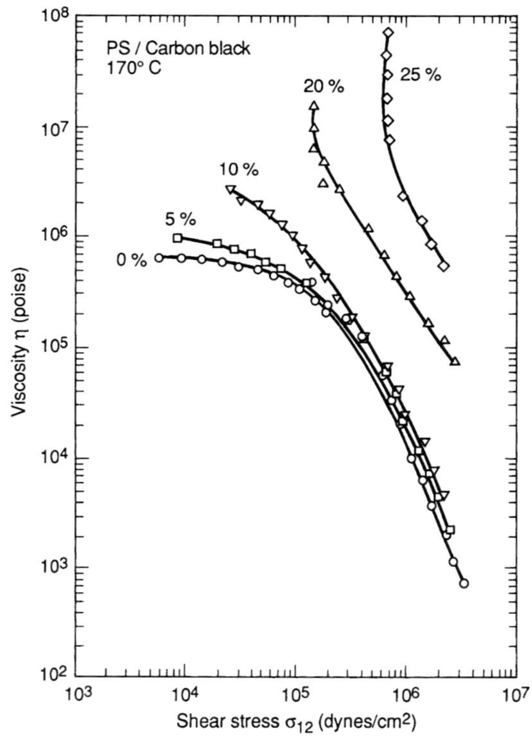

Figure 2.6: Viscosity-shear stress behavior of particle filled compounds based on polystyrene-carbon black compounds [36]

In polymer solutions, there is a very strong concentration dependence of normal stresses. For concentrated solutions of flexible chain polymers, the principal normal stress difference coefficient may be expressed as [28]:

$$\Psi_1 \sim M_w^{3.5}\, M_z^{3.5}\, c^{10} \tag{2.39}$$

Polymer melts filled with small particles exhibit an increased viscosity, especially at very low shear rates. At high levels of loading, yield values Y develop [34–38] (see Figure 2.6). This effectively means that at shear stresses lower than Y, flow ceases. This leads to equations of form [38]:

$$\begin{aligned}\sigma_{12} &= Y + K\,\dot{\gamma}^n \\ &= Y + \frac{A\,\dot{\gamma}}{1+(A/K)\,\dot{\gamma}^{1-n}}\end{aligned} \tag{2.40}$$

which have been used to fit shear viscosity data.

2.6 Constitutive Equations

Three dimensional constitutive equations have been developed for the stress tensor in polymer fluid systems [14–18]. These constitutive equations focus on representing the deviations exhibited by these fluids for Newtonian fluid behavior.

The stress tensor, $\underset{\sim}{\sigma}$ in a Newtonian fluid may, following Navier and Stokes, can be expressed as:

$$\underset{\sim}{\sigma} = -p\,\underset{\sim}{I} + 2\,\eta\,\underset{\sim}{d} \tag{2.41}$$

Here, $\underset{\sim}{I}$ is the unit tensor and $\underset{\sim}{d}$ is the rate of deformation tensor. Specifically:

$$\underset{\sim}{I} = \begin{vmatrix} 1 & 0 & 0 \\ 0 & 1 & 0 \\ 0 & 0 & 1 \end{vmatrix} \tag{2.42}$$

$$\underset{\sim}{d} = \frac{1}{2}\left[\nabla \underset{\sim}{v} + (\nabla \underset{\sim}{v})^{\mathrm{T}}\right] \tag{2.43}$$

where the superposed 'T' represents transpose.

The simplest constitutive equation expressing the non-Newtonian character found in polymer systems is to use Eq. (2.41) with η dependent on deformation rate. However, we must now express deformation rate, not only as a shear rate as in the previous section, but as a 3-dimensional quantity in terms of $\underset{\sim}{d}$. To express, η, a scalar, as a function of the rate of deformation tensor, $\underset{\sim}{d}$, we introduce invariants. These represent combinations of components of the tensor that are independent of coordinate system [14–18]. The most important of the invariants is the second invariant $t\,r\,\underset{\sim}{d}^2$, i.e., the trace of the square of the second rate of deformation invariant.

Specifically, we write:

$$\eta = \eta\,(t\,r\,\underset{\sim}{d}^2) \tag{2.44}$$

where:

$$t\,r\,\underset{\sim}{d}^2 = \sum_i \sum_j d_{ij}\, d_{ij} \tag{2.45}$$

The non-Newtonian shear viscosities of Eqs. (2.25) and (2.26) may be represented in terms of Eq. (2.44). For the power law fluid of Eq. (2.25) we may write:

$$\eta = K\,(2\,t\,r\,\underset{\sim}{d}^2)^{(n-1)/2} \tag{2.46}$$

For Eq. (2.26) we may write:

$$\eta = \frac{\eta_0}{1+(\eta_0\,/\,K)\,(2\,t\,r\,\underset{\sim}{d}d^2)^{(1-n)/2}} \tag{2.47}$$

The constitutive equations described above do not predict normal stresses or memory effects. There has been a very extensive development of constitutive equations by different researchers that predict these effects [16–18]. For long duration shear flows it has been shown, rather generally, that viscoelastic fluids may be expressed through:

$$\underset{\sim}{\sigma} = -p\,\underset{\sim}{I} + 2\,\eta\,\underset{\sim}{d} - \Psi_1\,\frac{\delta \underset{\sim}{d}}{\delta t} + \Psi_2\left(\frac{\delta \underset{\sim}{d}}{\delta t} + 4\,\underset{\sim}{d}^2\right) \tag{2.48}$$

This equation is known for Criminale, Ericksen, and Filbey [39]. Here $\delta \underset{\sim}{d}/\delta t$ is defined through:

$$\frac{\delta \underset{\sim}{d}}{\delta t} = \frac{\partial}{\partial t}\underset{\sim}{d} + (\underset{\sim}{v}\cdot\nabla)\,\underset{\sim}{d} - \nabla \underset{\sim}{v}\cdot \underset{\sim}{d} - \underset{\sim}{d}\cdot\nabla \underset{\sim}{v} \tag{2.49}$$

The quantities η, Ψ_1, and Ψ_2 are the experimental shear viscosity and normal stress coefficients that depend on $t\,r\,\underset{\sim}{d}^2$, as suggested by the non-linear forms of Eqs. (2.26), (2.27), (2.35), and (2.36). Thus, the non-linear forms of Eqs. (2.46) and (2.47) could be used for the shear viscosity in Eq. (2.48).

The weakness of Eq. (2.49) is that it cannot represent 'strong' transients such as those giving rise to wave propagation effects.

More general constitutive equations which may be used for a broad range of different flow fields have not developed the same consensus. We will not discuss these more general constitutive equations here but rather refer to the books of Middleman [17], Vinogradov and Malkin [18], and White [16].

2.7 Conservation of Energy and Heat Conduction

Generally, twin screw extruders have heating and cooling sections. Thus, it is necessary to consider energetic as well as dynamic phenomena when describing their behavior [14]. The basis of any such consideration must be the first law of thermodynamics, which equates the change in internal energy of the process fluid to the sum of the heat and work input. They may be expressed as:

$$\frac{\mathrm{d}U}{\mathrm{d}t} = Q + W \tag{2.50}$$

where U is internal energy, Q is total heat flow, and W the power input.

The heat flow Q into a body may be expressed in terms of a local heat flux q, as:

$$Q = -\oint \underset{\sim}{q} \cdot \underset{\sim}{n}\, \mathrm{d}a \tag{2.51}$$

where $\underset{\sim}{n}$ is a unit normal to surface area da.

The transport of heat in materials is governed by Fourier's law of heat conduction which relates local heat flux $\underset{\sim}{q}$ to temperature gradient, through [14]:

$$q = -k\,\frac{\mathrm{d}T}{\mathrm{d}x} \tag{2.52a}$$

$$\underset{\sim}{q} = -k\,\nabla T \tag{2.52b}$$

where k is the thermal conductivity. This quantity is represented here as a scalar. The thermal conductivity of molten polyethylene is of order 0.25 J/(m-s-°C) and may vary with pressure and temperature. In anisotropic media $\underset{\sim}{q}$ and ∇T may be in different directions. This will make thermal conductivity a second order tensor with components k_{ij}.

We must reduce Eq. (2.50) to a differential equation in a deforming fluid. We proceed as follows. The change in internal energy of a material may be expressed through the differential:

$$\mathrm{d}U = \int [\rho\, c_\mathrm{v}\, \mathrm{d}V]\, \mathrm{d}T + \left(\frac{\partial U}{\partial V}\right)_\mathrm{T} \mathrm{d}V \tag{2.53}$$

where V refers to the total volume.

The rate of work done on a body by external forces W may be expressed a:

$$\begin{aligned} W &= \oint \underset{\sim}{t} \cdot \underset{\sim}{v} \, da \\ &= \oint \underset{\sim}{\sigma} \cdot \underset{\sim}{v} \cdot \underset{\sim}{n} \, da \end{aligned} \tag{2.54}$$

Combining together Eqs. (2.51) through (2.54) for Q, dU, and W leads after some manipulation to equations of form [14]:

$$\rho c_v \left[\frac{\partial T}{\partial t} + (\underset{\sim}{v} \cdot \nabla) T \right] = k \nabla^2 T + \rho^2 \left(\frac{\partial u}{\partial \rho} \right) \nabla \cdot \underset{\sim}{v} + \sum_i \sum_j \sigma_{ij} \frac{\partial v_i}{\partial x_j} \tag{2.55}$$

where u is internal energy per unit mass. For incompressible fluids, this simplifies to:

$$\rho c_v \left[\frac{\partial T}{\partial t} + (\underset{\sim}{v} \cdot \nabla) T \right] = k \nabla^2 T + \sum_i \sum_j \sigma_{ij} \frac{\partial v_i}{\partial x_j} \tag{2.56}$$

In Cartesian coordinates, Eq. (2.56) has the form:

$$\rho c \left(\frac{\partial T}{\partial t} + v_1 \frac{\partial T}{\partial x_1} + v_2 \frac{\partial T}{\partial x_2} + v_3 \frac{\partial T}{\partial x_3} \right) = k \left(\frac{\partial^2 T}{\partial x_1^2} + \frac{\partial^2 T}{\partial x_2^2} + \frac{\partial^2 T}{\partial x_3^2} \right) + \sum_i \sum_j \sigma_{ij} \frac{\partial v_i}{\partial x_j} \tag{2.57}$$

2.8 Hydrodynamic Lubrication Theory

In the treatment of the flow of highly viscous fluids through narrow channels including those with moving members, the approximation known as 'Hydrodynamic Lubrication Theory' [16, 40–43] is extremely useful. It was originally developed by Osborne Reynolds [40] to apply to the flow of Newtonian lubricating oils in machines with moving parts. This approximation is based on making 'order of magnitude' analysis on viscous fluids moving through narrow channels. It presumes inertia and normal stresses may be neglected. Consider flow in the 1–3 plane of a Cartesian coordinate system, where shearing is in the '2' direction (Figure 2.7). The equations of motion are taken to be:

$$\begin{aligned} 0 &= \frac{\partial \sigma_{11}}{\partial x_1} + \frac{\partial \sigma_{12}}{\partial x_2} = -\frac{\partial p}{\partial x_1} + \frac{\partial \sigma_{12}}{\partial x_2} = -\frac{\partial p}{\partial x_1} + \frac{\partial}{\partial x_2} \eta \frac{\partial v_1}{\partial x_2} \\ &= \frac{\partial \sigma_{33}}{\partial x_3} + \frac{\partial \sigma_{32}}{\partial x_2} = -\frac{\partial p}{\partial x_3} + \frac{\partial \sigma_{32}}{\partial x_2} = -\frac{\partial p}{\partial x_3} + \frac{\partial}{\partial x_2} \eta \frac{\partial v_3}{\partial x_2} \end{aligned} \tag{2.58}$$

The change of σ_{11} and σ_{33} along the 1 and 3 coordinate directions is attributed to being solely due to pressure.

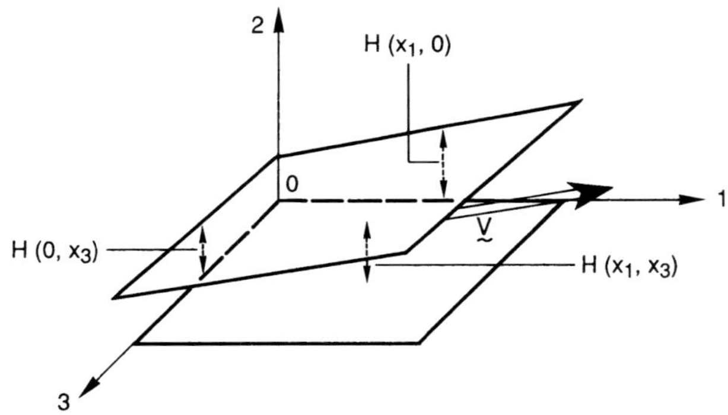

Figure 2.7: Shearing lubrication flow

The approximation of neglecting inertia is, of course, a better one for highly viscous polymer melts than it is for lower viscosity oils. The approximation of neglect of normal stress variations in the direction of flow must be much poorer than for Newtonian oils. N_1, i.e., $\sigma_{11} - \sigma_{22}$ is quite large, generally greater than the shear stresses, σ_{12}, in polymer melts. The inherent presumption of:

$$\frac{\partial \sigma_{12}}{\partial x_2} \gg \frac{\partial N_1}{\partial x_1} \tag{2.59}$$

is critical for the application of the Hydrodynamic Lubrication Theory.

Accepting now Eq. (2.58), the basic premise of the Hydrodynamic Lubrication Theory, we will proceed with the development of this formulation. Let us presume that the fluid is Newtonian and the boundary conditions on the upper and lower surface bounding the fluid are:

$$\left.\begin{aligned} v_1(x_1,H,x_3) &= U_1 \\ v_3(x_1,H,x_3) &= U_3 \\ \\ v_1(x_1,0,x_3) &= 0 \\ v_3(x_1,0,x_3) &= 0 \end{aligned}\right\} \tag{2.60}$$

where:

$$H = H(x_1,x_3) \tag{2.61}$$

We may solve Eqs. (2.58) for the velocity field:

$$v_1(x_1,x_2,x_3) = U_1\left(\frac{x_2}{H}\right) - \frac{H^2}{2\eta}\left(\frac{\partial p}{\partial x_1}\right)\left[\left(\frac{x_2}{H}\right) - \left(\frac{x_2}{H}\right)^2\right] \tag{2.62a}$$

$$v_3(x_1,x_2,x_3) = U_3\left(\frac{x_2}{H}\right) - \frac{H^2}{2\eta}\left(\frac{\partial p}{\partial x_3}\right)\left[\left(\frac{x_2}{H}\right) - \left(\frac{x_2}{H}\right)^2\right] \tag{2.62b}$$

We may define mean velocities $\bar{v}_1$ and $\bar{v}_3$, and fluxes q_1 and q_2:

$$q_1 = H\,\bar{v}_1 = \int_0^H v_1(x_2)\,\mathrm{d}x_2 \tag{2.63a}$$

$$q_3 = H\,\bar{v}_3 = \int_0^H v_3(x_2)\,\mathrm{d}x_2 \tag{2.63b}$$

We may rewrite the incompressible continuity equation of Eq. (2.10) in terms of the mean velocities through the channel thickness *H*. This may be expressed as:

$$\frac{\partial q_1}{\partial x_1} + \frac{\partial q_3}{\partial x_3} = 0 \tag{2.64a}$$

$$\frac{\partial H\,\bar{v}_1}{\partial x_1} + \frac{\partial H\,\bar{v}_3}{\partial x_3} = 0 \tag{2.64b}$$

For a Newtonian fluid, the values of q_1 and q_3 are from Eqs. (2.62) and (2.63) of form:

$$\begin{aligned} q_1 &= \frac{1}{2} U_1\,H - \frac{H^3}{12\,\eta}\left(\frac{\partial p}{\partial x_1}\right) \\ q_3 &= \frac{1}{2} U_3\,H - \frac{H^3}{12\,\eta}\left(\frac{\partial p}{\partial x_3}\right) \end{aligned} \tag{2.65}$$

The continuity equation of Eq. (2.64) takes on the form:

$$\frac{\partial}{\partial x_1}\left(H^3 \frac{\partial p}{\partial x_1}\right) + \frac{\partial}{\partial x_3}\left(H^3 \frac{\partial p}{\partial x_3}\right) = 6\,\eta\left[\frac{\partial}{\partial x_1}(U_1\,H) + \frac{\partial}{\partial x_3}(U_3\,H)\right] \tag{2.66}$$

This expression was obtained by Reynolds [40] in 1886, and is called Reynolds equation. Certain special cases of Eq. (2.66) are worthy of attention. Most notable of these is the case where:

$$\begin{aligned} &H \cong \text{constant} \\ &U_1 = U_3 = 0 \end{aligned} \tag{2.67}$$

Here, Eq. (2.66) reduces to:

$$\bar{v}_1 = -\frac{H^2}{12\,\eta}\frac{\partial p}{\partial x_1} \qquad \bar{v}_3 = -\frac{H^2}{12\,\eta}\frac{\partial p}{\partial x_3}$$

or:

$$\bar{v} = -\frac{H^2}{12\,\eta}\nabla p \tag{2.68}$$

Reynolds equation is:

$$\frac{\partial^2 p}{\partial x_1^2} + \frac{\partial^2 p}{\partial x_3^2} = 0 \tag{2.69}$$

or:

$$\nabla^2 p = 0 \qquad (2.69)$$

We see that Reynolds equation becomes equivalent to Laplaces' equation [41]. The mean velocity is give by a velocity potential function as shown in Eq. (2.69). This is the same situation that arises in the two-dimensional flow of a non-viscous (inviscid fluid). This special form of hydrodynamic lubrication theory is known as Hele-Shaw flow theory [41]. The formulation of this paragraph applies to flows similar to the filling of injection molds by thermoplastics. It has been discussed by various researchers in this light [16, 42, 43].

The formulation of lubrication theory may also be applied to non-Newtonian fluid models. An extensive literature has been developed along these lines, especially as applied to simulation of polymer processing operations [42]. One proceeds by returning to Eq. (2.58) and introducing forms of the now non-Newtonian shear viscosity η. This would have the form for a power law fluid of:

$$\eta = K\left[\left(\frac{\partial v_1}{\partial x_2}\right)^2 + \left(\frac{\partial v_3}{\partial x_2}\right)^2\right]^{(n-1)/2} \tag{2.70}$$

Analytical solutions are no longer possible.

References

1. P. J. Flory, Principles of Polymer Chemistry, Ithaca, N.Y. (1953).
2. F. W. Billmeyer, Textbook of Polymer Science, Interscience, N.Y. (1962).
3. P. Meares, Polymers: Structure and Bulk Properties, Van Nostrand-Reinhold, London (1965).
4. J. M. Schultz, Polymer Material Science, Prentice-Hall, Englewood Cliffs (1974).
5. W. Minoshima, J. L. White, and J. E. Spruiell, *J. Appl. Polym. Sci.*, **25**, 287 (1980).
6. A. Ghijsels and J. J. S. M. Ente in Rheology (Proc 8th Int. Rheo. Conj.), edited by G. Astarita, G. Marrucci, and L. Nicolais, p. 15, Plenum, N.Y. (1980).

7. H. Yamane and J. L. White, *Polym. Eng. Sci.*, **15**, 131 (1987).
8. W. Minoshima and J. L. White, *J. Non-Newt. Fluid Mech.*, **19**, 275 (1986).
9. J. L. White and H. Yamane, *Pure Appl. Chem.*, **59**, 193 (1987); H. Yamane and J. L. White, *Nihon Reoroji Gakkaishi*, **15**, 131 (1987).
10. J. C. Staton, J. P. Keller, R. C. Kowalski, and J. W. Harrison, U.S. Patent 3,551,943 (1971).
11. R. C. Kowalski, U.S. Patent 3,563,972 (1972).
12. T. Kanai and J. L. White, *Polym. Eng. Sci.*, **24**, 1185 (1984).
13. P. Schlack, *Pure Appl. Chem.*, **15**, 507 (1967).
14. R. B. Bird, W. E. Stewart, and E. N. Lightfoot, *Transport Phenomena*, Wiley, N.Y. (1960).
15. A. C. Eringen, Non-Linear Theory of Continuous Media, McGraw-Hill, N.Y. (1962).
16. J. L. White, Principles of Polymer Engineering Rheology, Wiley, N.Y. (1990).
17. S. Middleman, The Flow of High Polymers, Wiley, N.Y. (1967).
18. G. V. Vinogradov and A. Y. Malkin, Rheology of Polymers, Mir, Moscow (1980).
19. K. Weissenberg, *Nature*, **159**, 310 (1947).
20. M. Reiner, Deformation and Flow, Lewis, London (1948).
21. G. V. Vinogradov and A. Y. Malkin, *J. Polym. Sci.*, **4**, 135 (1966).
22. W. Minoshima, J. L. White, and J. E. Spruiell, *Polym. Eng. Sci.*, **20**, 1166 (1980).
23. W. W. Graessley, *J. Chem. Phys.*, **47**, 1942 (1967).
24. S. Onogi, T. Masuda, I. Shiga and F. M. Costaschuk, *Appl. Polym. Symp.*, **20**, 37 (1973).
25. H. Yamane and J. L. White, *Polym. Eng. Rev.*, **2**, 167 (1982).
26. M. L. Williams, R. F. Landel, and J. D. Ferry, *J. Amer. Chem. Soc.*, **77**, 3701 (1955).
27. M. F. Johnson, W. W. Evans, I. Jordan, and J. D. Ferry, *J. Colloid Sci.*, **7**, 493 (1952).
28. Y. Ide and J. L. White, *J. Appl. Polym. Sci.*, **18**, 2997 (1974).
29. J. Hermans, *J. Colloid Sci.*, **17**, 638 (1962).
30. H. Aoki, J. L. White, and J. F. Fellers, *J. Appl. Polym. Sci.*, **23**, 2293 (1979).
31. S. P. Papkov, V. G. Kulichikhin, V. D. Kalmykova, and A. Y. Malkin, *J. Polym. Sci. Polym. Phys.*, **12**, 1753 (1974).
32. B. L. Lee and J. L. White, *Trans. Soc. Rheology*, **18**, 467 (1974).
33. K. Oda, J. L. White, and E. S. Clark, *Polym. Eng. Sci.*, **18**, 25 (1978).
34. F. M. Chapman and T. S. Lee, *Soc. Plast. Eng. J.*, **26**, (1)37 (1970).
35. G. V. Vinogradov, A. Y. Malkin, E. P. Plotnikova, O. Y. Sabsai, and N. E. Nikolayeva, *Int. J. Polym. Mat.*, **2**, 1 (1972).
36. V. M. Lobe and J. L. White, *Polym. Eng. Sci.*, **19**, 617 (1979).
37. Y. Suetsugu and J. L. White, *J. Appl. Polym. Sci.*, **28**, 1481 (1983).
38. J. L. White, Y. Wang, A. I. Isayev, N. Nakajima, F. C. Weissert, and K. Min, *Rubber Chem. Technol.*, **60**, 337 (1987).
39. W. O. Criminale, J. L. Ericksen, and G. L. Filbey, *Arch. Rat. Mech. Anal.*, **1**, 410 (1958).
40. O. Reynolds, *Phil. Trans. Roy. Soc.*, **A 177**, 157 (1886).
41. H. Lamb, Hydrodynamics, 6th ed., Cambridge U Press (1932).
42. Z. Tadmor and C. G. Gogos, Principles of Polymer Processing, 2nd ed., Wiley, N.Y. (2006).
43. J. L. White, *Polym. Eng. Sci.*, **15**, 44 (1975).

3 Single Screw Extrusion

3.1 Introduction

In order to understand the mechanisms of operation, applications, and performance of multiple screw extrusion systems, one must begin with single screw extruders. It is the purpose of this chapter to concisely describe the technology and science of single screw extrusion. Our intention is not to be exhaustive, but rather to be thorough enough that a proper perspective will be developed from which the reader may appreciate the remainder of this book.

We begin by a discussion of the development of single screw extrusion technology. There is an enormous amount of patent literature and it is necessary to be selective in dealing with it. In the remainder of the chapter we will discuss the mechanisms of flow melting and mixing in single screw extruders. From a historical perspective we note the article of Paton, Squires, Darnell, Cash and Carley [1] and the early treatise of Schenkel [2]. The more recent volumes of Tadmor and Klein [3], Rauwendaal [4], and White and Potente [5] deserve citation.

3.2 Development of Single Screw Extruders

The idea of screw pumps is very old, often being attributed to Archimedes. The development of single screw extrusion machines seems to occur in the latter years of the 19th century, roughly contemporary as we will see to non-intermeshing counter-rotating twin screw extruders. An 1871 U.S. patent to Sturges [6] has a spirally arranged flange on a shaft or screw for pumping soap (Figure 3.1). An 1877 U.S. patent to Higbie [7] describes the use of a screw for conveying and drying grain (Figure 3.2). The first patent on screw extrusion of polymers is an 1879 patent of M. Gray [8] on extrusion of gutta percha with specific application to wire coating (Figure 3.3). There may well have been many applications of single screw extrusion by industrial firms without patents or publications. Some authors [2] indicate the Hamburger Gummiwerke (now Phoenix Gummiwerke AG) used screw extruders for rubber compounds in 1873. In any case, it is clear that screw pumps became common to the process industries in the last 30 years of the 19th century.

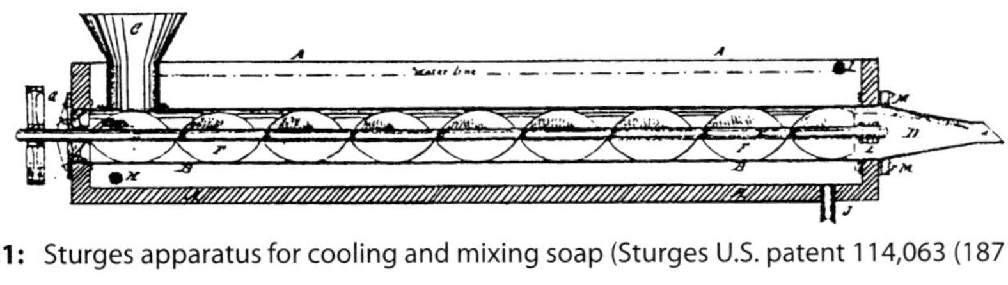

Figure 3.1: Sturges apparatus for cooling and mixing soap (Sturges U.S. patent 114,063 (1871))

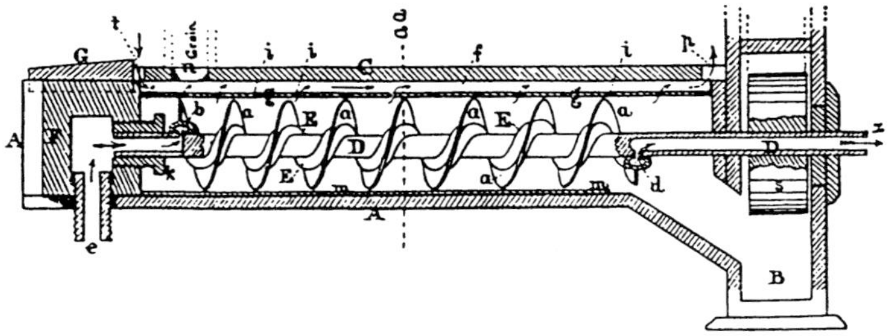

Figure 3.2: Higbie's grain conveyor and drier (Higbie U.S. patent 192,069 (1877))

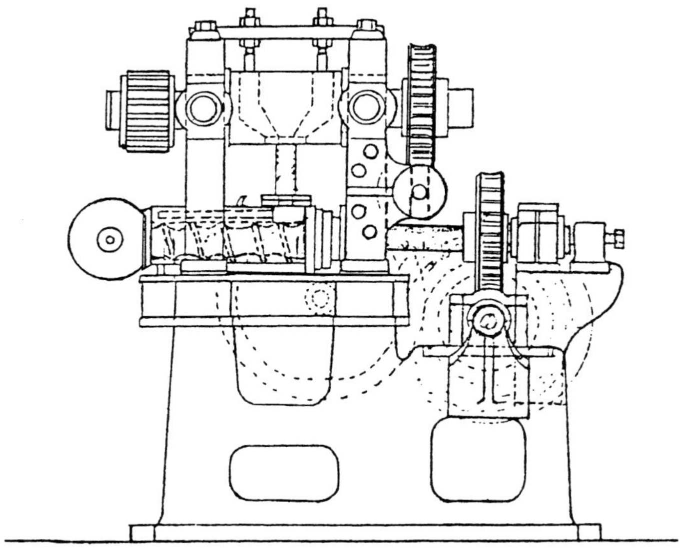

Figure 3.3: Gray's wire coating extruder (Gray British patent 5,056 (1879))

Variations in screw and screw extruder design have a long history. Desgoffe and DiGiorgio [9], within a few years of Higbie's and Gray's patent, described a single screw extruder with decreasing screw pitch in the machine direction. This extruder was supposed to provide both mixing and kneading (Figure 3.4) to the material. The rubber industry was an early user of screw extrusion equipment, and it was here that the first devolatilizing extruder was developed (Figure 3.5), achieved by placing an aperture between two screw sections [8, 9] the second of which contained a port for existing volatiles. There was considerable activity to develop the capability for continuous compounding of rubber. As early as 1919, single screw extruders were developed with multiple entry ports; in front of each these ports there would be a pressure reducing aperture [12] (Figure 3.6). The necessity of kneading and mixing rubber and early plastics led in the 1920s to segmented screw designs [13] (Figure 3.7) and in the 1930s to threaded and grooved barrel liners [14] (Figure 3.8). Pin barrel designs have a long history [15, 16] (Figure 3.9). The basic ideas were developed though they were to be often forgotten and subsequently re-invented.

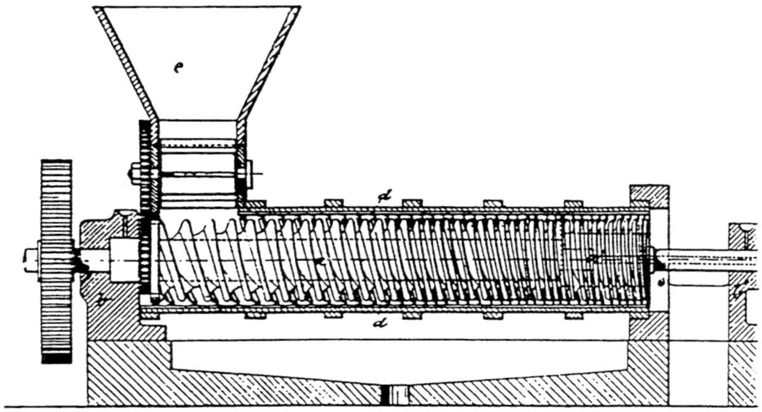

Figure 3.4: Early screw extruder with varying pitch
(Desgoffe and DiGiorgio German patent 26,177 (1884))

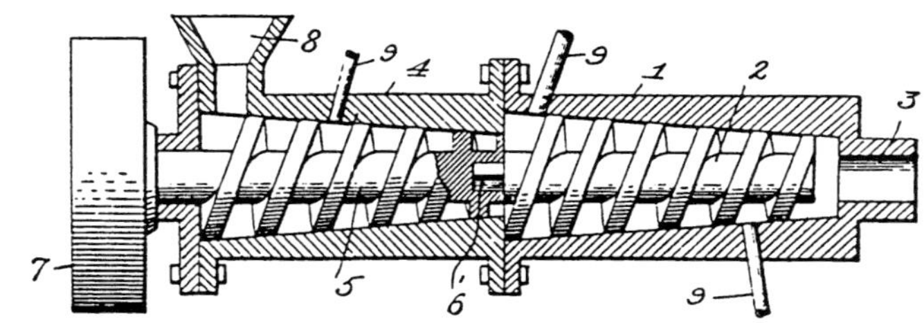

Figure 3.5: Early devolatilizing screw extruder (Price U.S. patent 1,156,096 (1915))

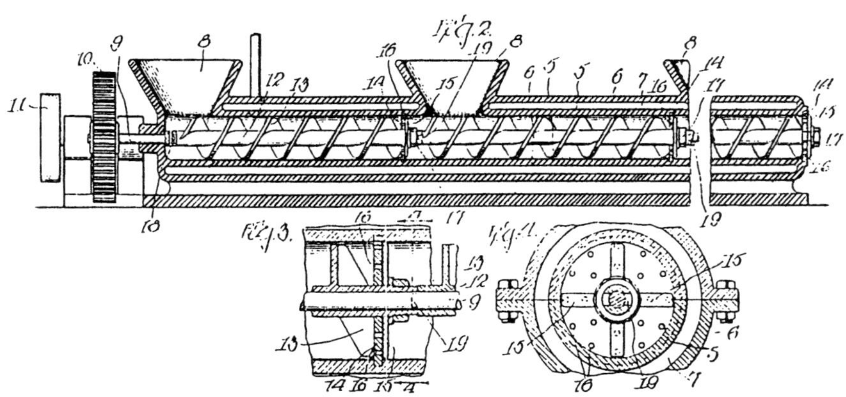

Figure 3.6: Multiple entry point screw extruders (Felix U.S. patent 1,320,128 (1919))

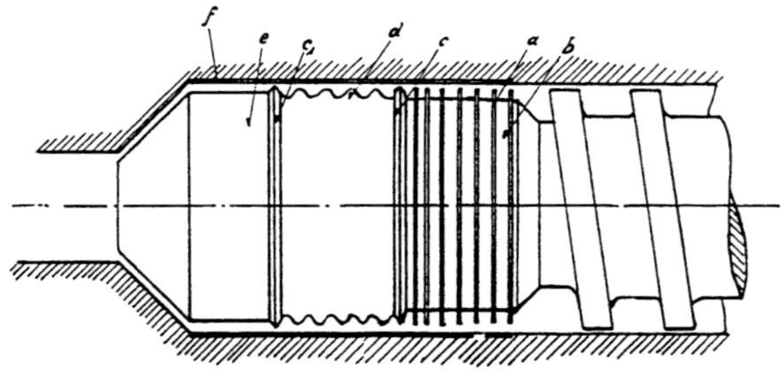

Figure 3.7: Segmented screw designs (Stich German Patent 476,748 (1929))

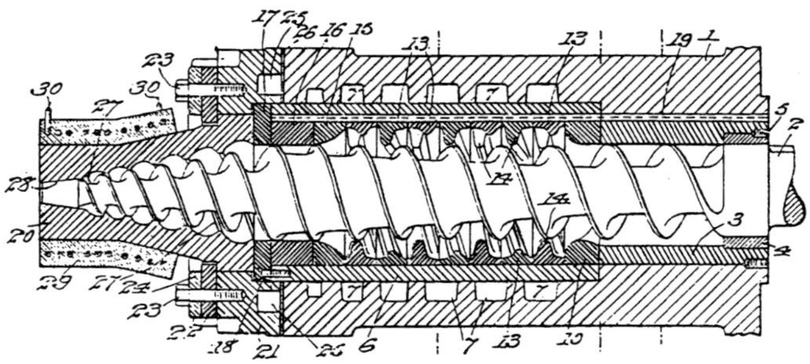

Figure 3.8: Screw extruder with thread and grooved barrel (Royle U.S. patent 1,904,884 (1933))

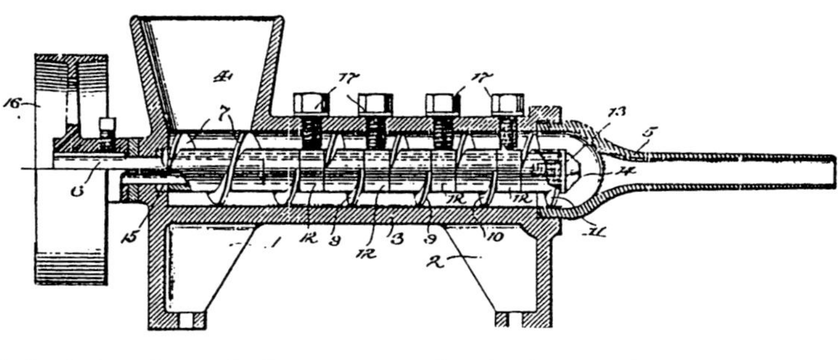

Figure 3.9: Early pin barrel extruder for sausage stuffing (Anderson U.S. patent 1,848,236 (1932))

The patents described earlier roughly cover the period between 1880 and 1940 during which many companies around the world became deeply involved in the manufacture of screw extruders. This involved firms such as Royle and Farrel-Birmingham in the United States, Francis Shaw in Manchester, England, and Paul Troester Maschinenfabrik in Hannover, Germany. The earliest screw extruders intended for rubber were steam-heated. In 1935, Paul Troester Maschinenfabrik manufactured a combined steam and electrically heated extruder [2, 17–19]. In 1939, Paul Troester Maschinenfabrik began to offer completely electrically heated plastics extruders with compressed air cooling [2, 17–19].

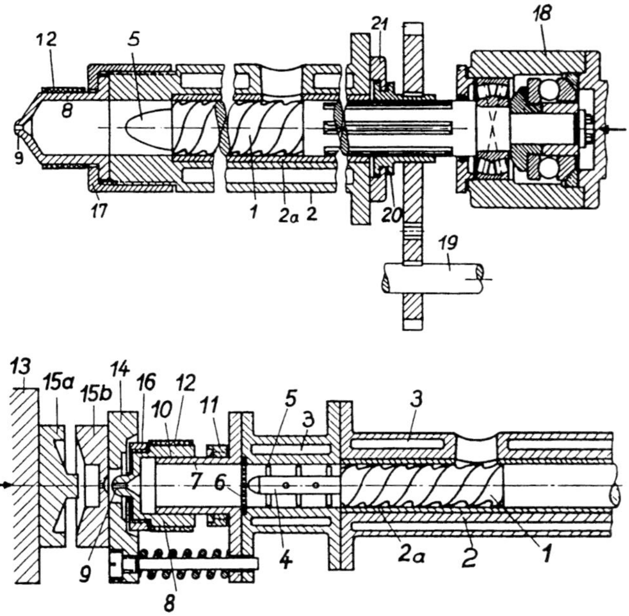

Figure 3.10: Beck's screw injection molding machine [20]

The 1940s saw the introduction of two remarkable machines based on reciprocating screws. The most important of these is the reciprocating screw injection technology, probably best attributed to H. Beck (Figure 3.10) of the I.G. Farbenindustrie [20]. Here, polymer is added to the feed section and subsequently pumped forward by the rotating screw. Melt is accumulated in front of the gate to a mold by the pumping action. Periodically the screw moves forward, injecting polymer melt into the mold. The screw injection step allows the transformation of the inherently batch molding process into a semi-continuous process. A second machine was the Ko-Kneader of H. List [21], where a screw with sliced flights reciprocated in a pin barrel extruder to produce mixing and self wiping characteristics (see Chapter 16). This machine was pioneered commercially by Buss AG in Pratteln, near Basel, Switzerland, and Baker-Perkins in North America and is discussed in Chapter 16.

The post World War II period has been a major expansion of these activities. Special attention has been given to the development of homogenization and mixing sections on screws [22–29] (Figure 3.11). Extruders became larger and larger. Screw machines with diameters of 250 to 600 mm began to be produced and used in industry. A 1977 machine produced by Hermann Berstorff Maschinenfabrik of Hannover had a drive power of 400 kW and outputs of 17–22 tons/hour.

In the 1960s, Kautex introduced grooves into the barrel of the feed zone in thermoplastic extruders [30]. The value of grooves in increasing the output of single screw extruders was clearly shown in a joint program of Kautex and BASF on extrusion blow molding of high density polyethylene.

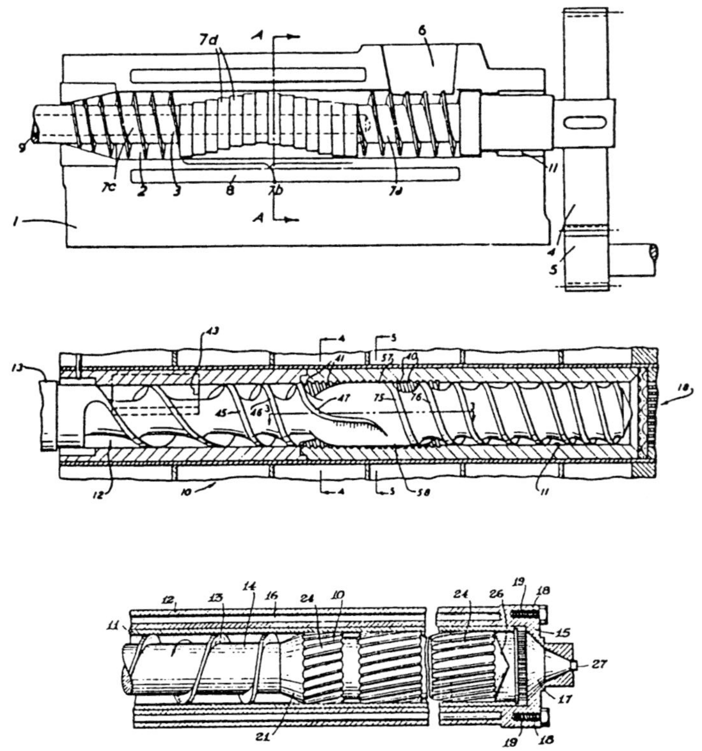

(a)

(b)

(c)

Figure 3.11: Mixing and homogenizing screw designs,
(a) Zona [21], (b) Henning [22], (c) Dulmage [24]

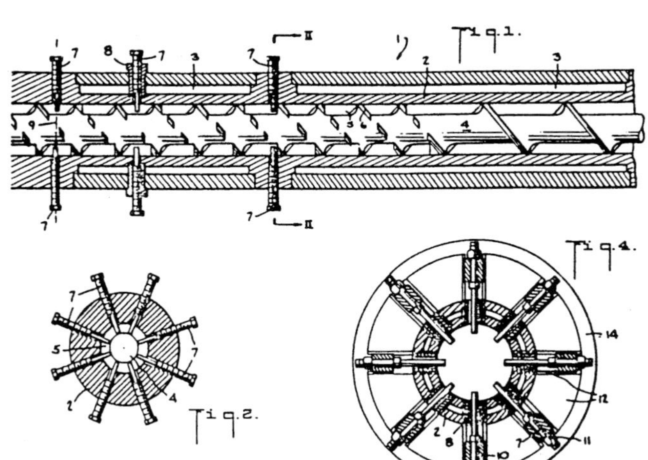

Figure 3.12: Pin barrel rubber extruder
(Menges, Harms, and Hegele, U.S. patents 4,178,104 and 4,199,263)

Pin barrel extruders (Figure 3.12) became of great importance to the rubber industry and provided high output rates of homogenized product [31]. Paul Troester Maschinenfabrik was the first licensee and took the lead in their manufacture.

Efforts at improving single screw extruders continue. This may be seen in the patent and engineering literature. Typically, recent developments include Grünschloss 'Helibar Screw Extruder' [32].

Tadmor [33] has devised a continuous processing machine based on discs placed on a rotating shaft inside of a stationary casing. This disc pack processor acts as a continuous pumping device but does not use a screw principle. This machine is being developed by the Farrel Corporation (successor to Farrel-Birmingham).

3.3 Mechanisms of Metering Flow

The basic analysis for simulation of flow in a single screw extruder dates back to a 1922 paper on flow of oils [34]. This was subsequently considered in papers by Pearsall [35], Rowell and Finlayson [36], Pigott [37], Meskat [38], Carley, Mallouk and McKelvey [39–41], Maillefer [42], and Mohr and Mallouk [43]. The theory was reviewed by Paton et al. [1] and Schenkel [2] in 1959, and subsequently in the 1962 book by McKelvey [44]. It is treated in great detail in the more recent volumes of Tadmor and Klein [3] and Rauwendaal [4].

The analysis is based on embedding a coordinate system in the root of an extruder screw which is flattened out (Figure 3.13). The barrel is then considered to move relative to the screw at an absolute velocity:

$$U = \pi D N \tag{3.1}$$

where D is the outer screw diameter and N is the rotation speed. The coordinate system is set up so that direction '1' is along the screw channel, '2' is perpendicular to the screw root, and '3' is perpendicular to the flights. The barrel velocity is then of form:

$$\begin{aligned} \underset{\sim}{U} &= U_1\, \underline{e}_1 + 0\, \underline{e}_2 + U_3\, \underline{e}_3 = U \cos\phi\, \underline{e}_1 - U \sin\phi\, \underline{e}_3 \\ &= \pi D N \cos\phi\, \underline{e}_1 - \pi D N \sin\phi\, \underline{e}_3 \end{aligned} \tag{3.2}$$

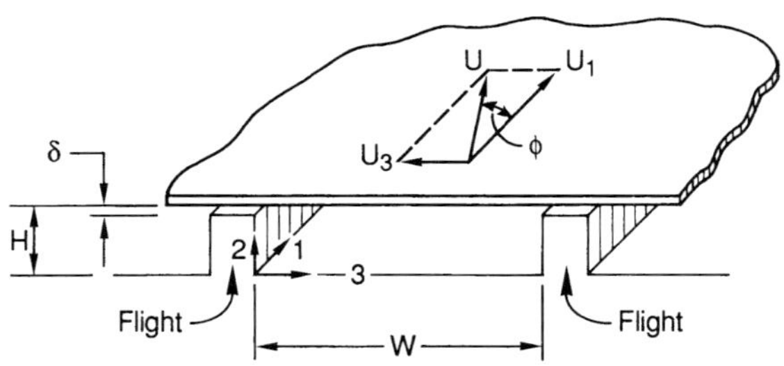

Figure 3.13: Coordinate and model system for analysis of flow in single screw extruder

The modeling is based on applying hydrodynamic lubrication theory. This leads to the equations of motion:

$$0=-\frac{\partial p}{\partial x_1}+\frac{\partial \sigma_{12}}{\partial x_2}+\frac{\partial \sigma_{13}}{\partial x_3} \tag{3.3a}$$

$$0=-\frac{\partial p}{\partial x_3}+\frac{\partial \sigma_{32}}{\partial x_2} \tag{3.3b}$$

If one further presumes the fluid is Newtonian, Eq. (3.3) becomes:

$$0=-\frac{\partial p}{\partial x_1}+\eta\left(\frac{\partial^2 v_1}{\partial x_2^2}+\frac{\partial^2 v_1}{\partial x_3^2}\right) \tag{3.4a}$$

$$0=-\frac{\partial p}{\partial x_3}+\eta\,\frac{\partial^2 v_3}{\partial x_2^2} \tag{3.4b}$$

Generally, shearing by the flight walls is considered a second order effect and Eq. (3.4a) simplifies to:

$$0=-\frac{\partial p}{\partial x_1}+\eta\,\frac{\partial^2 v_1}{\partial x_2^2} \tag{3.5}$$

The velocity field is:

$$v_1(x_2)=U_1\left(\frac{x_2}{H}\right)-\frac{H^2}{2\,\eta}\frac{\partial p}{\partial x_1}\left[\left(\frac{x_2}{H}\right)-\left(\frac{x_2}{H}\right)^2\right] \tag{3.6}$$

The velocity field is illustrated in Figure 3.14. The relative motion of the screw and barrel pushes the fluid forward. There is a backward pressure flow caused by the die.

The net flow rate along the screw channel is:

$$Q=W\int_0^H v_1\,\mathrm{d}x_2=\frac{W\,H\,U_1}{2}-\frac{W\,H^3}{12\,\eta}\frac{\mathrm{d}p}{\mathrm{d}x_1} \tag{3.7}$$

where U_1 is $\pi\,D\,N\cos\phi$.

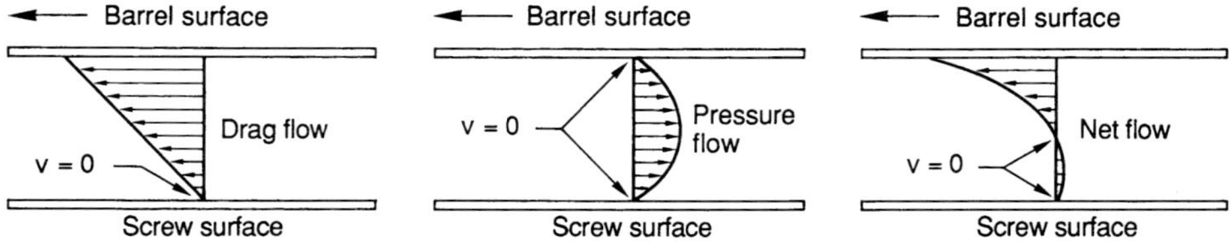

Figure 3.14: Velocity profiles along screw channel determined from model

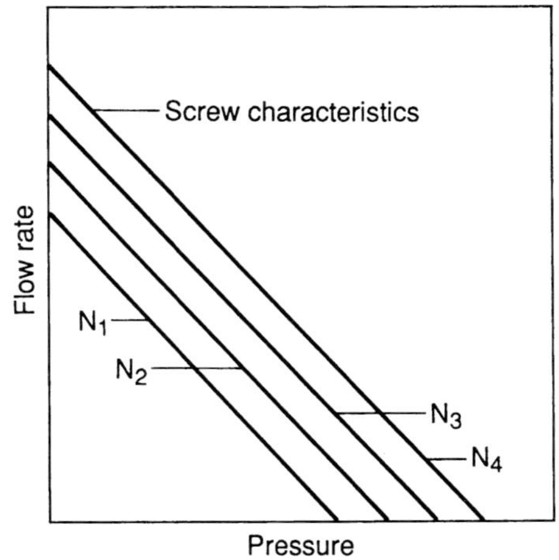

Figure 3.15: Screw characteristic curves

This simple result is due to Pearsall [35]. Equation (3.7) is the basis of *screw characteristic curve* representations which relate throughout Q to pressure rise Δp at specified screw speed U or N. This is shown in Figure 3.15.

Let us now turn to the transverse flow, the '3' component. This has a form analogous to Eq. (3.6):

$$v_3(x_2) = U_3\left(\frac{x_2}{H}\right) - \frac{1}{2\eta}\frac{\partial p}{\partial x_3}\left[\left(\frac{x_2}{H}\right) - \left(\frac{x_2}{H}\right)^2\right] \tag{3.8}$$

where $U_3 = -\pi D N \sin\phi$.

Velocity components $v_2(x_3)$ must also exist, which give rise to the velocity field of Figure 3.16.

The circulating flows implicit in the $v_1(x_2)$ and $v_3(x_2)$ components of Eqs. (3.6) and (3.7) were observed in an experimental study by Eccher and Valentinotti [45] published in 1958.

The formulation as described contains the essentials of the velocity fields. We have, however, presumed the fluid in the screw flight as Newtonian. More generally we must consider the variation of shear viscosity with shear rate and replace Eq. (3.4a) with:

$$0 = -\frac{\partial p}{\partial x_1} + \frac{\partial}{\partial x_2}\left(\eta\frac{\partial v_1}{\partial x_2}\right) + \frac{\partial}{\partial x_3}\left(\eta\frac{\partial v_1}{\partial x_3}\right) \tag{3.9a}$$

which may be simplified to:

$$0 = -\frac{\partial p}{\partial x_1} + \frac{\partial}{\partial x_2}\left(\eta\frac{\partial v_1}{\partial x_2}\right) \tag{3.9b}$$

if we may neglect shearing by the screw flights.

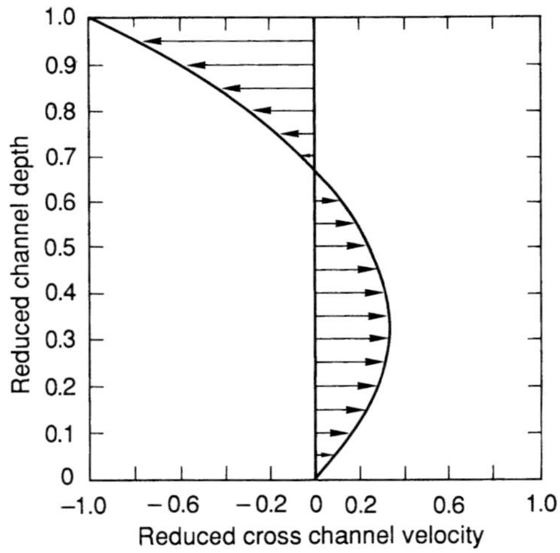

Figure 3.16: Transverse velocity field in extruder screw channel

The first analyses of non-Newtonian flow in screw channels were carried out by Mori, Otatake and Igarashi [46, 47] for a Bingham plastic in 1954/1955. Of more importance over the long run were analyses for power law fluids by Rotem and Shinnar [48], Griffith [49], and Kroesser and Middleman [50]. For a power law fluid:

$$\eta = K\left(\frac{\partial v_1}{\partial x_2}\right)^{n-1} \tag{3.10}$$

which gives rise in place of Eq. (3.9b) to:

$$0 = -\frac{\partial p}{\partial x_1} + \frac{\partial}{\partial x_2}\left[K\left(\frac{\partial v_1}{\partial x_2}\right)^n\right] \tag{3.11}$$

This has the solution:

$$v_1(x_2) = \frac{n}{n+1}\frac{1}{K^{1/n}}\frac{1}{(\partial p/\partial x_1)}\left[\left(x_2\frac{\partial p}{\partial x_1} + C_1\right)^{(n+1)/n} - C_1^{(n+1)/n}\right] \tag{3.12}$$

where C_1 is to be determined from:

$$U_1 = \frac{n}{n+1}\frac{1}{K^{1/n}}\frac{1}{(\partial p/\partial x_1)}\left[\left(H\frac{\partial p}{\partial x_1} + C_1\right)^{(n+1)/n} - C_1^{(n+1)/n}\right] \tag{3.13}$$

A numerical solution is required to determine C_1. This solution is dealt with in some detail in the monographs of Tadmor and Klein [3] and Rauwendaal [4].

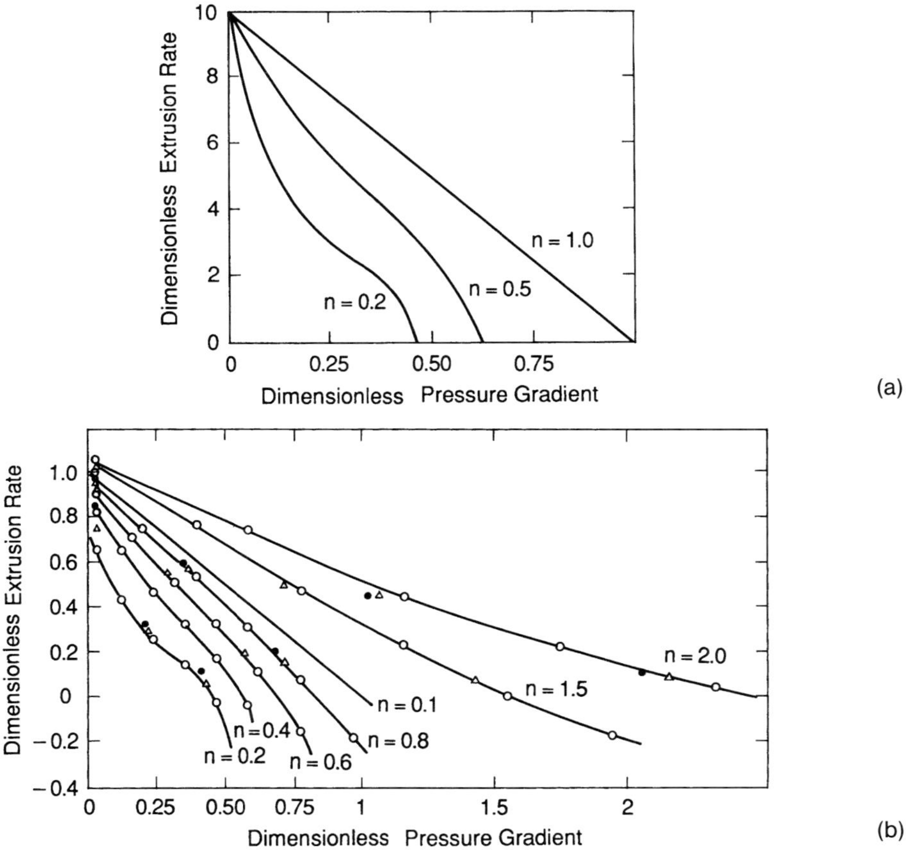

Figure 3.17: Screw characteristic curves for a power law fluid, (a) Kroesser and Middleman [50], (b) Griffith [49] based on Eq. (3.16)

Equation (3.13) may be integrated to give:

$$
\begin{aligned}
Q &= W \int_0^H v_1(x_2)\,\mathrm{d}x_2 \\
&= \frac{n}{n+1}\,\frac{W}{K^{1/n}\,(\partial p/\partial x_1)}\left\{\frac{[H(\partial p/\partial x_1)+C_1]^{(2n+1)/n}-C_1^{(2n+1)/n}}{[(2\,n+1)/n]\,(\partial p/\partial x_1)}-C_1^{(n+1)/n}\,H\right\}
\end{aligned}
\tag{3.14}
$$

where C_1 again comes from Eq. (3.13). We plot calculations based on Eq. (3.14). In Figure 3.17a, we see that the linear $Q-\Delta p$ behavior of Newtonian fluids is lost and the pumping ability of the extruder is reduced.

Griffith [49] considers transverse flow as well longitudinal flow along the screw channel. This leads to a replacement of Eqs. (3.10) and (3.11):

$$\eta = K\left[\left(\frac{\partial v_1}{\partial x_2}\right)^2 + \left(\frac{\partial v_3}{\partial x_2}\right)^2\right]^{(n-1)/2} \tag{3.15}$$

with:

$$0 = -\frac{\partial p}{\partial x_1} + \frac{\partial}{\partial x_2} K\left[\left(\frac{\partial v_1}{\partial x_2}\right)^2 + \left(\frac{\partial v_3}{\partial x_2}\right)^2\right]^{(n-1)/2} \frac{\partial v_1}{\partial x_2} \tag{3.16a}$$

$$0 = -\frac{\partial p}{\partial x_3} + \frac{\partial}{\partial x_2} K\left[\left(\frac{\partial v_1}{\partial x_2}\right)^2 + \left(\frac{\partial v_3}{\partial x_2}\right)^2\right]^{(n-1)/2} \frac{\partial v_3}{\partial x_2} \tag{3.16b}$$

Griffith presents a numerical solution of this case which we may summarize through screw characteristic curves in Figure 3.17b.

Clearly, increasing non-Newtonian character through decreasing 'n' decreases pumping capacity (i.e., Q at fixed $\partial p/\partial x_1$).

Thus far, we have not discussed viscous heating in screw extruders. This has been considered by Griffith [49], Zamodits and Pearson [51] and Martin [52]. One must solve:

$$\rho\, c\left(v_1 \frac{\partial T}{\partial x_1} + v_2 \frac{\partial T}{\partial x_2} + v_3 \frac{\partial T}{\partial x_3}\right) = k\frac{\partial^2 T}{\partial x_2^2} + \eta\left[\left(\frac{\partial v_1}{\partial x_2}\right)^2 + \left(\frac{\partial v_3}{\partial x_2}\right)^2 + \left(\frac{\partial v_1}{\partial x_3}\right)^2\right] \tag{3.17}$$

together with the force balances. In addition the temperature of viscosity is considered, as well as non-Newtonian viscosity. Griffith [49] and Zamodits and Pearson [51] simplify Eq. (3.17) to:

$$0 = k\frac{\partial^2 T}{\partial x_2^2} + \eta\left[\left(\frac{\partial v_1}{\partial x_2}\right)^2 + \left(\frac{\partial v_3}{\partial x_2}\right)^2\right] \tag{3.18}$$

and use Eq. (3.15) with:

$$K = A\, e^{-a(T-T_0)} \tag{3.19}$$

Martin [52] argues the importance of the convective term and sought to solve:

$$\rho\, c\, v_1 \frac{\partial T}{\partial x_1} = k\frac{\partial^2 T}{\partial x_2^2} + \eta\left[\left(\frac{\partial v_1}{\partial x_2}\right)^2 + \left(\frac{\partial v_3}{\partial x_2}\right)^2\right] \tag{3.20}$$

The simulations described above are summarized in more detail in the monographs of Tadmor and Klein [3] and Rauwendaal [4].

3.4 Metering Flow: Screw Flight and Clearance Effects

The analyses of the previous section neglect the influence of the screw flights. Flowing melts adhere to screw flights just as they do to the screw surface. A second effect is that the narrow gap between the flight tips and barrel define a clearance which connects two regions in different pressures. This pressure difference gives rise to a backward leakage flow. It is our purpose in this section to describe these two flow effects and present analyses and interpretations of them as they represent corrections.

The flow of a Newtonian fluid in a 2-dimensional channel is defined by Eq. (3.4a) with boundary conditions:

$$\begin{aligned} v_1(0, x_3) &= 0 \\ v_1(x_2, 0) &= 0 \\ v_1(x_2, W) &= 0 \\ v_1(H, x_3) &= U \end{aligned} \tag{3.21}$$

Solutions of Eq. (3.3a) for screw extrusion using the above boundary conditions have been discussed since the 1920s. An early critical discussion of these solutions was given by Meskat [38].

These solutions have a typical form:

$$v_1(x_2, x_3) = \left[\frac{4}{\pi} \sum_{n=1,3,5} \frac{\sinh(n\pi x_2/W)}{n \sinh(n\pi H/W)} \sin\frac{n\pi x_3}{W}\right] U - \frac{H^2}{2\eta}\frac{\partial p}{\partial x_1} \cdot \left[\frac{x_2}{H} - \left(\frac{x_2}{H}\right)^2 + \frac{8}{\pi^3} \sum_{n=1,3,5} \frac{\cosh\{n\pi(W/H)[(x_3/W)-(1/2)]\}}{\cosh[n\pi(W/2H)]} \sin\left(\frac{n\pi x_2}{H}\right)\right] \tag{3.22}$$

Equation (3.22) represents a generalization of Eq. (3.6) allowing the velocity field $v_1(x_2, x_3)$ to go to zero at the flight walls located at $x_3 = 0$ and W.

The flow rate Q of Eq. (3.7) through the screw channel may be expressed now as:

$$Q = \frac{H\,W\,U}{2} F_d - \frac{H^3\,W}{12\,\eta} \frac{dp}{dx_1} F_p \tag{3.23}$$

where F_d and F_p are shape factors which express the effects of the flight walls. The specific forms of F_d and F_p are:

$$F_d = \frac{16\,W}{\pi^3\,H} \sum_{n=1,3,5}^{\infty} \frac{1}{n^3} \tanh\frac{n\pi H}{2\,W} \tag{3.24a}$$

$$F_p = 1 - \frac{192}{\pi^5\,W} \sum_{n=1,3,5}^{\infty} \frac{1}{n^5} \tanh\frac{n\pi H}{2\,W} \tag{3.24b}$$

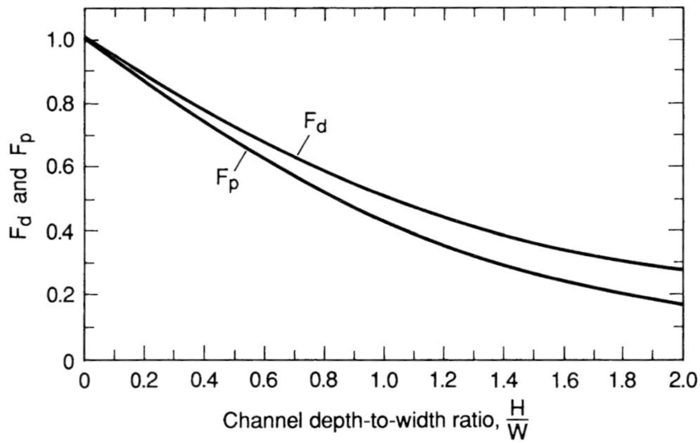

Figure 3.18: F_d and F_p shape factors of Eqs. (3.23) and (3.24)

F_d and F_p are plotted in Figure 3.18 as a function of H/W. If H/W is less than 0.6 then to a good approximation

$$F_d = 1 - 0.571 \frac{H}{W} \tag{3.25a}$$

$$F_p = 1 - 0.625 \frac{H}{W} \tag{3.25b}$$

For many extruder screws H/W ranges between 0.03 and 0.10. This makes F_d and F_p in the range 0.94 to 0.98.

Similar developments may be carried out with non-Newtonian models such as power law fluids. These are based on solutions of Eq. (3.6a) where we have written the shear viscosity:

$$\eta = K\left[\left(\frac{\partial v_1}{\partial x_2}\right)^2 + \left(\frac{\partial v_1}{\partial x_3}\right)^2\right]^{(n-1)/2} \tag{3.26}$$

This gives:

$$0 = -\frac{\partial p}{\partial x_1} + K \cdot \left\{\frac{\partial}{\partial x_2}\left[\left(\frac{\partial v_1}{\partial x_2}\right)^2 + \left(\frac{\partial v_1}{\partial x_3}\right)^2\right]^{(n-1)/2} \frac{\partial v_1}{\partial x_2} + \frac{\partial}{\partial x_3}\left[\left(\frac{\partial v_1}{\partial x_2}\right)^2 + \left(\frac{\partial v_1}{\partial x_3}\right)^2\right]^{(n-1)/2} \frac{\partial v_1}{\partial x_3}\right\} \tag{3.27}$$

If transverse flow were considered as well we would have to introduce $\partial v_3 / \partial x_2$ into Eqs. (3.26) and (3.27). Equation (3.16b) would also need to be introduced.

The occurrence of leakage flows over screw flights has been considered almost since the beginning of research on the fluid mechanics of screw extrusion [35, 39]. It was realized by Carley et al. [39] that the pressure differential from flight to flight will produce a backwards flow to lower pressure regions. The leakage flow for a Newtonian fluid will have the form:

$$Q_L = \frac{\pi D_b \delta^3 \Delta p_f}{12 \eta b} E \tag{3.28}$$

where δ is the flight clearance, b the flight thickness, D_b the barrel diameter, E an eccentricity factor, which is unity if the screw is centered, and Δp_f is the pressure drop over a flight. Q_L would be subtracted from the throughput through the screw channel, i.e., from Eq. (3.23). Later authors noted that the drag flow is also reduced by the presence of the leakage. This suggests a form (neglecting F_d and F_p):

$$Q = \frac{H W U}{2}\left(1 - \frac{\delta}{H}\right) - \frac{H^3 W}{12 \eta} \frac{\Delta p}{L}(1 + f) \tag{3.29a}$$

with:

$$f_L = \frac{\pi D_b \delta^3 L}{H^3 W b} \tag{3.29b}$$

Mohr and Mallock [43] have developed a more general form of leakage flow in extruder screws than that presented here.

3.5 Melting in a Screw Extruder

Often the key region of a plasticating screw extruder is where the pellets are melted. This has long been realized by extrusion technologists and led to basic studies of the mechanisms of the melting process. The key studies were made by Maddock [53], who fed mixtures of natural and pigmented plastic pellets to the hopper of a screw extruder. When steady state was achieved, a sudden freeze of the polymer within the extruder was achieved by suddenly cooling the barrel. The extruder screw was then pushed out of the barrel and the solidified plastic unwound and sectioned. The sections allow interpretation of the state of the polymer prior to the freezing. In regions where the plastic was not brought into the state of a melt, the pellets of pigmented and unpigmented plastic retain their identity. Where the melt state was achieved, the polymer takes on the homogeneous coloring of the pigment. The mechanism of melting is surmised by following the state of those two regions along the length of the extruder screw. Subsequent experiments of this type have been reported by Street [54], Tadmor, Duvdevani and Klein [55], Menges and Klenk [56], Martin [57], and Decker [58].

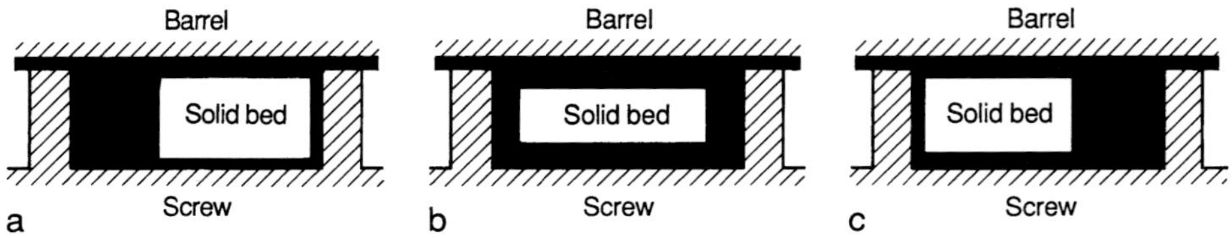

Figure 3.19: (a) Melting observations of Maddock [53], (b) Melting observations of Dekker [58], (c) Melting observations of Menges and Klenk [56]

Maddock [53] and Tadmor et al. [55] clearly saw a melt layer along the barrel which was scraped off by the transverse flow in the screw channel and accumulated at the active flight. Menges and Klenk [56] found the melt pool formed on the passive flight. Dekker's [58] work clearly showed melt forming completely around the solid bed, around the screw flights and base, as well as the barrel (see Figure 3.19 a, b, c).

The first modeling study of melting of a solid bed in a screw channel was presented by Tadmor [59] in 1966. He considered Maddock's mechanism melting along the barrel and accumulation of melt by the leading flight. A decade later, a new model was presented by Lindt [60] based on the experiments of Dekker [58]. A similar model was presented by Shapiro, Halmos and Pearson [61] at about the same time.

In the original model of Tadmor [59], the solid bed is considered to melt as in Figure 3.19a. Melt is generated in a thin film of thickness δ between the screw and barrel and accumulates in a melt pool beside the pushing flight. The heat flux from the film into the solid bed is taken as the sum of heat conduction from the barrel and viscous dissipation heating in the film:

$$q = \frac{k_m}{\delta}(T_b - T_m) + \frac{\eta(\pi D N - U_s)^2}{2\delta} \tag{3.30}$$

where U_s is the solid bed velocity in the 1–3 plane it use the geometry defined in Section 3.4 and 3.5. This heat flux goes into heating and melting the solid bed:

$$q = \rho_s [c_s (T_m - T_s) + \lambda] v_{s2} \tag{3.31}$$

Here, c_s is the solid heat capacity, ρ_s the solid density, λ the heat of fusion, and v_{s2} the solid bed interface velocity. We may obtain v_{s2}, the rate of solid into the melt phase, by combining Eqs. (3.30) and (3.31). The rate of melting is:

$$\omega = \rho_s v_{s2} W_s = \rho_m \frac{U_3}{2}\delta \tag{3.32}$$

where W_s is the width of the solid bed. This leads to the expression for the rate of disappearance of the solid bed of:

$$\rho_s v_{s1} \frac{d}{dx_1}(W_s H) = -\omega = -\rho_s v_{s2} X \tag{3.33}$$

The solution of Eq. (3.33) leads to, $W_s(x_1)$, which specified the melting process. The above model was upgraded by Tadmor and his co-workers to consider non-Newtonian shear viscosity and temperature dependence of the viscosity [3].

In the model of Lindt [60] it is considered that a melt film develops around and encapsulates the solid bed. Lindt formulates force balances of from:

$$0 = -\frac{\partial p}{\partial x_3} + \frac{\partial}{\partial x_2}\eta\frac{\partial v_3}{\partial x_2} \tag{3.34a}$$

coupled with the energy balance:

$$0 = k\frac{\partial^2 T}{\partial x_2^2} + \eta\left[\left(\frac{\partial v_1}{\partial x_2}\right)^2 + \left(\frac{\partial v_3}{\partial x_2}\right)^2\right] \tag{3.34b}$$

in both the upper and lower film layers of Figure 3.19b. For an isothermal Newtonian fluid, Eq. (3.34a) may be solved to give the form:

Barrel film layer:

$$q_b = \frac{1}{2}\delta_b U_3 - \frac{\delta_b^3}{12\eta}\frac{\partial p_b}{\partial x_3} \tag{3.35a}$$

Screw film layer:

$$q_s = -\frac{\delta_s^3}{12\eta}\frac{\partial p_s}{\partial x_3} \tag{3.35b}$$

where:

$$q_b = q_s + q_l \tag{3.35c}$$

Here, δ is the film thickness and subscripts 'b' and 's' represent the barrel and screw surfaces. Equation (3.35c) indicates that the flux at the barrel q_b divides a leakage flow q_l over the screw flights and a flux q_s which circulates around the bed.

Lindt [60] argues that the position of the solid bed is determined by the relative magnitudes of the pressure profiles and gradients. The cross-channel force on the bed in the 3-direction is:

$$F_3 = W_B\left(\eta\frac{U_3}{\delta_b} - \frac{\delta_b}{2}\frac{\partial p_b}{\partial x_3} - \frac{\delta_s}{2}\frac{\partial p_s}{\partial x_3} - H_{bed}\frac{\partial p_b}{\partial x_3}\right) \tag{3.36}$$

where W_B is the width of the bed.

The pressure gradient in the '1' direction along the channel is related to the transverse pressure gradient. Roughly:

$$W_B\frac{\partial p_b}{\partial x_3} \cong \pi D\frac{\partial p}{\partial x_1} \tag{3.37}$$

Figure 3.20: Maillefer screw

The magnitude of F_3 determines the position of the solid bed. Lindt reasons that the varying observations of Maddock [53], Menges and Klenk [56], and Dekker [56] are associated with the relative value of F_3 as determined by variables such as screw rotation speed and pressure gradient.

It would seem reasonable to presume that Tadmor's [59] model would be valid in the initial stages of melting with Lindt's [60] model being operative in later stages. In these later stages, the Tadmor model would be an asymptote of the Lindt model when the solid bed is pressed against the passive flight.

More recently, Viriyathorkorn and Kassahon [62] have approached this problem by numerically simulating the flow, considering all the material through the cross-section. Different phases are not distinguished. The rheological properties are considered to vary with temperature and the latent heat of fusion is lumped into the temperature dependence of the heat capacity. The predictions generally agree with the assessment of the above paragraphs.

One of the striking developments in response to the various fundamental studies of melting was Maillefer's [27] invention of the barrier screw (Figure 3.20). Here, a new flight begins to develop in the melting region, which seeks to separate the melt from the solid beds and to control it along the length of the screw. This led to a new generation of screw designs for single screw extruders, as has been discussed by Rauwendaal [4, 63].

3.6 Solid Conveying in a Plastics Screw Extruder

In a plasticating screw extruder, pellets are added to a hopper, and then are pumped forward by the relative motion of the screw and barrel. Subsequently, the pellets are melted by heaters in the barrel and by friction against the barrel. It is our purpose in this section to discuss this solid conveying zone of a single screw extruder.

Experimental studies of the solid conveying region of screw extruders have been described by Darnell and Mol [64]. Pellets appear to move through the barrel as a solid plug. Significant pressure fluctuations are observed.

Studies of flow in the solid conveying region of a screw extruder begin with the work of Maillefer [42] and later Darnell and Mol [64]. They consider the bed of pellets behaving like a continuum, as it is dragged forward by the relative motion of the barrel and screw. This is resisted by frictional forces exerted on the root of the screw and the flanks of the flights.

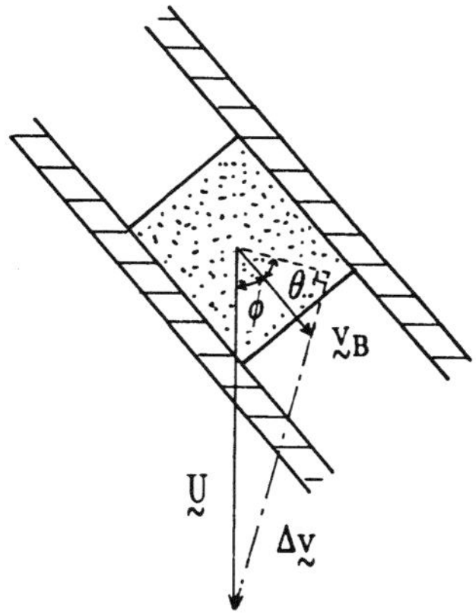

Figure 3.21: Movement of solid bed in screw channel

Pressure develops in the bed along the direction of the screw channel. The solid bed moves at a velocity $\underline{v}_B$, which is less than the velocity component of the barrel along the screw channel. We may write (see Figure 3.21):

$$\underline{v}_B = \underline{U} - \Delta \underline{v} \tag{3.38}$$

where the quantity $\Delta \underline{v}$ determines the direction of the drag force $\underline{F}_d$ on the barrel on the bed.

The development of the pressure field along the length of the screw is associated with the difference between the friction of the barrel and the friction of the screw.

Maillefer expresses this simply as:

$$v_B = U \frac{\cos(\phi + \theta)}{\cos\phi} \tag{3.39}$$

$$\frac{dp}{dx_1} = (\mu_b \sin\theta - k\,\mu_s) \frac{p}{b \sin\theta} \tag{3.40}$$

where θ is the angle between the velocity vector of the slipping solid polymer plug $\underline{v}_B$, and the perpendicular to $\underline{U} - \underline{v}_B$. μ_b is the coefficient of friction of the barrel and μ_s that of the screw. The pressure buildup depends primarily on μ_b. Equations (3.38) and (3.39) were generalized by Darnell and Mol [64] to include screw and barrel curvature.

The above paragraph strongly suggests that to build up large pressures in a screw extruder, one should have large barrel coefficients of friction μ_b. This leads to the idea of a grooved barrel extruder [31]. Grooved barrel extruders develop very high pressures and as a result develop very high outputs. Pressure flow in the extruder is in the same direction as the drag flow by the barrel. Critical discussions of grooved barrel extruders are given by Boes et al. [31], Menges and Hegele [65], Fritz [66], Potente [67], and Koch [68].

3.7 Residence Time Distributions in Single Screw Extruders

The length of mean time of residence in a screw extruder as well as the range or distribution of times of residence of individual material particles is of great importance. Residence time distributions in screw extruders seem to have first been considered by Pinto and Tadmor [3, 69] and later by Lidor and Tadmor [70]. They developed out of early studies of laminar shear mixing by Mohr, Saxton, and Jepson [71], and by McKelvey [44].

One proceeds by tracking the helical motion of fluid particles as they translate along the screw channel. For a Newtonian fluid, these are defined by Eqs. (3.6) and (3.8) near the barrel; the velocity component normal to the screw flights $v_3(x_2)$ is in the negative 3 direction as the barrel velocity U_3 is $(-\pi\, D\, N \sin\phi)$ in the neighborhood of the screw $v_3(x_2)$ becomes positive. The division point between positive and negative velocities may be determined from:

$$\int_0^H v_3(x_2)\,\mathrm{d}x_2 = 0 \tag{3.41}$$

The solution to this is essentially given by Eq. (3.8b). When x_2 is greater than 2 $H/3$, $v_3(x_2)$ is negative and when x_2 is less than 2 $H/3$, $v_3(x_2)$ is positive. To trace individual particles, we use

$$\int_{x_2}^H v_3'(x_2)\,\mathrm{d}x_2 + \int_0^{x_2} v_3(x_2)\,\mathrm{d}x_2 = 0 \tag{3.42}$$

This leads to:

$$\left(\frac{x_2}{H}\right)^2 - \left(\frac{x_2}{H}\right)^3 = \left(\frac{x_2'}{H}\right)^2 - \left(\frac{x_2'}{H}\right)^3 \tag{3.43}$$

A particle with position x_2 will move along the screw channel with velocity components $v_1(x_2)$ and $v_3(x_2)$ for a period of time, roughly $W(x_2)/v_3(x_2)$, after which it will move to position x_2' as governed by Eq. (3.43). It will then move with velocity components $v_1(x_2')$ and $v_3(x_2')$ along the screw channel for a subsequent period of $W(x_2')/v_3(x_2')$.

The residence time, t_{res}, of a fluid particle is the axial length of the extruder channel L divided by the average axial fluid velocity i.e.:

$$t_{\text{res}} = \frac{L}{\overline{v}_1} \tag{3.44}$$

Here, $\overline{v}_1$ is the axial velocity of a particle which alternates its position between x_L and x_2'. This has the form:

$$\overline{v}_1 = v_1(x_2)\,y + v_1(x_2')\,(1-y) \tag{3.45}$$

where y represents the fraction of time that the particle spends in each of these positions clearly:

$$y = \frac{|v_3(x_2')|}{|v_3(x_2)| + |v_3(x_2')|} \tag{3.46a}$$

$$= \frac{1}{1 + \left|\dfrac{v_3(x_2)}{v_3(x_2')}\right|} \tag{3.46b}$$

where it is assumed that W is the same at x_2 and x_2'.

We may rewrite t_{res} now as:

$$t_{res} = \frac{L}{U_1}\left\{\frac{1}{y\left[\dfrac{x_2}{H} - \dfrac{H^2}{2\eta U_1}\dfrac{\partial p}{\partial x_1}\left[\dfrac{x_2}{H} - \left(\dfrac{x_2}{H}\right)^2\right]\right] + (1-y)\left[\dfrac{x_2'}{H} - \dfrac{H^2}{2\eta U_1}\dfrac{\partial p}{\partial x_1}\left[\dfrac{x_2'}{H} - \left(\dfrac{x_2'}{H}\right)^2\right]\right]}\right\} \tag{3.47}$$

Equation (3.47) may be rearranged to:

$$t_{res} = \frac{L}{3\pi D N \cos\phi(1+\Phi)}\left(\frac{3a - 1 + 3\sqrt{1 + 2a + 3a^2}}{a(1-a) + \sqrt{1 + 2a + 3a^2}}\right) \tag{3.48a}$$

where a is x_2/H, and Φ is:

$$\Phi = \frac{H^2}{\delta \eta U_1}\frac{\partial p}{\partial x_1} \tag{3.48b}$$

Equation (3.48) indicates that the minimum residence time occurs at the position of x_2 equal to $2H/3$ as we should expect.

The residence time distribution function $f(t_{res})\,dt_{res}$ is related to the throughput dQ associated with residence time between t_{res} and $t_{res} + dt_{res}$. It will thus be associated with two positions in the channel cross-section, x_2 and x_2'. We have then:

$$f(t_{res})\,dt_{res} = \frac{dQ(x_2) + dQ'(x_2')}{Q} \tag{3.49}$$

now:

$$\begin{aligned} dQ &= v_1 W\,dx_2 = v_1 H W\,da \\ &= W H U_1\,a(1 + 3\Phi - 3a\Phi)\,da \end{aligned} \tag{3.50}$$

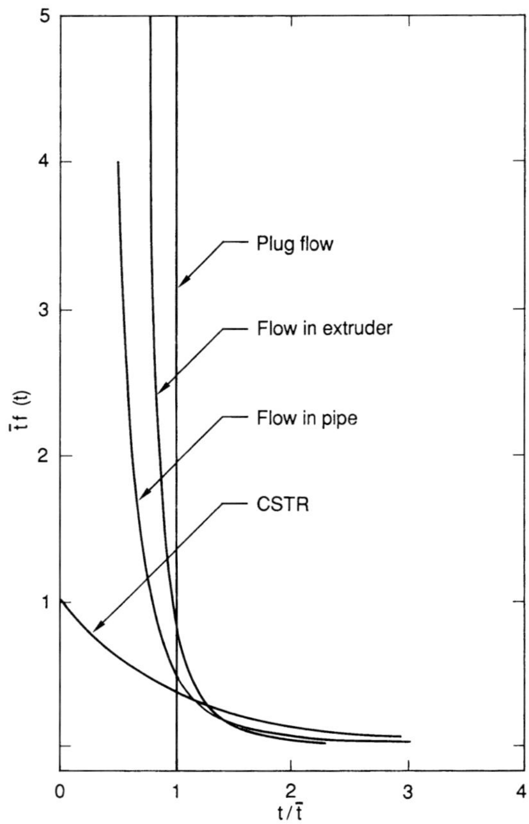

Figure 3.22: $\bar{t}_{res}\, f(t_{res})$ as a function of $t_{res}/\bar{t}_{res}$ for a Newtonian fluid in a single screw extruder compared to plug flow, Poiseuille flow and a CSTR from Pinto and Tadmor [69] (reprinted with permission of the Society of Plastics Engineers)

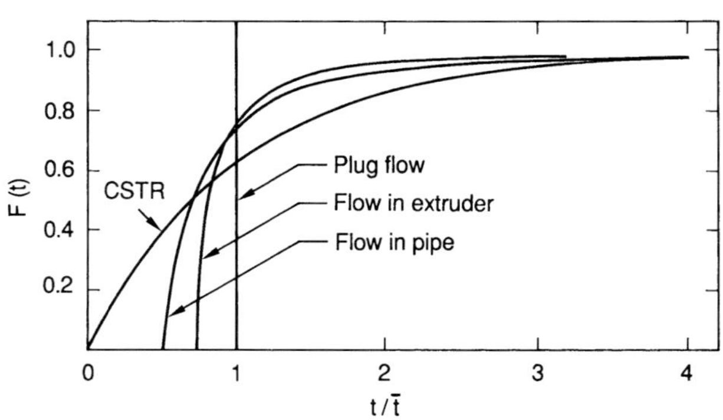

Figure 3.23: $F(t_{res})$ as a function of $t_{res}/\bar{t}_{res}$ for a Newtonian fluid in a single screw extruder compared to plug flow, Poiseuille flow and a CSTR from Pinto and Tadmor [69] (reprinted with permission of the Society of Plastics Engineers)

This leads to:

$$f(t)\,\mathrm{d}t = \frac{3\,a}{\sqrt{1+2\,a-3\,a^2}}\left(1-a+\sqrt{1+2\,a-3\,a^2}\right)\mathrm{d}a \tag{3.51}$$

We may determine from this a mean residence time $\bar{t}_{\mathrm{res}}$. This has the form:

$$\bar{t}_{\mathrm{res}} = \frac{\int_0^\infty t_{\mathrm{res}}\, f(t_{\mathrm{res}})\,\mathrm{d}t_{\mathrm{res}}}{\int_0^\infty f(t_{\mathrm{res}})\,\mathrm{d}t_{\mathrm{res}}} = \int_0^\infty t_{\mathrm{res}}\, f(t_{\mathrm{res}})\,\mathrm{d}t_{\mathrm{res}} \tag{3.52}$$

We may integrate Eq. (3.52) with Eq. (3.51), which yields:

$$\bar{t}_{\mathrm{res}} = \frac{2\,L}{\pi\,D\,N\,\cos\theta}\left(1+\Phi\right) \tag{3.53}$$

In Figure 3.22, we plot $\bar{t}_{\mathrm{res}}\, f(t_{\mathrm{res}})$ as a function of $t_{\mathrm{res}}/\bar{t}_{\mathrm{res}}$ and in Figure 3.23 the cumulative residence time distribution $F(t_{\mathrm{res}})$ as a function of $t_{\mathrm{res}}/\bar{t}_{\mathrm{res}}$:

$$F(t_{\mathrm{res}}) = \int_0^{t_{\mathrm{res}}} f(t_{\mathrm{res}})\,\mathrm{d}t_{\mathrm{res}} \tag{3.54}$$

It can be seen that the distribution of residence times for an idealized extruder screw is much broader than for plug flow but not as broad as for Poiseuille flow in a pipe or an idealized continuously stirred tank reactor (CSTR).

3.8 Scale-Up

The problem of scale-up in single screw extruders has long attracted serious attention [44, 72–80]. The earliest study by Carley and McKelvey [72] was based on an isothermal Newtonian fluid model which had been exploited by DuPont researchers [39–41, 43, 44] in this period to represent flow in single screw extruders. They begin with Eq. (3.6), which we write:

$$Q = \frac{1}{2}\,\pi\,D\,W\,H\,N\,\cos\phi - \frac{W\,H^3}{12\,\eta}\,\frac{\Delta p}{L} \tag{3.55}$$

They scale up the dimensions uniformly so that:

$$D_{\mathrm{L}} = \xi\,D \qquad H_{\mathrm{L}} = \xi\,H \qquad W_{\mathrm{L}} = \xi\,W \qquad \phi_{\mathrm{L}} = \phi \tag{3.56}$$

They then have:

$$Q_L = \xi^3 \left(\frac{1}{2} \pi D W H \cos\phi \right) N - \xi^4 \frac{W H^3}{12 \eta} \frac{dp}{dx_1} \tag{3.57}$$

If one scales up all dimensions uniformly, i.e., D, H, W, and, L and maintains screw rotation speed N, then the open discharge Q will increase with ξ^3.

The problem with this approach is that it neglects heat build-up due to viscous dissipation. This can be roughly described as:

$$\frac{\text{Heat generation}}{\text{Heat conduction losses}} = \frac{\eta \left(\frac{U}{H} \right)^2}{k \left(\frac{\Delta T}{H^2} \right)} = \frac{\eta \left(\frac{\pi D N}{H} \right)^2}{k \left(\frac{\Delta T}{H^2} \right)} = \frac{\eta \pi^2 D^2 N^2}{k \Delta T} \tag{3.58}$$

This leads to the large extruder generating more heat relative to conductive heat losses at a rate of increase of ξ_2, i.e., the square of the scale factor. The polymer melt may literally burn up in a large extruder.

A thoughtful view of extruder scale-up was subsequently given by Maddock [73], who expressed concern over residence time. Maddock considered various cases. First, D is increased by a factor ξ, while channel depth H and screw rotation speed N are constant. The open discharge rate Q_L increases as ξ^2. Maddock concluded that this scale-up would cause excessive heating as can be seen by Eq. (3.58). Maddock then considered the case where screw diameter D and W increase with ξ, while H is constant and N decreases inversely with ξ. Open discharge Q would now increase with ξ. Equation (3.58) indicates the rate of heat generation balances heat conduction in the same manner as the small extruder. There would be larger residence times. This would improve distributive mixing but lead to larger heat build-up. Maddock prefers the case where D and W are increased by ξ, H by $\sqrt{\xi}$, and N is decreased as $1/\sqrt{\xi}$. Here, Q at open discharge increases with ξ^2. From Eq. (3.58), heat generation relative to heat conduction does not change. The residence time increases, but not as much as in the previous case, and the extent of heating is not as bad.

Maddock [74] returned to the subject some years later and studied scale up using numerical simulation of the melting and metering region, probably based on the melting model of Tadmor [59] (Section 3.5) and the models of Section 3.3. This led Maddock to propose to increase H by the 0.7 or 0.75 power rather than the 0.5 power of ξ or D in his scale-up.

A similar modeling approach to scale-up was subsequently taken by Pearson [76]. However, Pearson uses as a criteria maintaining Eq. (3.58) as a constant and also the ratio of convective to conductive heat flux.

Models of scale-up in screw extruders have also been developed by Schenkel [77] and by Potente [78–80], who has written a monograph [80] on the topic. These models are also based on making the energy equation independent of scale.

Schenkel [77], and Potente and Fischer [78–80], developed the modeling of scale-up in the same period. Following the latter authors we begin with the mechanico-thermal energy balance

$$P_s = \rho\, C_p\, Q\, \Delta T + \Delta p_d\, Q \tag{3.59a}$$

where P_s is the driving power on the screw and Δp_d the die pressure. Division by $\rho\, C_p\, Q\, \Delta T$ gives:

$$\frac{P_s}{\rho\, C_p\, Q\, \Delta T} = 1 + \frac{\Delta p_d}{\rho\, C_p\, \Delta T} \tag{3.59b}$$

If we maintain the melt temperature constant, it follows that:

$$P_s \sim Q \tag{3.60a}$$

$$\Delta p_d = \text{constant} \tag{3.60b}$$

where P_s is:

$$P_s = N \cdot M_s \tag{3.61}$$

where M_s is the screw torque.

The channel depth H of the screw will vary with its diameter. This may be taken through:

$$\left(\frac{H}{H_0}\right) = \left(\frac{D}{D_0}\right)^{\psi} \tag{3.62}$$

The drag flow in the screw may be written:

$$Q_d \sim H\, W\, D\, N \tag{3.63a}$$

and if W scales with D:

$$Q_d \sim D^{2+\psi}\, N \tag{3.63b}$$

The pressure flow may be written (for a Newtonian fluid) as:

$$Q_p \sim \frac{H^3\, W\, \Delta p_d}{L} \tag{3.64a}$$

and if L scales with D:

$$Q_p \sim D^{3\psi} \tag{3.64b}$$

The ratio of pressure flow to drag flow is:

$$a = \frac{Q_p}{Q_d} \sim \frac{D^{2\psi-2}}{N} \tag{3.65}$$

The volume throughput is:

$$Q = (1-a)\,Q_\mathrm{d} \sim D^{3\psi} \tag{3.66}$$

Potente and Fischer also worry about the associated problems of feeding and non-Newtonian behavior.

References

1. J. B. Paton, P. H. Squires, W. H. Darnell, F. M. Cash and J. F. Carley in Processing of Thermoplastic Materials, edited by E. C. Bernhardt, Reinhold, N.Y. (1959).
2. G. Schenkel, Schneckenpressen fur Kunststoffe, Hanser, Munich (1959) and Kunststoff Extruder-Technik, Hanser, Munich (1963).
3. Z. Tadmor and I. Klein, Engineering Principles of Plasticating Extrusion, Van Nostrand, Reinhold, N.Y. (1970).
4. C. Rauwendaal, Polymer Extrusion, 4th ed. Hanser, Munich (2001).
5. J. L. White and H. Potente (Eds), Screw Extrusion, PPS Progress in Polymer Processing Series, Hanser, Munich (2003).
6. J. D. Sturges, U.S. Patent 114,063 (1877).
7. W. H. Higbie, U.S. Patent 192,069 (1877).
8. M. Gray, British Patent 5,056 (1879).
9. A. Desgoffe and L. A. DiGiorgio, German Patent 26,177 (1884).
10. R. B. Price and W. J. Steinle, U.S. Patent 1,211,370 (1917).
11. W. J. Steinle, U.S. Patent 1,283,947 (1918).
12. B. B. Felix, U.S. Patent 1,320,128 (1919).
13. E. Stich, German Patent 476,748 (1929).
14. V. Royle, U.S. Patent 1,904,884 (1933) and U.S. Patent 2,200,997 (1940).
15. C. Wurster, German Patent 137,813 (1901).
16. F. B. Anderson, U.S. Patent 1,848,236 (1932).
17. G. Schenkel in Kunststoffe, Ein Werkstoff Macht Karriere, edited by W. Glenz, Hanser, Munich (1985).
18. G. Schenkel, *Int. Polym. Proc.*, **3**, 3 (1988).
19. H. Decker, German Patent 735,000 (1943).
20. H. Beck, German Patent (filed 1943) 858,310 (1952).
21. H. List, Swiss Patent 247,704 (1947) and U.S. Patent 2,505,125 (1950).
22. V. Zona, U.S. Patent 2,485,854 (1949).
23. G. E. Henning, U.S. Patent 2,496,625 (1950).
24. L. E. Bankey, U.S. Patent 2,519,614 (1950).
25. F. E. Dulmage, U.S. Patent 2,607,077 (1952) and U.S. Patent 2,765,490 (1956).
26. V. Zona, U.S. Patent 2,765,490 (1956).

27. C. Maillefer, Swiss Patent 363,149 (1961) and German Auslegeschrift 1,207,079 (1965).
28. R. B. Gregory and W. S. McCormick, U.S. Patent 3,300,810 (1967).
29. P. Geyer, U.S. Patent 3,375,549 (1968).
30. A. Boes, A. Kramer, V. Lohrbacher and A. Schneiders, *Kunststoffe*, **80**, 659 (1990).
31. G. Menges, E. G. Harms and R. Hegele, German Offenlegungschrift 2,235,784 (1974), U.S. Patent 4,178,104 (1979) and U.S. Patent 4,199,263 (1980).
32. E. Grünschloss, *Int. Polym. Process.*, **17**, 291 (2002), *ibid* **18**, 235 (2003).
33. Z. Tadmor, U.S. Patent 4,142,805 (1979).
34. Anonymous, *Engineering*, **114**, 606 (1922).
35. R. H. Pearsall, *Automobile Engineer*, 145 (1924).
36. H. S. Rowell and D. Finlayson, *Engineering*, **126**, 249 (1928).
37. W. T. Pigott, *Trans. ASME*, **73**, 947 (1951).
38. W. Meskat, *Kunststoffe*, **41**, 417 (1951); *ibid* **45**, 87 (1955).
39. J. F. Carley, R. M. Mallouk and J. M. McKelvey, *Ind. Eng. Chem.*, **45**, 974 (1953).
40. J. F. Carley and R. A. Strub, *Ind. Eng. Chem.*, **45**, 970 (1953).
41. J. M. McKelvey, *Ind. Eng. Chem.*, **45**, 982, (1953).
42. C. Maillefer, *British Plastics*, 394 (1954).
43. W. D. Mohr and R. S. Mallouk, *Ind. Eng. Chem.*, **51**, 765 (1959).
44. J. M. McKelvey, Polymer Processing, Wiley, N.Y. (1962).
45. S. Eccher and A. Valentinotti, *Ind. Eng. Chem.*, **50**, 829 (1958).
46. Y. Mori, N. Otatake and H. Igaroshi, *Kogaku Kagaku*, **18**, 227 (1954).
47. Y. Mori and N. Otatake, *Kogaku Kagaku*, **19**, 9 (1955).
48. Z. Rotem and R. Shinnar, *Chem. Eng. Sci.*, **15**, 130 (1961).
49. R. M. Griffith, *IEC Fund*, **1**, 180 (1962).
50. F. W. Kroesser and S. Middleman, *Polym. Eng. Sci.*, **5**, 231 (1965).
51. H. J. Zamodits and J. R. A. Pearson, *Trans. Soc. Rheology*, **13**, 357 (1969).
52. B. Martin, *Plastics and Polymers*, (April) 113 (1970).
53. B. H. Maddock, *SPE Journal*, **15** (May) 383 (1959).
54. L. F. Street, *Int. Plastics Eng.*, **1**, 289 (1961).
55. Z. Tadmor, I. J. Duvdevani and I. Klein, *Polym. Eng. Sci.*, **7**, 198 (1967).
56. G. Menges and K. P. Klenk, *Kunststoffe*, **51**, 598 (1967).
57. G. Martin, *Kunststofftechnik*, **7**, 238 (1969).
58. J. Dekker, *Kunststoffe*, **66**, 130 (1976).
59. Z. Tadmor, *Polym. Eng. Sci.*, **6**, 185 (1966).
60. J. T. Lindt, *Polym. Eng. Sci.*, **16**, 284 (1976).
61. J. Shapiro, A. L. Halmos and J. R. A. Pearson, *Polymer*, **17**, 905 (1976); *ibid*, **24**, 1199 (1978).
62. M. Viriyathakorn and B. Kassahon, *SPE ANTEC Tech Papers*, **30**, 81 (1984).
63. C. Rauwendaal, *Polym. Eng. Sci.*, **26**, 1245 (1986).
64. W. H. Darnell and E. A. J. Mol, *SPE Journal*, **12**, (April), 21 (1956).
65. G. Menges and R. Hegele, *Kunststoffberater*, **11**, 1071 (1970).
66. H. G. Fritz, *Chem. Ing. Tech.*, **55**, 256 (1983); *Polym. Proc. Eng.*, **5**, 209 (1988).

67. H. Potente, *Kunststoffe*, **75**, 439 (1985).
68. H. Potente and M. Koch, *Int. Polym. Process.*, **4**, 208 (1989).
69. G. Pinto and Z. Tadmor, *Polym. Eng. Sci.*, **10**, 279 (1970).
70. G. Lidor and Z. Tadmor, *Polym. Eng. Sci.*, **16**, 450 (1976).
71. W. D. Mohr, R. L. Saxton and C. H. Jepson, *Ind. Eng. Chem.*, **49**, 1857 (1957).
72. J. F. Carley and J. M. McKelvey, *Ind. Eng. Chem.*, **45**, 989 (1953).
73. B. H. Maddock, *SPE Journal*, **15**, (November), 983 (1959).
74. B. H. Maddock, *Polym. Eng. Sci.*, **14**, 853 (1974).
75. R. T. Fenner, *Polymer*, **16**, 298 (1975).
76. J. R. A. Pearson, *Plastics and Rubber Processing*, 113 (1976).
77. G. Schenkel, *Ind. Anzeiger*, **94**,113 (1972); *Kunststofftechnik*, **12**, 171, 203 (1973); *Kunstotoffe*, **68**, 155 (1978).
78. H. Potente and P. Fischer, *Kunststoffe*, **67**, 242 (1977).
79. H. Potente, *Rheol. Acta.*, **17**, 406 (1978).
80. H. Potente, Auslegen von Schneckenmaschinen Baureihen, Hanser, Munich (1981).

4 Technology of Intermeshing Co-Rotating Twin Screw Extruders

4.1 Introduction

The modular intermeshing co-rotating twin screw extruder has become the most important twin screw machine. It is also probably the most complex of the commercial machines, involving several different types of elements which are closely intermeshing and self-wiping. The patent technology on intermeshing co-rotating twin screw extruders dates to the beginning of the century. Commercial machines date to 1939 and the modern modular machine to the late 1950s.

We shall use a roughly historical perspective in this chapter to describe the development of the technology of intermeshing co-rotating twin screw extruders. Section 4.2 summarizes early technology. The first commercial machine, the LMP Colombo extruder, is described in Section 4.3. The I.G. Farbenindustrie-Bayer A.G. development program, which led to kneading disc blocks and modular twin screw machines, is described in Sections 4.4 to 4.6. The development of the Werner and Pfleiderer ZSK machine is summarized in Section 4.7. Section 4.8 describes a machine with a different type of mixing element devised by Kraffe de Laubarede. The Erdmenger-Bayer-Werner-Pfleiderer Four-Screw VSD-V machine is discussed in Section 4.9. Later developments by LMP and its R. H. Windsor licensee, including a twin screw injection molding machine, are described in Section 4.10. The origin of the Baker-Perkins modular twin screw machine is reviewed in Section 4.11. We move on to a discussion of more recent Bayer A.G. activities in Section 4.12. The development of the Berstorff machines is considered in Section 4.13. More recent Werner and Pfleiderer machines and developments are considered in Section 4.14. Newer machine manufacturers are summarized in Section 4.15. The application of these machines to rubber compounding is considered in Section 4.16.

The developments of this chapter are described (through in somewhat less detail) in earlier reviews by Herrmann [1], White et al. [2], White and Wang [3], and by Ullrich [4,5] as well as in the first edition of this book.

4.2 Origins

The origin of co-rotating twin screw extrusion technology appears to be in the 1860s. An 1869 U.S. patent by F. Coignet [6] of Paris, France, describes a machine he calls a 'malaxator", which pumps and "works" artificial stone paste. This is shown in Figure 4.1.

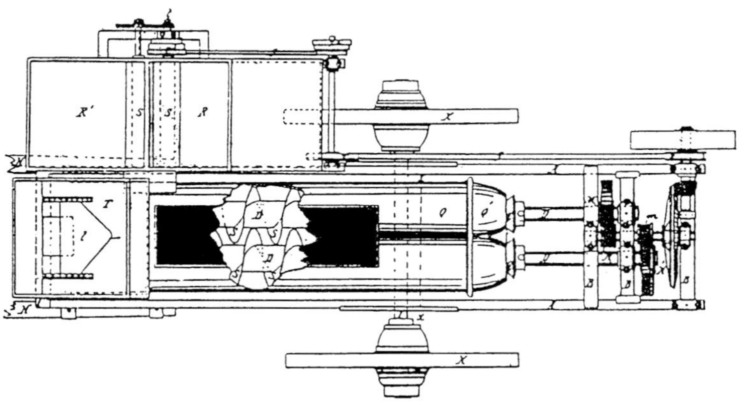

Figure 4.1: Coignet (1869) "malaxator" twin screw machine [6]

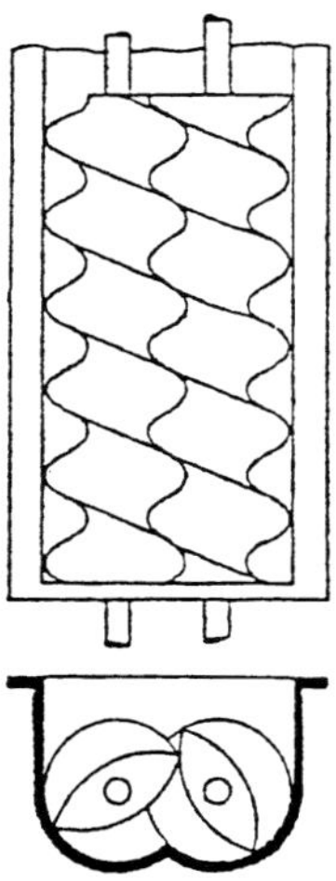

Figure 4.2: Wunsche (1901) self-wiping intermeshing co-rotating twin screw extruder [7]

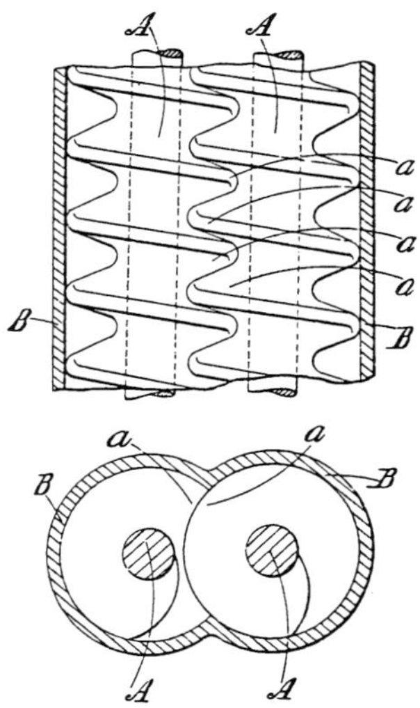

Figure 4.3: Easton (1916, 1920) self-wiping intermeshing co-rotating twin screw extruder [8]

A 1901 German patent by A. Wunsche [7] of Charlottenburg (Berlin) (Figure 4.2), describes an intermeshing co-rotating twin screw as well as multiple screw extrusion. This contains the first discussion and claim for a self-wiping twin screw machine.

A 1916 British patent application and a subsequent 1920 U.S. application [8] of R. W. Easton of London, England (Figure 4.3), apparently unaware of Wunsche [7],also proposed a fully intermeshing self-wiping twin screw machines for pumping gaseous liquid, viscous, or granular materials. Easton specifically states that his screw profiles are identical and complementary and fit closely into each other. In Easton's words [8]:

"In all these applications a great advantage arises from the fact that the surfaces of the two screws at the point of intermeshing are moving in opposite directions and thus they keep each other clean."

The new machine was viewed by Easton as an improved screw pump. Easton discusses applications in coke ovens and retorts.

A screw extrusion machine with six co-rotating screws arranged in a circle is described in a 1920 U.S. patent by H. La Casse [9].

A 1932 U.S. patent by W. K. Nelson [10] of Universal Gypsum and Lime of Chicago describes a process for manufacturing gypsum board using a continuous mixer consisting of a series of co-rotating double tipped self-wiping 'paddles'. In Nelson's words [10]:

"Two shafts are revolubly mounted in subtable bearings in the end castings in parallel relation to each other being equipped with mixer paddles fixedly mounted on the shafts and so shaped and arranged that when the shafts are revolved at the same speed and in the same direction, each paddle on one shaft has its entire face scraped clean by a cooperating paddle on the other shaft."

This is shown in Figure 4.4. The machine does not contain screws, but the staggering of the paddles gives a forward pumping action. Again, Nelson [10] says:

"The paddles are arranged in progressively varied angular positions so as to have a tendency on rotation in counter-clockwise direction to forward the contents of the mixer."

A 1935 German patent by the industrial firm Baur & Cie. [11] of Brugg, Switzerland describes a partially-intermeshing co-rotating twin screw extruder.

A 1936 U.S. patent of F. F. Pease [12] of Lever Brothers of Cambridge, Mass., describes a similar co-rotating intermeshing twin screw extruder for pumping soap. This machine is also self-wiping. The Pease patent gives considerably more engineering detail than the Coignet, Wunsche, Easton, or Baur & Cie. patents.

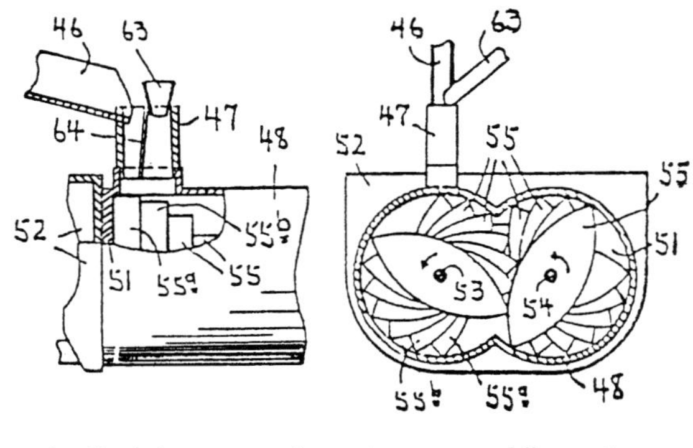

Figure 4.4: Nelson (1931) self-wiping co-rotating twin screw paddle continuous mixer [10]

4.3 The Colombo LMP RC Extruder

The first commercial development of a co-rotating twin screw extruder can be contributed to Roberto Colombo and Lavorazione Materie Plastiche (LMP) of Turin, Italy. LMP was created in 1937 for handling thermosetting compounds and manufacturing thermoset products for the Italian automotive industry [13]. Colombo, one of the principals of the firm, developed a co-rotating intermeshing twin screw extruder in 1938, which was applied by LMP to the production of rigid polyvinyl chloride tubes, previously produced by hydraulic presses. He filed a patent in 1939 in Italy and in 1940/1941 in France and Switzerland [14]. The machine is shown in Figure 4.5. Colombo also described co-rotating extruders involving three, four, five, and eight screws (Figure 4.6). LMP states that they sold and delivered their first twin screw extruders to the I.G. Farbenindustrie in Germany in 1939.

The I.G. Farbenindustrie (Interessengemeinschaft der Farbenindustrie) was founded in 1925, combining essentially all of the major chemical firms of Germany. This was made necessary by the hostility of the allies (especially France) following World War I. The companies involved included Friederich Bayer and Company, BASF, and Farbwerke Hoechst. This was the company that received the first LMP co-rotating twin screw extruders. At least some of the LMP twin screw extruders were used as continuous polymerization reactors. A patent filed in 1943 by the I.G. Farbenindustrie [15] describes the application of such twin screw extruders to polymerization of butadiene with sodium. This reaction would be:

$$CH_2{=}CH-CH{=}CH_2 \xrightarrow{Na} \left[\begin{array}{l} CH_2-CH \\ \qquad\;\; | \\ \qquad\; CH \\ \qquad\;\; \| \\ \qquad\; CH_2 \end{array} \right] \qquad \text{(Structure 4.1)}$$

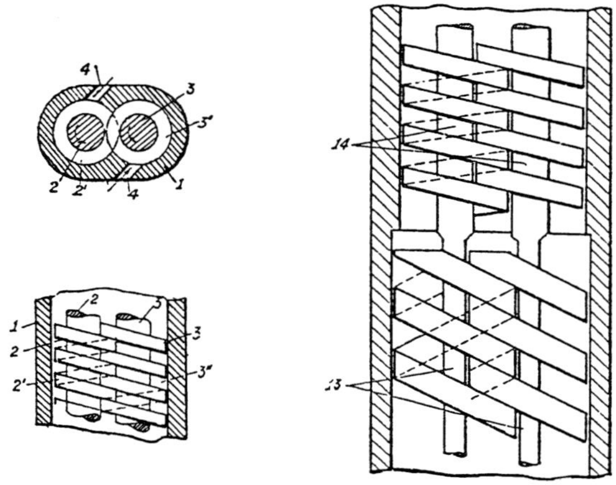

Figure 4.5: Colombo (1939) intermeshing co-rotating twin screw extruder [14]

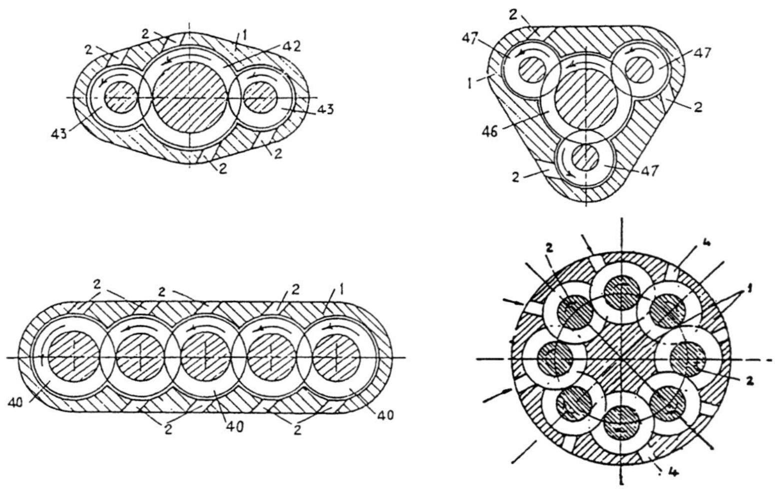

Figure 4.6: Colombo (1939) intermeshing co-rotating multiple screw extruders [14]

which largely produces the vinyl-1,2 microstructure as indicated. They also reported the condensation polymerization:

$$ONC-(CH_2)_6-CNO + OH-(CH_2)_4-OH \longrightarrow \left[O-\overset{\overset{\displaystyle O}{\|}}{C}-NH-(CH_2)_6-NH-\overset{\overset{\displaystyle O}{\|}}{C}-O-(CH_2)_4 \right]$$

(Structure 4.2)

which produces a polyurethane. These appear to be the first examples of polymerization carried out continuously using twin screw extruders as reactors.

Colombo filed patents in Britain [16], the US [17], Canada [18], and elsewhere in the post-war period. These emphasize the application to extrusion of structural members and elimination of porosity. Following the war, LMP sold manufacturing licenses for the Colombo intermeshing twin screw machine in other European countries and Japan. The Colombo patent was licensed to R. H. Windsor in England, CAFL (Creusot-Loire) in France, and the Ikegai Iron Works in Japan. R. H. Windsor machines were sold in the USA through the R. J. Stokes Machine Company of Philadelphia, PA. They were designated as RC (Roberto Colombo?) extruders. The Colombo screws in these patents involve non-threaded sections between intermeshing screw regions. As such they are not completely intermeshing and self-wiping. As one approached the die, the screw thread sections had a small diameter and a smaller pitch. The barrel would also decrease in its inner diameter as one moved towards the die (Figure 4.7). This compression was presumably to eliminate porosity in extrudates. Greenwood [19], in a 1953 paper, speaks of Stokes-Windsor extruders with models RC-65, RC-100, and RC-200. The RC 100 was a 90 mm (3.6 inch) twin screw machine that delivered 100–160 lbs/hour. The RC-200 was a 3-screw machine with 108 mm diameter screws that produced 200–300 lbs/hour. The machines appear to have been primarily used for extrusion of rigid polyvinyl chloride pipe.

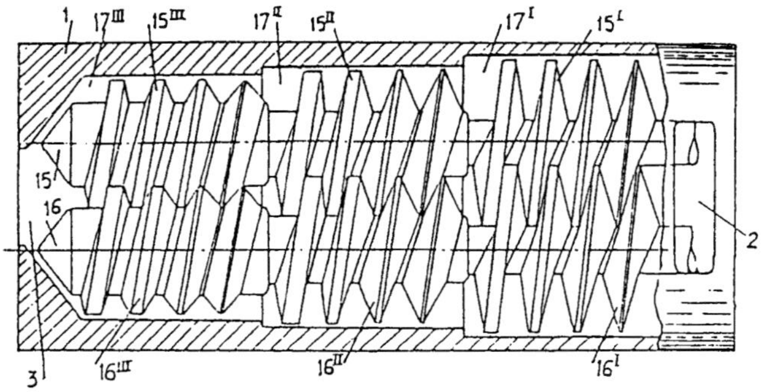

Figure 4.7: Detail of screw-barrel character of Colombo (1947) intermeshing co-rotating twin screw extruder [17]

Detailed characteristics of the Colombo machines of the 1950s and 1960s were discussed by Schaerer [20], Baigent [21], Beyer [22], Schutz [23], and Martelli [24]. Schutz [23] estimated about 40 R. H. Windsor-Colombo machines were sold in the US by 1962.

4.4 The Meskat-Erdmenger-Geberg, I.G. Farbenindustrie Wolfen Works Twin Screw Development Program

A separate development of the intermeshing co-rotating twin screw extruder took place in the I.G. Farbenindustrie in their Wolfen Works near Bitterfeld in the 1940s. This was associated with Walter Meskat, Rudolf Erdmenger, and A. Geberg. These activities are best followed through the career of Erdmenger [25], who recorded his views and experiences. The son of the chief engineer of a Farbwerke Hoechst chemical factory, Erdmenger received a Dipl. Ing. degree from the Techniche Hochschule of Munich in 1935. He later worked (1937–1939) with Maschinenfabrik Paul Leistritz in Nuremburg on screw pumps and screw kneading machines. Here he knew, and perhaps worked with, F. Burghauser, P. Leistritz, and the joint program with the I.G. Farbenindustrie-Hoechst on intermeshing counter-rotating twin screw extruders with decreasing C-chamber volumes in the machine direction. The Leistritz-I.G. Farbenindustrie-Hoechst Program will be described in Section 7.3.

In 1939, Erdmenger left Leistritz and joined the I.G. Farbenindustrie at their Wolfen location. After World War II, this area would be in the USSR zone and later the German Democratic Republic. Here, he worked with A. Geberg in Walter Meskat's group under the direction of Kurt Riess. The I.G. Farbenindustrie Wolfen Works activities of Meskat et al. were concerned with cellulose derivatives and man-made fibers. At first, Erdmenger worked with Meskat in developing a continuous process for removing the liquor from alkali celluloses. Meskat had been involved in these areas since at least the mid 1930s. Later, these activities were summarized in a paper by Riess and Meskat [26].

In the 1940s, Meskat and Erdmenger, with A. Geberg, devised fully-intermeshing, self-cleaning twin- and triple-screw screw extruders (see Sections 18.3 and 19.6). The first patent application by this group describing an intermeshing co-rotating machine was filed by Meskat [27] in October 1943 for an apparatus for the devolatilization of water from fiber suspensions. This patent, which contains no detail on the twin screw machine, was issued in 1952. A patent applied for in July 1944 by Meskat and Erdmenger [28] (but not issued until 1953) describes in detail a self-wiping twin screw machine for kneading, gelatinizing, and squeezing of plastic masses, intended for post-reactor processing of polymerization and polycondensation reactions. They describe 3-screw as well as twin screw machines. The twin screw machine is shown in Figure 4.8. Riess and Meskat [26] described how Kaufmann, Buhmann, and Meskat used this machine for the production of green carbon electrodes. Meskat and Erdmenger [29] filed a second patent in July 1944 for a self-wiping intermeshing 3-screw co-rotating extruder, which was used for the removal of water from fiber suspensions (Figure 4.9). This patent was intended to follow up the above cited October 1943 application of Meskat [27].

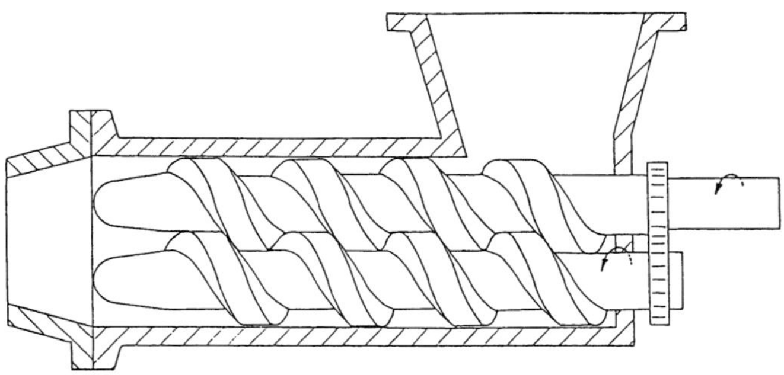

Figure 4.8: Meskat-Erdmenger (1944) intermeshing co-rotating twin screw extruder [28]

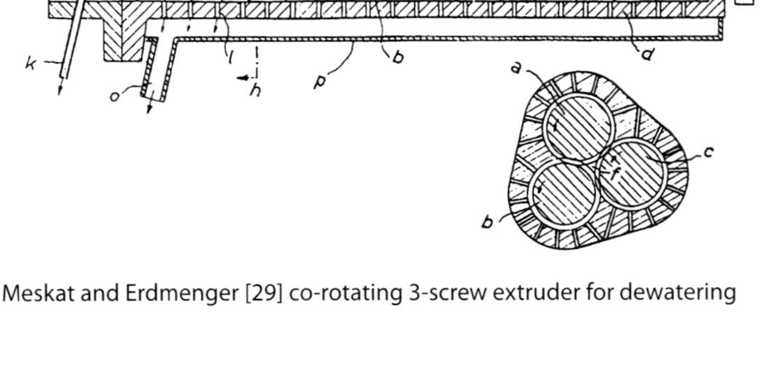

Figure 4.9: Meskat and Erdmenger [29] co-rotating 3-screw extruder for dewatering

4.5 Meskat Bayer (Dormagen) Program

When World War II ended in 1945, the I.G. Farbenindustrie Wolfen Works and Bitterfeld were occupied by the American Army. The occupation of this region was agreed to be turned over to the Soviet Union (in exchange for a part of Berlin). The Americans gave an opportunity to the I.G. Farbenindustrie scientists, engineers, and managers to move to the West. Riess, Meskat, and Erdmenger all moved to what became West Germany, the German Federal Republic (Bundesrepublik Deutschland). The I.G. Farbenindustrie was now broken up by the allies into the original industrial firms that had founded it. This was largely Friederich Bayer and Company, which became Farbenfabriken Bayer (later Bayer AG), Farbwerke Hoechst, BASF, and a new company Chemische Werke Hüls. Riess was able to obtain an executive position in Farbenfabriken Bayer in Leverkusen. He was also able to hire both Meskat and Erdmenger, with Meskat stationed in Dormagen and Erdmenger in Leverkusen.

Farbenfabriken vorm. Friederich Bayer and Company was founded in 1863 by Friederich Bayer and Friederich Weskott in Elberfeld to manufacture dyestuffs from coal tar [30]. In succeeding years, the company developed outstanding research laboratories which led to the many new dyestuffs and later pharmaceuticals, including aspirin. At the turn of the century, Bayer came under the leadership of Carl Duisberg, an innovative research chemist. Duisberg became General Director in 1912, the same year Bayer moved its headquarters to Leverkusen on the northeastern outskirts of Cologne. In the years preceding World War I, Bayer chemists developed synthetic rubber, which they subsequently manufactured during the war. In 1925, Duisberg and Bayer took the lead in establishing the I.G. Farbenindustrie. The succeeding years saw the Bayer wing of the I.G. Farbenindustrie develop the modern emulsion polymerized Buna S and Buna N synthetic rubber, sulfa drugs, and polyurethane. In the post World War II period, Bayer held strong to its research tradition, which was possible in the British zone of occupation where they received encouragement unlike the big chemical companies in the American and French zones [31].

In the post-war period, Meskat and Erdmenger carried out separate research and development programs on twin screw extrusion, one in Dormagen and the other in Leverkusen. Meskat, in Dormagen, worked jointly with J. Pawlowski. Their joint efforts on twin screw extruders led to a major development. This was a December 1950 patent application [32] for a modular intermeshing co-rotating twin screw machine with left-handed as well as right-handed screw elements (Figure 4.10). This is described as a mixing and kneading machine. In 1951, Riess and Meskat [26] published a paper on screw machines for viscous and plastic fluids which includes a description of the Meskat-Erdmenger twin screw machine and the Meskat-Pawlowski modular twin screw machine. In a 1955 plenary lecture to the Congress of the European Federation of Chemical Engineering Methods (published in *Chemie Ingenieur Technik*), Riess [33] described how the Meskat-Pawlowski apparatus was used for the continuous production of spinning solutions and esterification of cellulose. This is another very early example of reactive twin screw extrusion. It appears to have involved six separate interconnected twin screw machines (Figure 4.11).

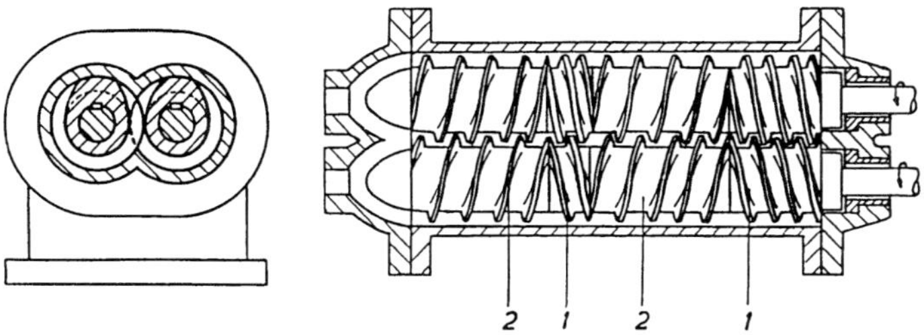

Figure 4.10: Meskat-Pawlowski [32] modular intermeshing co-rotating twin screw extruder with right-handed and left-handed screw elements

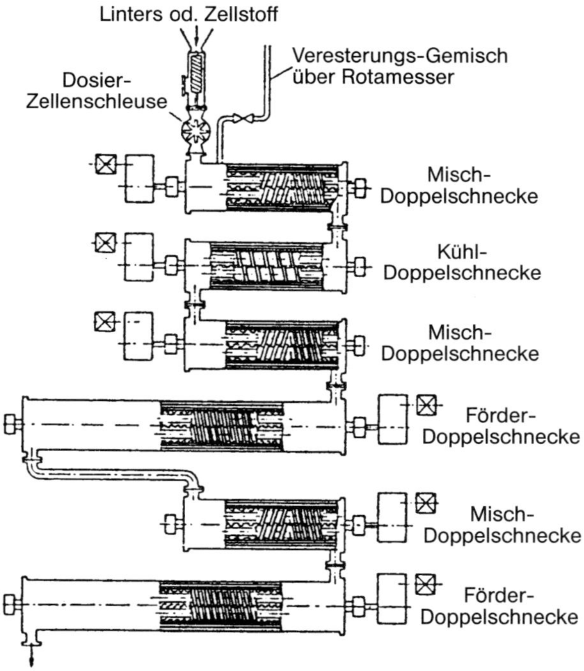

Figure 4.11: Bayer AG application of Meskat-Pawlowski modular intermeshing co-rotating twin screw extruder for the esterification of cellulose and linters Reported by Riess [33] in 1955 (reprinted with permission of Chemie Ingenieur Technik)

In 1951, Meskat seems to have moved from Dormagen to Leverkusen. Pawlowski also moved perhaps at the same time. They worked, however, separately from Erdmenger, and seem not to be involved in twin screw extrusion development. Meskat and Pawlowski were actively involved in single screw extrusion (see Chapter 3). When the 3rd International Rheological Congress was held in West Germany in 1958, Meskat took on the position of founder and editor of a new journal, *Rheologica Acta*, which published the proceedings and then became a permanent fixture in the rheological world.

4.6 Erdmenger Bayer (Leverkusen) Program

In Leverkusen in the late 1940s, Erdmenger sought to develop twin screw machines with superior mixing ability. First came the labyrinth twin screw machine that contained barriers [34]. The patent was applied for in September 1949 and issued in 1951.

In the same period, Erdmenger [35, 36] developed self-wiping co-rotating kneading disc machines for mixing and kneading. Single-tipped and double-tipped kneading disc blocks are included in the September 1949 German patent application [35] (Figure 4.12). The kneading discs in Erdmenger's patents were keyed onto shafts. It is to be noted that the kneading discs are staggered into configurations (which resemble screws) in order to pump the material forward. Some experimental machines were built by Leistritz in Nuremberg [37] for Bayer AG. Erdmenger's U.S. Patent [36], when issued, claimed only single-tipped or cam kneading disc blocks, presumably because the U.S. patent examiner cited Nelson's early work [10]. However, it should be noted that Erdmenger was an independent inventor.

A 1951 paper by Riess and Erdmenger [37] on continuous mixing machinery notes the Meskat-Erdmenger intermeshing co-rotating twin screw extruder and the co-rotating kneading disc machines.

In a 1953 patent application, Telle and Erdmenger [38] described the use of the kneading disc block machine to pulverize agglomerated powders. A 1953 patent application by Erdmenger [39] which, issued in both Germany and the U.S., describes staggered intermeshing co-rotating triple tipped kneading disc blocks keyed onto shafts (Figure 4.13).

Erdmenger moved in other directions as well. On December 31, 1951, W. Winkelmuller, Erdmenger, S. Neidhardt, G. Hirschberg, and B. Fortuna [40] filed a patent in which a pair of intermeshing co-rotating twin screw extruders feed transversely in a second pair of twin screws. This is carried out in such a way as to produce devolatilzation, e.g., by having the intersection near the feed hopper in the second twin screw machine (see Section 18.3).

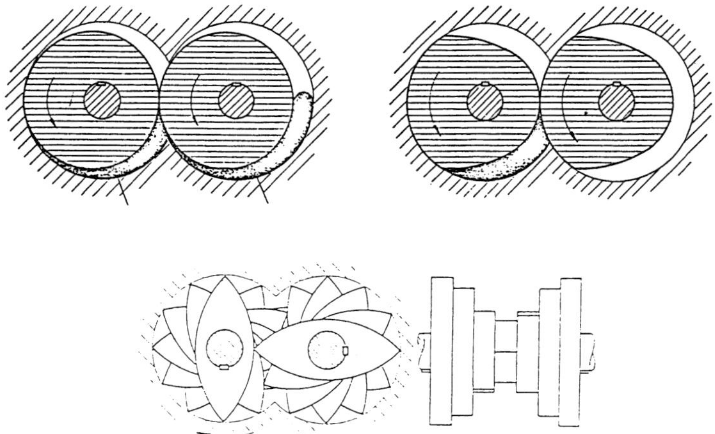

Figure 4.12: Erdmenger's single- and double-tipped kneading disc block machine [35]

In these years, Erdmenger devoted himself, as we shall describe in more detail in the next section, to modular machines involving screw and kneading disc block elements (Figure 4.14). He eventually filed patents for such a machine in 1958/1959 [41]. Erdmenger also considered four screw machines which we will discuss later.

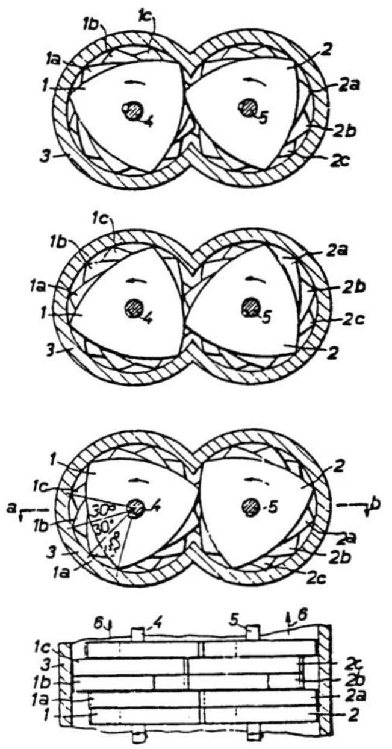

Figure 4.13: Erdmenger's triple-tipped kneading disc block machine [39]

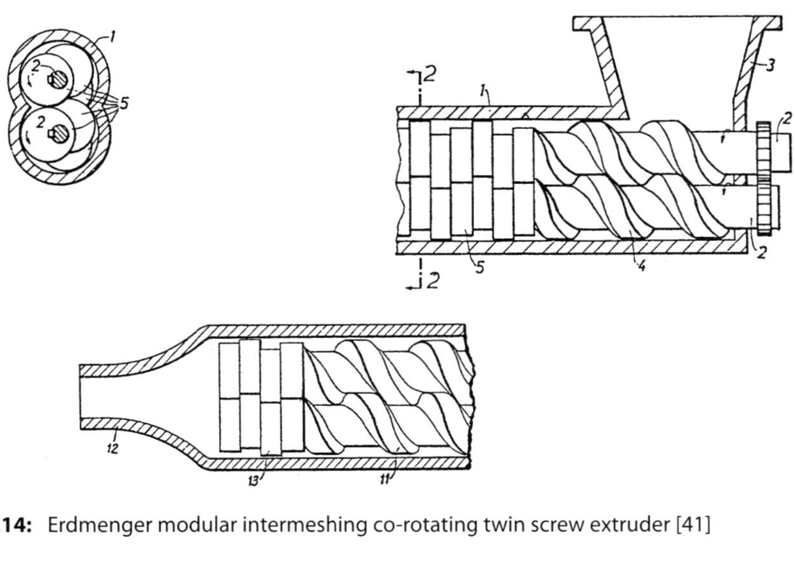

Figure 4.14: Erdmenger modular intermeshing co-rotating twin screw extruder [41]

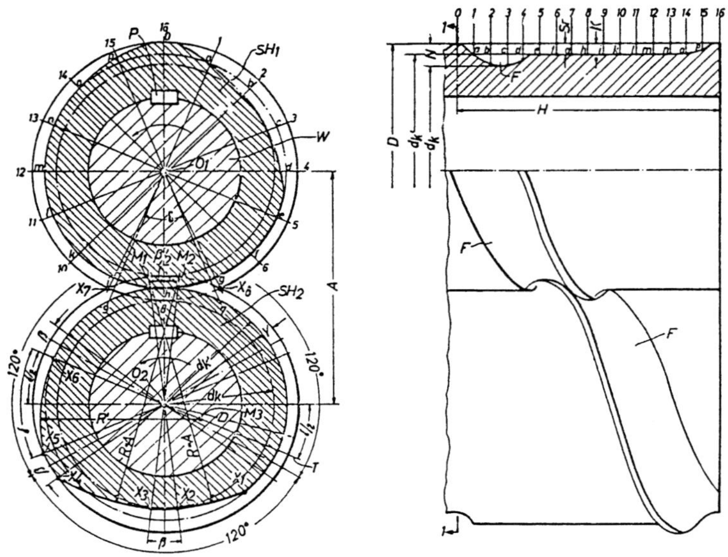

Figure 4.15: Erdmenger intermeshing co-rotating twin screw element with spiral groove [45]

Riess retired from Bayer in 1962. In that year, Meskat and Pawlowski [42], and Erdmenger [41] published papers in *Chemie Ingenieur Technik* dedicated to Riess. Only the paper by Erdmenger treats twin screw extruders. Here, Erdmenger [43] gave detailed consideration to devolatilization, intermeshing co-rotating twin screw and kneading disc machines, and the Winkelmuller et al. [40] transverse twin screw machine is discussed.

Two years later (1964) Erdmenger [44] published a broader ranging paper on intermeshing co-rotating twin screw extrusion which included significant experimental studies of flow mechanisms.

In the 1960s, Erdmenger gave attention to the development of new self-wiping twin screw element designs. Patents filed in 1962/1963 describe a new design in which a spiral groove, whose axial dimension is much less than the pitch, is cut at the leading flight [45]. This had a scraping contact with the opposite flight (Figure 4.15). This acts to reduce the thickness of the material being pumped through the screw element.

4.7 Kraffe de Laubarede's Homogenizing Screw

Between 1949 and 1953, Leonce Kraffe de Laubarede [46, 47] of Paris, France, devised a modified intermeshing co-rotating twin screw extruder in which the final section in front of the die consists of intermeshing discs containing slots. This is shown in Figure 4.16. This was intended for homogenizing of the thermoplastic melt.

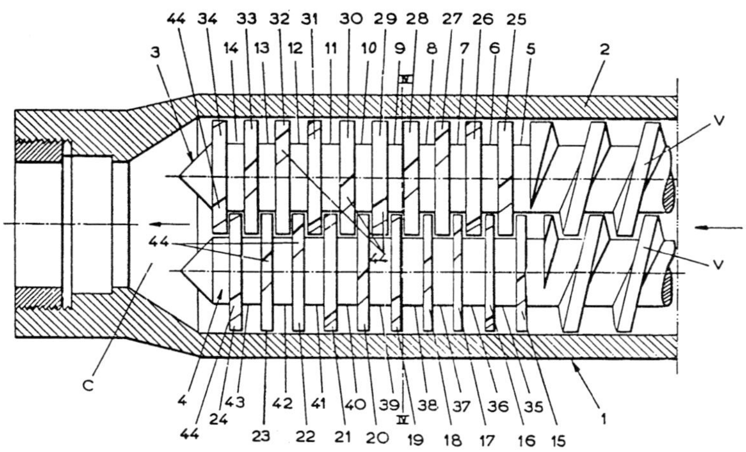

Figure 4.16: Kraffe de Laubarede's mixing screw element containing slots in a disc [46, 47]

4.8 The Werner-Pfleiderer ZSK System Erdmenger

In 1953, Bayer AG joined forces with the German mixing machinery manufacturer Werner and Pfleiderer of Stuttgart to commercially develop intermeshing self-wiping co-rotating twin screw extruder which would be manufactured and sold by Werner and Pfleiderer. The Werner and Pfleiderer firm had its origins in Stuttgart in 1880 and had first manufactured batch mixers for the bakery industry [1]. They later manufactured internal mixers for the rubber industry (Section 13.2) and in the post World War II period manufactured the counter-rotating continuous mixers named the Kammerkneter (see Section 13.7). The Werner and Pfleiderer team was led by G. Fahr and H. Ocker, and included R. Fritsch and H. Herrmann. A range of machines were produced in this program. The first prototype was built in 1955. Erdmenger had, in 1958/1959, filed patents for a modular twin screw extruder with screw and kneading disc block elements. A patent was issued in the United States in 1964 [41]. The combination of kneading disc modules with screw modules was the key to the new co-rotating twin screw machines. The right-handed screws transported, the left-handed screws controlled pressure fields, while the kneading disc block melted and mixed.

Werner and Pfleiderer began the manufacture of modular intermeshing co-rotating twin screw extruders at least as early as 1957. A photograph of a 'Type ZSK 83/700 System Erdmenger' (ZSK denotes Zwei Schnecken Kneter) is shown in Figure 4.17. In 1959, a paper by Fritsch and Fahr [48] described an updated ZSK 83/700 System Erdmenger. The machine contained screw (both left- and right-handed) elements and triple-tipped kneading discs. It had devolatilization ports and compounding ingredients could be added at different points along the screw (Figure 4.18). The screw had an 83 mm diameter and was 700 mm long, i.e., an L/D of 8.4.

Figure 4.17: Photograph of a 1957 Werner and Pfleiderer 83/700 System Erdmenger modular intermeshing co-rotating twin screw extruder (courtesy of Werner and Pfleiderer)

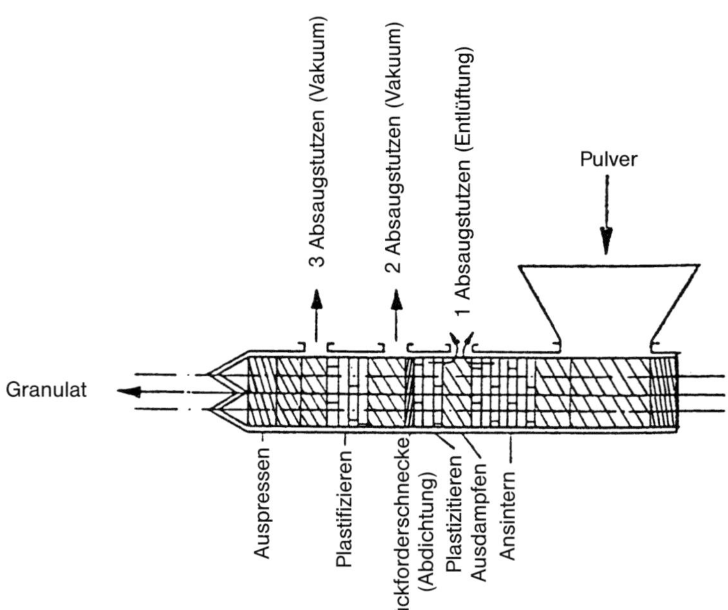

Figure 4.18: Early Werner and Pfleiderer modular intermeshing co-rotating twin screw extruder with devolatilization ports from Fritsch and Fahr [48]

Commercial modular intermeshing co-rotating twin screw extrusion machines were being sold by Werner and Pfleiderer in the late 1950s. The early Bayer and the Werner and Pfleiderer twin screw extruders had a vertical arrangement of the two co-rotating screws. In 1962, they began to fabricate machines with horizontal screws and variable modular screw lengths [49]. A 1964 paper by Herrmann [50] presents the Werner-Pfleiderer ZSK System Erdmenger machines in detail, indicating machines of varying scale were being produced, including the ZSK 160 with 160 mm diameter screws. Herrmann indicates applications to mixing, devolatilization, homogenization, and plastification of thermoplastics. He notes materials being processed included polyethylenes, polystyrene, ABS, and synthetic rubber. A 1966 paper by Herrmann [51] describes a basic study of mixing in ZSK machines.

Patents were filed from as early as 1963 by Werner and Pfleiderer engineers on apparatus design for process engineering [54–57], screw element manufacture [58], and applications [52, 56–58] of these machines (Figure 4.18). These included general processes such as blending of thermoplastics [55] and removal of volatiles, as well as devolatilization [57]. Ocker of Werner and Pfleiderer files for a patent application in 1968 [56] on removal of volatile components that involved screw configurations containing left-handed screw elements or truncated sections that lower pressure. The flowing material then enters a vacuum stage which draws off the volatiles. After leaving the vacuum region, there is a second obstruction which acts as a seal. Patents by Werner and Pfleiderer also treat specific applications, such as manufacture of chocolate [59], compounding thermosetting resins [60], and continuous soap production [61].

The development of the ZSK machine was described in a 1972 monograph on extrusion technology by Herrmann [1] of Werner and Pfleiderer.

Figure 4.19 shows a photograph dating to the 1970s showing screw and kneading disc modules mounted on a shaft and typical barrel sections of Werner and Pfleiderer twin screw extruder.

In 1964, Gerhard Illing [60] with the University of Erlangen filed a patent assigned to Werner and Pfleiderer on the addition polymerization and copolymerization of vinyl monomers such as styrene and acrylonitrile, as well as methyl methacrylate. Illing [61] later described the polymerization using modular intermeshing co-rotating twin screw machines of caprolactam to form polyamide-6. Illing briefly described machine designs to carry through this reaction.

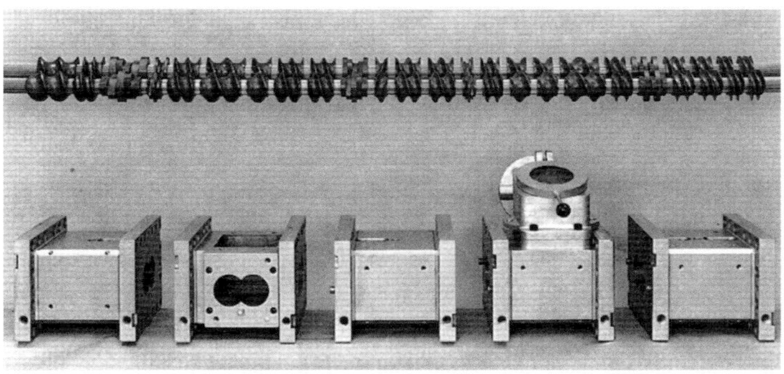

Figure 4.19: Photograph of more recent modular screw and barrel sections of a Werner and Pfleiderer ZSK (courtesy of Werner and Pfleiderer)

Catalytic agents are fed in from a side extruder. The kneading disc blocks are considered to be the position where polymerization occurs. Left-handed screw elements or valves are placed after the kneading disc blocks. The material then flows into a shear zone that controls molecular weight. This is followed by a devolatilization zone for degassing. In 1976, Mack and Herter [62] of Werner and Pfleiderer described efforts to apply these machines as polymerization reactors.

The early Werner and Pfleiderer modular co-rotating twin screw extruder had three thread starts or lobes. From 1973 they began to also manufacture a series of two-lobed machines [49].

4.9 Erdmenger-Oetke 4-Screw VDS-V Devolatilizer

Erdmenger and his co-workers at Bayer continued their efforts to develop new multi-screw machines. This involved not only twin screw machines but also machines with four screws, specifically two pairs of intermeshing co-rotating screws. This was described in a 1960 patent of Erdmenger and Oetke [63] and also in papers by Erdmenger [43, 44] and Erdmenger and Ullrich [64] in 1962/1970 (see Section 18.3 and Figure 18.4).

The machines built by Werner and Pfleiderer were designated VDS-V (VDS-V means Vierwellige Dichtprofilschnecke zum Verdampfen). This implies they were used for devolatilization with screws having deep channels. Such machines were later described in the above cited 1968 patent application by Ocker [54], which describes devolatilization in multi-screw machines. The major application of the VDS-V was devolatilization. Discussion can also be found in the monograph of Herrmann [1].

4.10 Further LMP-RH Windsor Developments

The LMP-RH Windsor machines were challenged by new technologies, especially that of Bayer/Werner and Pfleiderer. They did for a period respond with initiatives of their own.

In a unique development, K. H. Baigent [65] of R. H. Windsor in 1957 filed for a patent on a screw injection molding machine based on a co-rotating intermeshing twin screw machine. R. H. Windsor was a Colombo LMP licensee. Baigent notes that it was not possible to use the Colombo design screws and barrels. A uniform screw diameter and barrel diameter was introduced (Figure 4.20). The Baigent–R. H. Windsor technology was also discussed in a 1957 paper by H. Beck [66], the pioneer of screw injection molding technology. Schutz [23] briefly discussed the machine in his 1962 paper, indicating that Baigent had presented it at a December 1961 Society of Plastics Engineers Regional Technical Conference in Cleveland.

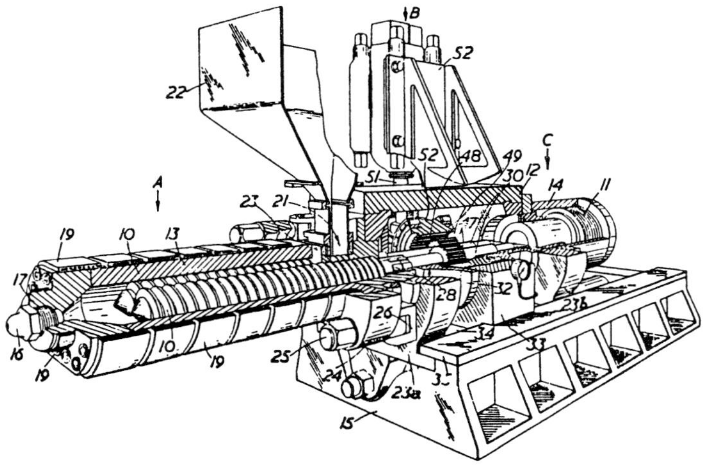

Figure 4.20: Baigent-RH Windsor twin screw injection molding machine [65]

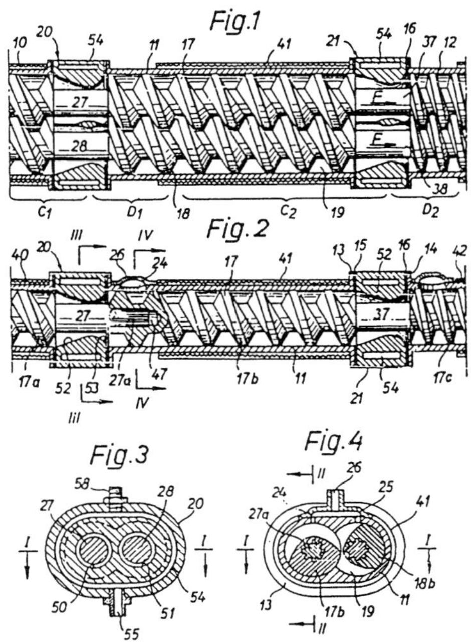

Figure 4.21: Colombo modular intermeshing co-rotating twin screw extruder with Venturi elements [67]

A patent filed in 1961 by R. Colombo [67] and licensed to LMP describes an intermeshing co-rotating twin screw machine with modular screws and Venturi-like barrel elements (Figure 4.21). The screws are right-handed and pump forward. The Venturi barrel elements, which coincide with small diameter positions on the modular shafts, produce pressure reductions and allow devolatilization.

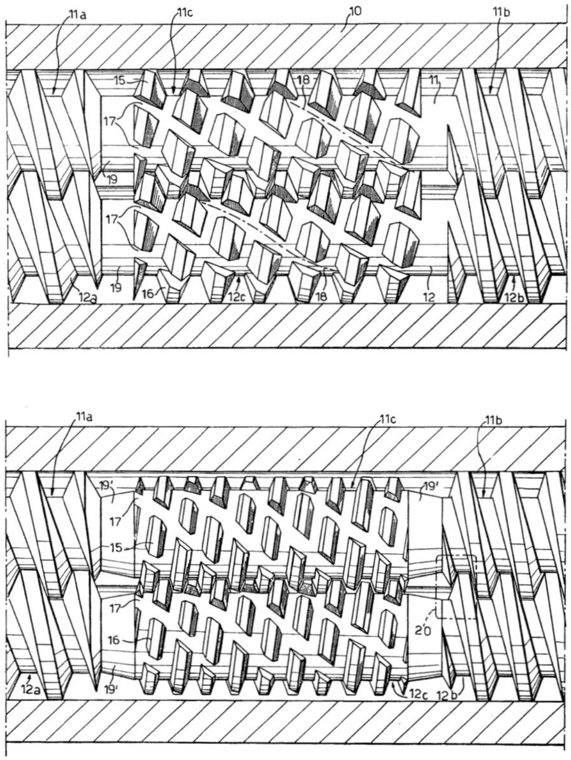

Figure 4.22: Colombo mixing element [68]

A second patent filed in 1963/1964 by Colombo [68] of LMP involved a new type mixing element, consisting of a left-handed backward pumping screw with slices in the flights (Figure 4.22). There is no mention of modularity of the screws or of mixing being the primary application of the machine.

4.11 Readco and Baker-Perkins Twin Screw Extruder

An independent development of a modular intermeshing co-rotating twin screw extruder took place at Readco (formally the Reading Machine Company of York, PA) in the 1960s. Beginning in 1962, Bernard A. Loomans and Ambrose K. Brennan, Jr. [69, 70] described a modular intermeshing twin screw machine in patents filed. The elements are screws and two-lobed kneading discs, called 'paddles', both of which are self-wiping (Figure 4.23). They presumably associated these elements with the paddles of Nelson's 1932 patent [9]. The Readco patent applications and technology were transferred to Baker Perkins (later APV Chemical

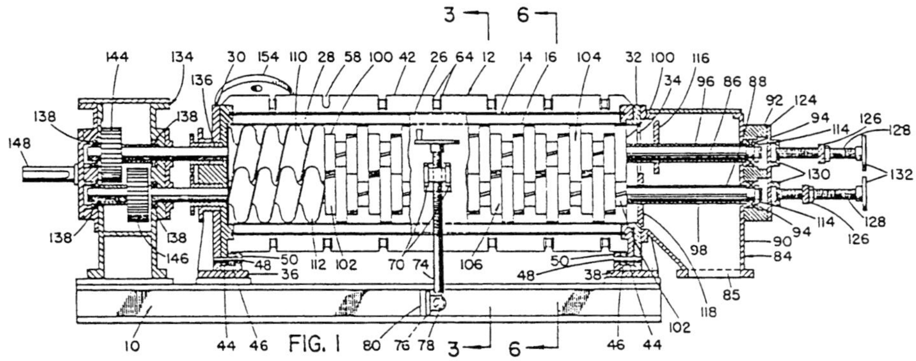

Figure 4.23: Loomans-Brennan modular twin screw machine [69, 70]

Machinery) of Saginaw, Michigan whose relationship to Werner and Pfleiderer is discussed in Section 13.2. When the patents issued, they were assigned to Baker Perkins. Several Readco employees, including Loomans, transferred to Baker Perkins in Saginaw. Baker Perkins began to manufacture machines of this type in the mid 1960s.

In the Baker Perkins machine, as distinct from the Werner and Pfleiderer machine, there are screws rotating clockwise rather than screws rotating counter-clockwise.

Among the applications of the Baker Perkins machine was reactive extrusion. A 1969 patent by Wheeler, Irving, and Todd [71] treats the continuous polycondensation and devolatilization of polyester and polyamide from pre-polymer in their machine.

There were many additional patents from the Baker Perkins group in Saginaw [72–74]. A 1975 patent of Loomans [74] describes a special new self-wiping profile including a radially-projecting hoop (Figure 4.24). Todd [75] of Baker Perkins, in a 1979 patent, described an adjustable saddle section which may be introduced between screw elements (Figure 4.25). This allows for flow and pressure control and better devolatilization.

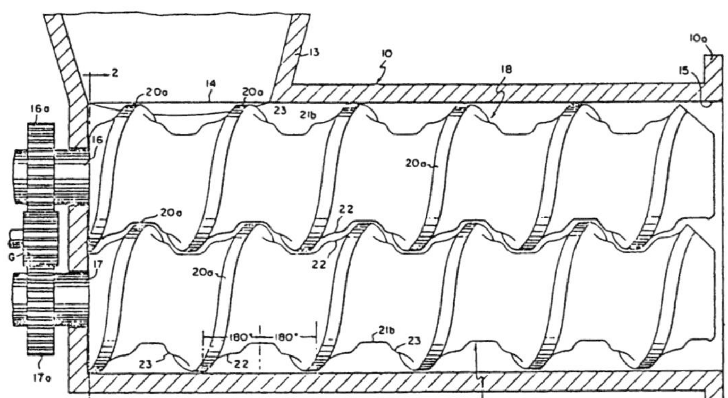

Figure 4.24: Loomans self-wiping profile [74]

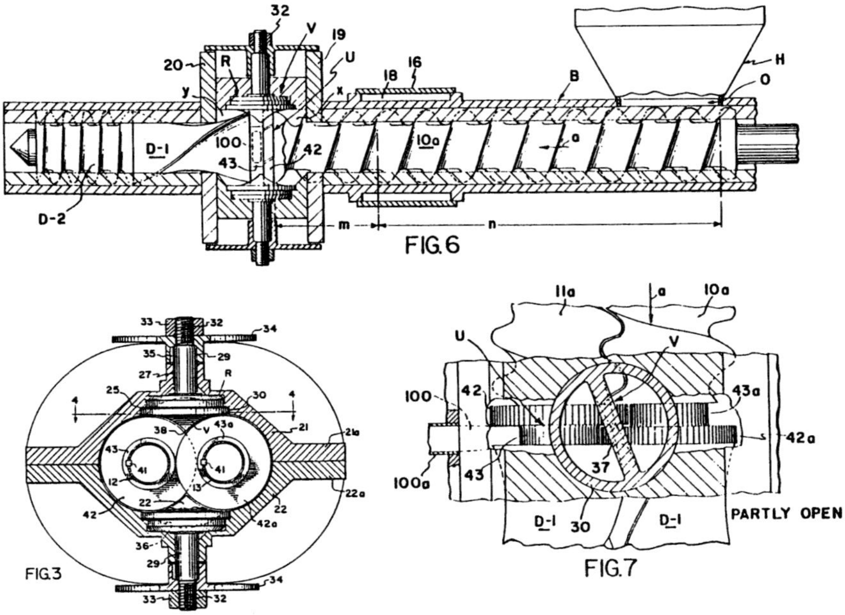

Figure 4.25: Todd's adjustable saddle section [75]

APV Chemical Machinery merged with Baker Perkins in 1986. In 1995, B and P Process Equipment and Systems, a successor to Baker Perkins, was developed in Saginaw, Michigan, US with marketing responsibility in North America and Japan. APV Baker based in Newcastle upon Tyne, England was charged with marketing to the rest of the world.

In 1969, Ambrose K. Brennan, Jr. [76], still with the Readco laboratories, which would become Teledyne, filed a new patent on a new modular intermeshing co-rotating twin screw extruder incorporating special paddle designs. These machines were marketed by Teledyne-Readco.

An interesting addition to these efforts is contained in a 1967 patent application by Brennan [77], which describes an internal mixer with intermeshing co-rotating rotors containing flights twisted into helices for longitudinal pumping. This was intended to be applied to highly viscous substances such as thermoplastics and elastomers.

4.12 Later Bayer Developments

Bayer AG has continued to make creative use of modular intermeshing co-rotating twin screw extruders [78–83]. Much of this involves reactive extrusion. It should be remembered that as early as 1955, Riess [33] had described the use at Bayer of intermeshing co-rotating twin screw extruder to esterify cellulose. A 1971 patent application by Erdmenger, Ullrich, and their co-workers [78] developed a process to produce cross-linked powder lacquers.

Erdmenger retired from Bayer in 1976. Shortly thereafter, he authored a short booklet *Schneckenmaschinen für die Hochviskos-Verfahrenstechnik* [25], which describes his development of the modular twin screw extruder. It was published by Bayer in 1978. After Erdmenger's retirement, his group was under the direction of M. Ullrich.

The Bayer efforts continued. A 1976 patent by M. Ullrich, with E. Meisert and A. Eitel [79], describes the use of this type of machine to produce polyurethane elastomers. Korber [80] patented the use of twin screw extruders in the formulation of hydroperoxide groups on polymers. Later, Binsack, Rempel, Korber, and Neuray [81] patented the formation of certain graft copolymers in an intermeshing co-rotating machine and their use with polybutylene terephthalate or polyethylene terephthalate in blends. Goyert, Meisert, Eitel and others [82] described a process technology using chain lengthening agents in this type of machine to produce improved thermoplastics. Quiring et al. [83] described a reactive co-extrusion process to produce polyurethane ureas. A constant theme of the Bayer AG reactive extrusion effort was the continuing primary, if not sole, usage of intermeshing co-rotating machines.

In another development, Rathjen and Ullrich [84, 85] described a new screw design for intermeshing co-rotating twin screw extruders for crystallization processes (Figure 4.26). These machines have slices in the screw flights, which catch the crystals and break them up through shearing caused by the relative motion of the screw and barrel.

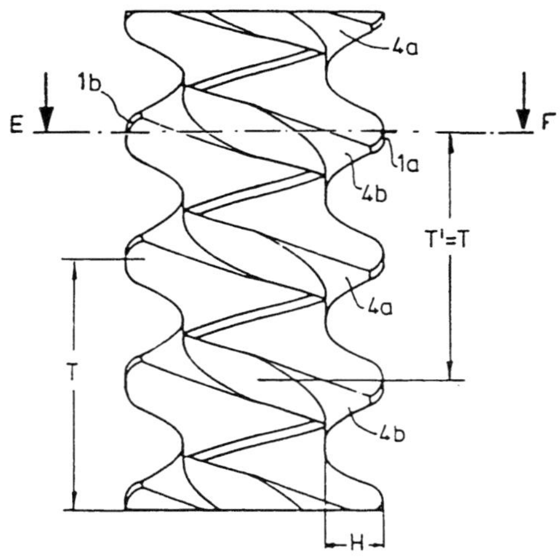

Figure 4.26: Rathjen and Ullrich modified screw elements for crystallization [84]

4.13 Later Werner and Pfleiderer Developments [49]

Werner and Pfleiderer continued to develop new generations of their ZSK modular co-rotating twin screw extruders. This generally meant (i) increases in the ratio of outer diameter to inner diameter of the screws (OD/ID ratio), and (ii) increases in specific torque (torque divided by centerline distance cubed (M/a^3). The fifth generation ZSK, called the 'Supercompounder', was introduced in 1983. It had the highest specific torque available among commercial modular co-rotating twin screw extruders at the time. In 1995, Werner and Pfleiderer introduced their sixth generation machine, called the 'Megacompounder', which had a specific torque that was 30% higher than the supercompounder. It again had the highest specific torque among commercial extruders of the type. The various OD/ID ratios and M/a^3 of the six generations of Werner and Pfleiderer ZSK are summarized in Table 4.1 [47].

Werner and Pfleiderer have also introduced new modular element designs. The older elements are the TME (Turbine Mixing) elements shown in Figure 4.27a [49]. More recent elements are described in a patent of Wobbe and Uhland [86]. These elements are also described by Grillo et al. [87] and Andersen [49]. They are identified as ZME (Zahn Misch Elemente). They are shown in Figure 4.27b and allow for better distributive mixing and milder dispersive mixing than kneading disc blocks. The ZME are generally self-wiping elements. They are recommended by Grillo et al. [87] for compounding glass fibers.

The late Heinz Herrmann of Werner and Pfleiderer always took an ambitious view of the future of compounding technology [88, 89]. The future of twin screw extrusion technology is described in a 1988 paper by Herrmann [88], whose English title is "Towards the Flexible and Intelligent Compounding Plants". This involves integration of polymerization, compounding (often reactive compounding), and converting into final shapes. This includes incorporation of new quality control sensors and control systems.

Table 4.1: Werner and Pfleiderer Twin Screw Extruder Generations

ZSK generation	Number of screw thread starts/lobes	$\left(\frac{\text{Outer diameter}}{\text{Inner diameter}}\right)$ of screw	$\frac{\text{Torque}}{C_L^3}$
ZSK Standard (1950s)	3	1.22	3.7–3.9
ZSK Variable	3	1.22	4.7–5.5
ZSK Variable	2	1.44	4.7–5.5
ZSK Compact	2 or 3	1.22 or 1.44	7.2–8.0
ZSK Super compounder	2	1.55	8.7
ZSK Mega compounder	2	1.55	11.3

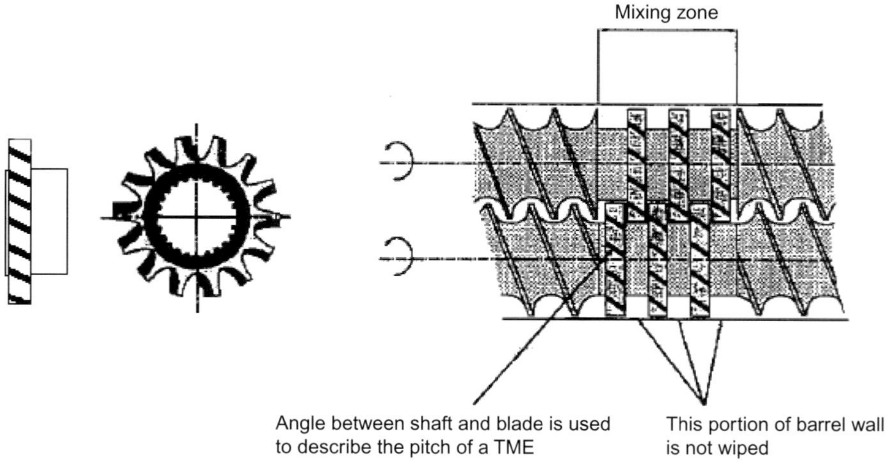

(a) Turbine Mixing Element (TME)

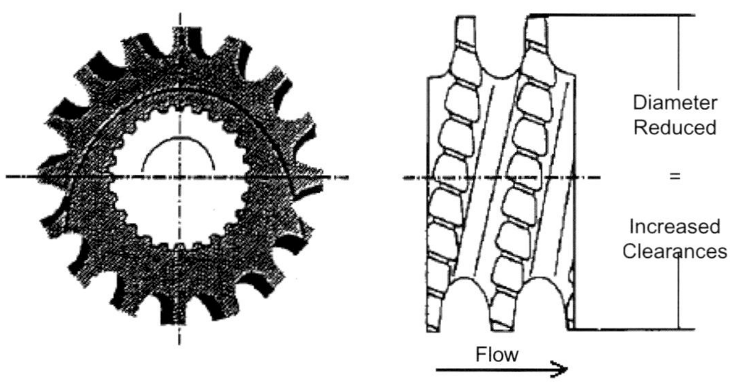

(b) Zahn Misch Element (ZME)

Figure 4.27: Werner and Pfleiderer special mixing elements (a) Turbine Mixing Elements (TME) (b) Zahn Misch Elements (ZME) (courtesy of Werner and Pfleiderer).

The ownership of Werner and Pfleiderer has varied in recent years. The Werner family and Baker Perkins exited as owners and were replaced by the German steel and machinery firm Krupp in the early 1990s. Werner and Pfleiderer then became Krupp, Werner and Pfleiderer. Subsequently, the German steel firm Thyssen acquired Krupp becoming Thyssen-Krupp. The new firm sold their Krupp Werner and Pfleiderer subsidiary to Georg Fischer of Switzerland. Werner and Pfleiderer then became Coperion Werner and Pfleiderer, a sister firm to Buss AG. Later, Buss AG was made independent.

4.14 Newer Machinery Manufacturers

Many firms entered the marketplace to manufacture modular co-rotating twin screw extruders when the Bayer patent system expired. We have already mentioned the contemporary early efforts of LMP, RH Windsor, Readco, and Baker Perkins. Various other companies have entered the field, including Maschinenfabrik Paul Leistritz GmbH (later Leistritz AG), Hermann Berstorff GmbH, Reifenhauser GmbH in Germany; Ikegai Corporation Ltd., Japan Steel Works, Kobe Steel, Mitsubishi Heavy Industries, and Toshiba Machine in Japan; Farrel Corporation, and Teledyne Readco in the US; Clextral in France; Pomini in Italy; Betol in England; and RH Windsor in India.

Berstorff and Leistritz joined their fellow German firm Werner and Pfleiderer as major factors in the American industry.

Kobe Steel and Farrel Corp. introduced rotor sections in the place of kneading disc blocks in their machines. Screw designs of the latter firm are described by Valsamis et al. [90] (Figure 4.28).

Ikegai Corporation, Ltd., of Japan, pioneer manufacturer of Colombo twin screw machines in Japan, now also makes modular twin screw extruders of the type pioneered by Bayer AG and Werner and Pfleiderer. A second original Colombo licensee, Clextral of France, today also manufactures modular co-rotating machines. Leistritz AG manufactures ZSE GL co-rotating modular machines.

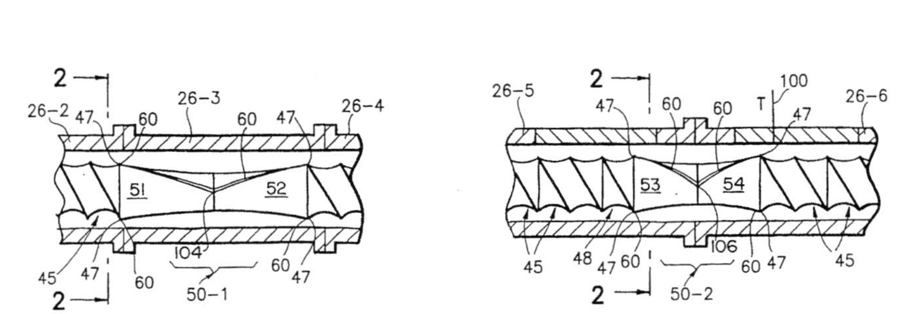

Figure 4.28: Farrel rotor-type modular mixing element [90]

In 1984, Hornsby and his co-workers [91, 92] described modular intermeshing co-rotating twin screw extruders developed at Brunel University in England. The initial machine was designed by Ivor Boyne at Brunel University and was manufactured by Gay's Ltd. (Hampton). It had 40 mm screws and was designated TS40 DV-L. In 1987, Hornsby [93], and Ess and Hornsby [94] discussed studies on this machine which they indicated had trapezoidal non-self-wiping flights. In their 1987 paper, Ess and Hornsby [94] indicated their machine to be the same as a Betol TS40 DV-L.

4.15 Berstorff Developments

Hermann Berstorff GmbH of Hannover, Germany [95–98] introduced a modular intermeshing co-rotating machine in 1981 after the Bayer patents expired. The machines were similar in design to those developed by Bayer and Werner and Pfleiderer described earlier. They included self-wiping screws and kneading disc blocks. Some machine designs have included throttling elements.

Berstorff introduced (in the 1980s) 'mixing gears' to compliment kneading disc blocks for the purpose of mixing. Berstorff's mixing gears provided improved distributive mixing without the severe shearing action of kneading disc blocks (Figure 4.29). The mixing gear design is reminiscent of the screws described by Kraffe de Laubarede [46, 47] in the early 1950s.

Another innovation of Berstorff is screw elements that are incompletely self-wiping. The leading screw flight is not self-wiping, but the neutral flight is. This also improves distributive mixing.

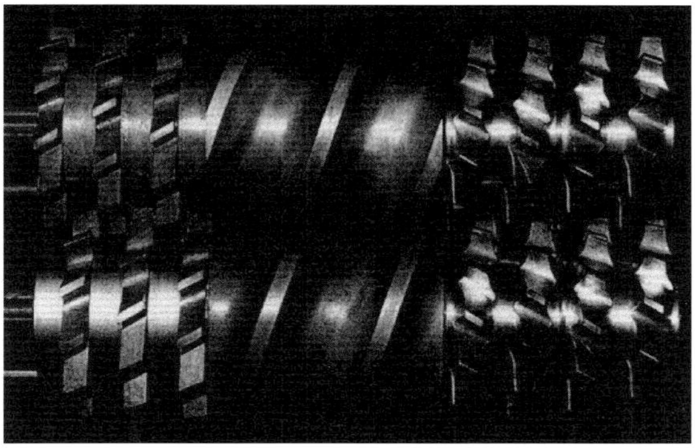

Figure 4.29: Typical arrangement of Berstorff gear mixing elements assembled on a ZE 90A twin screw extruder, (left). Parallel gear rings; (right) gear cut on a helix (courtesy of Bertorff)

4.16 Rubber Compounding

Efforts to continuously compound rubber have a long history. They at least date to the development of the Farrel Continuous Mixer (see Section 13.8) in the early 1960s. Efforts were made to apply this machine using powdered rubber such as the rubber developed in the 1960s by Chemische Werke Hüls [99–101] (B. F. Goodrich [102,103] and the Goodyear Tire and Rubber Company [104]).

In more recent years, there has been an extensive effort by various companies to apply the modular co-rotating twin screw extruder to rubber compounding. This has involved companies such as Goodyear [105, 106], Dow Corning Toray Silicone [107], and Pirelli [108]. An extensive program has been reported by Schuster and his co-workers [109–113] at the Deutsches Institut for Kautschuk Technologie in Hannover, Germany.

References

1. H. Herrmann, Schneckenmaschinen in der Verfahrenstechnik, Springer, Berlin (1972).
2. J. L. White, W. Szydlowski, K. Min, and M. Kim, *Adv. Polym. Technol.*, **7**, 295 (1987).
3. J. L. White and Y. Wang, *in* Progress in Polymer Processing, Vol. 1, edited by A. I. Isayev, Hanser, Munich (1990).
4. M. Ullrich, Historische Entwicklung der Gleichdrall Doppelschnecken, VDI Wissenforum, Leverkusen, June 2005.
5. M. Ullrich in Co-Rotating Twin Screw Extruders, edited by K. Kohlgrüber, Hanser, Munich (2007).
6. F. Coignet, U.S. Patent 93,035 (1869).
7. A. Wunsche, German Patent 131,392 (1901).
8. R. W. Easton, British Patent (filed September 25, 1916) 109,663 (1917); R. W. Easton, U.S. Patent (filed June 2, 1920) 1,468,379 (1923).
9. H. La Casse, U.S. Patent (filed August 25, 1919) 1,356,296 (1920).
10. W. K. Nelson, U.S. Patent (filed June 4, 1931) 1,868,671 (1932).
11. Baur and Cie, German Patent (filed May 6, 1933) 617,432 (1935).
12. F. F. Pease, U.S. Patent (filed August 17, 1933) 2,048,286 (1936).
13. Anonymous, *LMP and Polytal, The Extrusion Specialist*, Polytal and LMP, Turin (1986), (distributed at Dusseldorf K86 Show).
14. R. Colombo, Italian Patent (filed February 6, 1939) 370,578 (1939); Swiss Patent (filed January 3, 1941) 220,550 (1942).
15. Anonymous (BASF), German Patent (filed July 24, 1943) 895,058 (1953).
16. R. Colombo, British Patent (applied May 2, 1946) 629,109 (1949).
17. R. Colombo, U.S. Patent (filed August 7, 1947) 2,563,396 (1951).
18. R. Colombo, Canadian Patent (filed December 27, 1949) 517,911 (1955).
19. S. H. Greenwood, *Rubber World*, **129**, 73 (1953).

20. A. J. Schaerer, *Kunststoffe*, **44**, 105 (1954).
21. K. Baigent, *Trans. Plastics Inst.*, (April) 34 (1956).
22. S. Beyer, *Kunststoffe*, **48**, 157 (1958).
23. F. C. Schutz, *SPE Journal*, **18**, (September) 1147 (1962).
24. F. Martelli, *SPE Journal*, **27**, (1) 26 (1971).
25. R. Erdmenger, Schneckenmaschinen fur die Hochviskos-Verfahrenstechnik, Bayer AG, Leverkusen (1978).
26. K. Riess and W. Meskat, *Chem. Ing. Tech.*, **23**, 205 (1951).
27. W. Meskat, German Patent (filed October 17, 1943) 852,203 (1952).
28. W. Meskat and R. Erdmenger, German Patent (filed July 7, 1944) 862,668 (1953).
29. W. Meskat and R. Erdmenger, German Patent (filed July 28, 1944) 872,732 (1953).
30. E. Verg, G. Plumpe, and H. Schultheis, *Meilensteine,* Bayer AG, Leverkusen (1988).
31. R. G. Stokes, Divide and Prosper: The Heirs of I.G. Farben under Allied Authority 1945–1951, University of California Press, Berkeley (1988).
32. W. Meskat and J. Pawlowski, German Patent (filed December 10, 1950) 949,162 (1956).
33. K. Riess, *Chem. Ing. Tech.*, **27**, 457 (1955).
34. R. Erdmenger, German Patent (filed September 24, 1949) 815,641 (1951).
35. R. Erdmenger, German Patent (filed September 29, 1949) 813,154 (1951).
36. R. Erdmenger, U.S. Patent (filed September 20, 1950) 2,670,188 (1954).
37. K. Riess and R. Erdmenger, *VDI Zeitschr.*, **93**, 633 (1951).
38. O. Telle and R. Erdmenger, German Patent (filed July 22, 1953) 945,086 (1956).
39. R. Erdmenger, German Patent (filed July 28, 1953) 940,109 (1956); U.S. Patent (filed July 27, 1954) 2,814,472 (1957).
40. W. Winkelmuller, R. Erdmenger, S. Neidhardt, G. Hirschberg, and B. Fortuna, German Patent (filed December 10, 1951) 915,689 (1954).
41. R. Erdmenger, U.S. Patent (filed August 17, 1959) 3,122,356 (1964).
42. W. Meskat, *Chem. Ing. Tech.*, **34**, 742 (1962)
J. Pawlowski, *Chem. Ing. Tech.*, **34**, 749 (1962).
43. R. Erdmenger, *Chem. Ing. Tech.*, **34**, 751 (1962).
44. R. Erdmenger, *Chem. Ing. Tech.*, **36**, 175 (1964).
45. R. Erdmenger, U.S. Patent (filed March 26, 1963) 3,254,367 (1966).
46. L. Kraffe de Laubarede, U.S. Patent (filed July 10, 1950) 2,631,016 (1953).
47. L. Kraffe de Laubarede, British Patent (filed April 22, 1953) 738,784 (1955).
48. R. Fritsch and G. Fahr, *Kunststoffe*, **49**, 543 (1959).
49. P. G. Andersen, in Plastics Compounding: Equipment and Processing, edited by D. B. Todd, Hanser, Munich (1998).
50. H. Herrmann, *Kunststoff und Gummi*, **3**, 217 (1964).
51. H. Herrmann, *Chem. Ing. Tech.*, **38**, 25 (1966).
52. H. Boden, H. Ocker, G. Pfaff, and W. Worz, U.S. Patent (filed November 13, 1964) 3,305,894 (1967).
53. R. Fritsch and H. H. O. Kuhner, U.S. Patent (filed February 28, 1967) 3,392,962 (1968).
54. H. Ocker, U.S. Patent (filed November 15, 1968) 3,525,124 (1970).

55. H. Koch, U.S. Patent (filed November 20, 1968) 3,608,868 (1971).
56. R. Fritsch, U.S. Patent (filed February 24, 1966) 3,456,317 (1969).
57. H. Ocker, U.S. Patent (filed August 8, 1972) 3,682,086 (1972).
58. H. Herrmann and H. Ocker, U.S. Patent (filed February 7, 1972) 3,749,375 (1973).
59. H. Ocker, U.S. Patent (filed September 7, 1971) 3,764,114 (1973).
60. G. Illing, U.S. Patent (filed October 23, 1964) 3,536,680 (1970).
61. G. Illing, *Mod. Plastics*, (August) 72 (1969).
62. W. A. Mack and R. Herter, *Chem. Eng. Prog.*, **72** (1) 64 (1976).
63. R. Erdmenger and W. Oetke, German Auslegeschrift (filed March 16, 1960) 1,111,154 (1961).
64. R. Erdmenger and M. Ullrich, *Chem. Ing. Tech.*, **42**, 1 (1970).
65. K. Baigent, U.S. Patent (filed July 23, 1957) 2,916,769 (1959).
66. H. Beck, *Kunststoffe*, **47**, 469 (1957).
67. R. Colombo, U.S. Patent (filed December 27, 1961) 3,114,171 (1963).
68. R. Colombo, U.S. Patent (filed January 15, 1964) 3,252,182 (1966); German Offenlegungschrift (filed January 20, 1964) 1,554,761 (1970).
69. B. A. Loomans and A. K. Brennan, U.S. Patent (filed March 21, 1962) 3,195,868 (1965).
70. B. A. Loomans and A. K. Brennan, U.S. Patent (filed March 21, 1962) 3,198,491 (1965).
71. D. A. Wheeler, H. F. Irving, and D. B. Todd, U.S. Patent (filed October 30, 1969) 3,630,689 (1971).
72. B. A. Loomans, U.S. Patent (filed January 12, 1967) 3,423,074 (1969).
73. B. A. Loomans, U.S. Patent (filed March 6, 1973) 3,714,350 (1973).
74. B. A. Loomans, U.S. Patent (filed October 24, 1973) 3,900,187 (1975).
75. D. B. Todd, U.S. Patent (filed July 27, 1977) 4,136,968 (1979).
76. A. K. Brennan, U.S. Patent (filed November 14, 1969) 3,618,902 (1971).
77. A. K. Brennan, U.S. Patent (filed December 18, 1967) 3,490,750 (1970).
78. R. Erdmenger, M. Ullrich, R. Germadonk, J. Pedain, B. Quiring, and F. Wingler, U.S. Patent (filed January 21, 1971) 3,725,340 (1973).
79. M. Ullrich, E. Meisert, and A. Eitel, U.S. Patent (filed July 17, 1974) 3,963,679 (1976).
80. H. Korber, European Patent Application 0,002,760 (1979).
81. R. Binsack, D. Rempel, H. Korber, and D. Neuray, U.S. Patent (filed January 18, 1980) 4,260,690 (1981).
82. W. Goyert, E. Meisert, W. Grimm, A. Eitel, H. Wagner, G. Niederdellmann, and B. Quiring, U.S. Patent (filed December 13, 1979) 4,261,946 (1981).
83. B. Quiring, W. Wenzel, G. Niederdellmann, H. Wagner, and W. Goyert, U.S. Patent (filed June 20, 1980) 4,286,080 (1981).
84. C. Rathjen and M. Ullrich, European Patent (filed October 15, 1980) 0,049,835 (1982).
85. M. Ullrich and C. Rathjen, *Chem. Ing. Tech.*, **58**, 590 (1986).
86. H. Wobbe and E. Uhland, U.S. Patent (filed October 15, 1992) 5,318,358 (1994); German Patent 4,134,026 (1991).
87. J. Grillo, P. G. Andersen, and E. Papazoglou, *SPE ANTEC Tech. Papers*, **38**, 20 (1992).
88. H. Herrmann, *Kunststoffe*, **78**, 876 (1988).
89. H. Herrmann, in Plastics Extrusion Technology, edited by F. Hensen, Hanser, Munich (1988).

90. L. N. Valsamis, E. L. Canedo, J. M. Pereira, and D. U. Poscich, U.S. Patent (filed June 3, 1994) 5,487,602 (1996).
91. P. R. Hornsby, *Plastics Compounding*, (September/October) 65 (1983).
92. J. W. Ess, P. R. Hornsby, S. Y. Lin, and M. J. Beris, *Plastics Rubber Proc. Appl.*, **4**, 7 (1984).
93. P. R. Hornsby, *Plastics Rubber Proc. Appl.*, **7**, 237 (1987).
94. J. W. Ess and P. R. Hornsby, *Plastic Rubber Proc. Appl.*, **8**, 147 (1987).
95. D. Anders, U.S. Patent (filed July 7, 1981) 4,423,960 (1984).
96. D. Anders, *Kunststoffe*, **74**, 367 (1984).
97. D. Anders, *SPE ANTEC Tech. Papers*, **34**, 15 (1988).
98. M. H. Mack, in Plastics Compounding: Equipment and Processing, edited by D. B. Todd, Hanser, Munich (1998).
99. F. Zeppernick, *Kautsch Gummi Kunstst.*, **18**, 231 (1965).
100. F. Zeppernick, *Kautsch Gummi Kunstst.*, **18**, 313 (1965).
101. F. Zeppernick, *Kautsch Gummi Kunstst.*, **18**, 806 (1965).
102. T. R. Goshorn and F. R. Wolf, *Rubber Age*, (November) p. 77 (1965).
103. T. R. Goshorn, *J. Inst. Rubber Ind.*, (April) p. 77 (1972).
104. R. G. Bauer, J. L. Bush, H. K. Foley, D. C. Grimm, H. L. Gunnerson, G. E., Meyer, J. R. Purdon, and W. A. Wilson, in Science and Technology of Polymer Processing, edited by N. P. Suh and N. H. Sung, MIT Press, Cambridge, Mass., USA (1979).
105. P. K. Handa, C. M. Lansinger, V. R. Paramseran, and G. R. Schear, U.S. Patent (filed April 29, 1991) 5,158,725 (1992).
106. V. R. Eswaran, C. Kiehl, C. L. Magnus, and P. K. Handa, U.S. Patent (filed September 5, 1995) 5,711,904 (1998).
107. M. Hamada, T. Kinoshita, K. Kunimatsu, and S. Terashima, U.S. Patent (filed April 28, 1994) 5,409,978 (1995).
108. R. Caretta, R. Pessiwa, and A. Proni, European Patent Application (filed September 25, 1997) 0,911,359 A1 (1999).
109. R. Ophus, O. Skibba, R. H. Schuster, and U. Gorl, *Kautsch Gummi Kunstst.*, **53**, 279 (2000).
110. A. Amash, M. Bogun, R. H. Schuster, and M. Schmitt, *Plastics Rubber Comp.*, **30** 401 (2001).
111. U. Gorl, M. Schmitt, A. Amash, and M. Bogun, *Kautsch Gummi Kunstst.*, **55**, 23 (2002).
112. M. Bogun, F. Abraham, L. Muresan, R. H. Schuster, and H. J. Radasch, *Kautsch Gummi Kunstst.*, **57**, 363 (2004).
113. M. Bogun, S. Luther, R. H. Schuster, U. Gorl, and H. J. Radasch, *Kautsch Gummi Kunstst.*, **57**, 633 (2004).

5 Flow Mechanisms and Modeling of Intermeshing Co-Rotating Twin Screw Extruders

5.1 Introduction

In this chapter, we direct ourselves to the development of a fundamental understanding of the fluid mechanics of intermeshing co-rotating twin screw extruders. The basic understanding of the flow mechanisms was first revealed in the major patents which developed this technology, and are summarized in the previous chapter. Much of the early literature was confused because the authors did not properly distinguish between flow mechanisms in intermeshing co-rotating, and counter-rotating machines. Early patents generally realized the screw channels were fully open along their length. The first explicit discussion in the open scientific literature discussing this is given by Erdmenger [1] in 1964. Publications in scientific journals giving quantitative investigations of the fluid mechanics first appeared in the 1970s and continue into the present decade. In this chapter we will critically summarize this work.

We will begin in Section 5.2 with a discussion of screw geometry in intermeshing co-rotating twin screw systems. As in the case of counter-rotating machines, this is a topic requiring separate discussion. Section 5.3 summarizes fluid motions in forward pumping screw elements in these machines. The flow in backward pumping screw elements is considered in Section 5.4. Kneading disc blocks are described in Section 5.5. Non-isothermal behavior is considered in Section 5.6 and melting is described in Section 5.7. We then move on in Section 5.8 to considerations of global behavior of modular intermeshing co-rotating twin screw systems. The influence of the different modular elements on the behavior of the total system is described.

5.2 Screw Geometry

Intermeshing co-rotating twin screw extruders generally have self-wiping profiles for both their screw and kneading disc elements as may be seen in the Werner and Pfleiderer screw and kneading disc assemblies of Figure 5.1.

Early patents such as those of Wunsche [2], Easton [3], Pease [4], Colombo [5], and Meskat and Erdmenger [6] express the mechanisms of operation of intermeshing self-wiping co-rotating twin screw machines. Wunsche [2] first analyzed the detailed shape of the screw flights necessary. A. Geberg [7] at the Wolfen Works also later worked out the geometry of the self-wiping screw profile in 1944 and transmitted it to Meskat and Erdmenger.

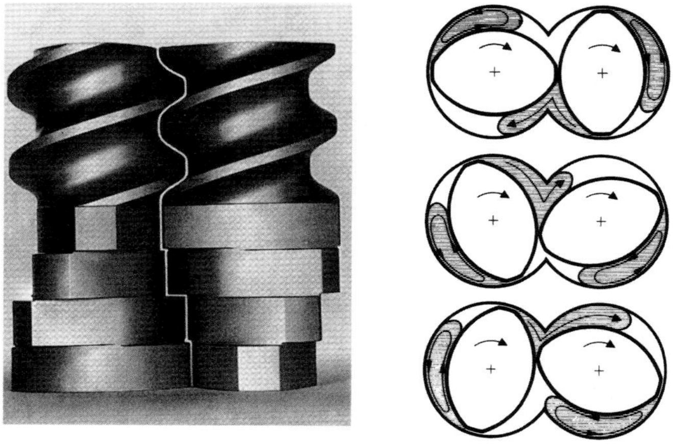

Figure 5.1: Self-wiping Werner and Pfleiderer screw and kneading disc assembly (courtesy of Werner and Pfleiderer)

A 1965 patent by Zimmermann [7] of Kalle later discussed the mathematical basis for the self-wiping profile. The problem was later taken up by Booy [8]. This author derived geometric relations for the shapes of the screws allowing them to become fully intermeshing and self-wiping. Booy showed that the channel depth in a self-wiping screw profile that is continuous and maintains a monotonic second increasing derivative must vary according to:

$$H(\theta) = R_s\,(1+\cos\theta) - \sqrt{C_L^2 - R_s^2 \sin^2\theta} \tag{5.1}$$

where R_s is the outer radius of the screw, θ is the angle of rotation of the screw and C_L is the centerline distance between the screws. For screw elements the angle θ may be expressed in forms of the axial distance along the length of the screw z:

$$\theta = 2\,\pi \frac{z}{S/p} \tag{5.2}$$

where S is the pitch and p is the number of thread starts.

Later patents of Erdmenger [9] and Loomans [10] have described different self-wiping screw profiles. These are shown in Figure 4.14 and 4.15. The screw profiles of Erdmenger and Loomans have 'grooves' and 'bumps' in them.

5.3 Modeling Flow in Forward Pumping Screw Elements

Mechanisms of flow are implicit in the patent literature. Indeed, Easton [3] fully recognized that the screw channels are fully open along the length of the screw, and it is implicit in all later patents. A reasonable discussion of the mechanisms of flow is given in a 1964 paper by Erdmenger [1] and a 1971 paper by Martelli [11]. Erdmenger notes that a fluid moves in a figure-8 motion through these screw channels along the length of screw. The mechanism is clarified in a series of papers by Armstroff and Zettler [12] of BASF and Herrmann and Burkhardt [13], Mack and Eise [14], Eise, Herrmann, Werner and Burkhardt [15] and Eise, Curry and Nangeroni [16] all of Werner and Pfleiderer.

Following these authors we note that in a fully intermeshing co-rotating twin screw extruder that in the screw elements, the channel is fully open as shown in Figure 5.2.

This is unlike the case of the intermeshing counter-rotating machine with its C-chambers and similar to a single screw extruder. In terms of the approximation that v_1, the velocity component along the screw channel, and v_3, the transverse velocity component, only depends upon the channel depth x_2, we have – as in Section 3.3:

$$0 = -\frac{\partial p}{\partial x_1} + \frac{\partial \sigma_{12}}{\partial x_2} = -\frac{\partial p}{\partial x_1} + \eta \frac{\partial^2 v_1}{\partial x_2^2} \tag{5.3a}$$

$$0 = -\frac{\partial p}{\partial x_3} + \frac{\partial \sigma_{32}}{\partial x_2} = -\frac{\partial p}{\partial x_3} + \eta \frac{\partial^2 v_3}{\partial x_2^2} \tag{5.3b}$$

with boundary conditions:

$$\begin{aligned} v_1(0) &= 0 \qquad & v_1(H) &= U_1 = U\cos\phi \\ v_3(0) &= 0 \qquad & v_3(H) &= U_3 = -U\sin\phi \end{aligned} \tag{5.4}$$

The velocity field in the channel direction is:

$$v_1(x_2) = U\cos\phi\left(\frac{x_2}{H}\right) - \frac{H^2}{2\eta}\frac{dp}{dx_1}\left[\left(\frac{x_2}{H}\right) - \left(\frac{x_2}{H}\right)^2\right] \tag{5.5}$$

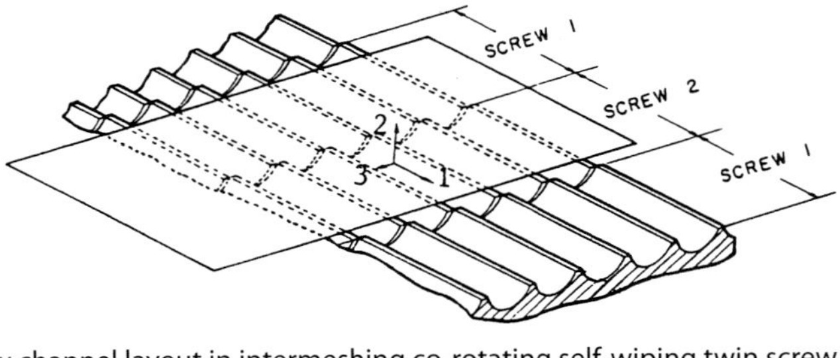

Figure 5.2: Screw channel layout in intermeshing co-rotating self-wiping twin screw extruder

This may be integrated to give for the flow rate:

$$Q = \frac{1}{2} W\, H\, U_1 - \frac{W\, H^3}{12\,\eta} \frac{\mathrm{d}p}{\mathrm{d}x_1} \tag{5.6}$$

The transverse flow, subject to the boundary condition of no flow over the screw flights, is:

$$v_3(x_2) = -U \sin\phi \left[2\left(\frac{x_2}{H}\right) - 3\left(\frac{x_2}{H}\right)^2 \right] \tag{5.7}$$

To proceed further, it was necessary to consider the geometry in detail, as done by Booy [8]. Analyses of flow through the screw channel considering the detailed geometry in a flattened configuration have been given by Denson and Hwang [12], Booy [18], and Szydlowski and White [19]. The former authors neglect the complications of the intermeshing regions and proceed by solving the equations:

$$0 = -\frac{\partial p}{\partial x_1} + \eta \left(\frac{\partial^2 v_1}{\partial x_2^2} + \frac{\partial^2 v_1}{\partial x_3^2} \right) \tag{5.8}$$

subjected to the boundary condition of no slip on the screw root defined by Eq. (5.4) and a velocity U on the barrel. They use a finite element method and apply Booy's screw channel shape as a boundary condition. The solution for the throughput is of form:

$$Q = \frac{1}{2} W\, H_0\, U_1\, F_D - \frac{W\, H^3}{12\,\eta} \frac{\mathrm{d}p}{\mathrm{d}x_1} F_p \tag{5.9a}$$

$$Q = \frac{1}{2} W\, H_0\, U \cos\phi\, F_D - \frac{W\, H^3}{12\,\eta} \frac{\mathrm{d}p}{\mathrm{d}x_1} F_p \tag{5.9b}$$

where H is a characteristic channel depth and F_D and F_p are shape factors. Denson and Hwang calculate and plot screw characteristic curves (Figure 5.3) in terms of the variables:

$$Q^* = \frac{Q}{2\,\pi\, R^3\, N} \cos\phi \qquad \frac{\mathrm{d}p^*}{\mathrm{d}x_1^*} = \frac{R}{2\,\pi\,\eta\, N \cos\phi} \frac{\mathrm{d}p}{\mathrm{d}x_1} \tag{5.10}$$

These have the linear decreasing form typical of calculations for Newtonian fluids in single screw extruders (Chapter 3) and tangential counter-rotating twin screw extruders (Chapters 11 and 12).

Szydlowski and White [19] in a subsequent study tried to accommodate the problems of the intermeshing region. As shown in Figure 5.4, the flow is diverted where the screws come together and this must add a pressure loss. They proceeded using hydrodynamic lubrication theory and considered shearing in the x_3 direction. They began with Eqs. (5.3 a, b) and integrated them to yield:

$$q_1 = \int_0^H v_1 \, dx_2 = \frac{1}{2} U_1 H - \frac{H^3}{12\,\eta} \frac{\partial p}{\partial x_1} \tag{5.11a}$$

$$q_3 = \int_0^H v_3 \, dx_2 = \frac{1}{2} U_3 H - \frac{H^3}{12\,\eta} \frac{\partial p}{\partial x_3} \tag{5.11b}$$

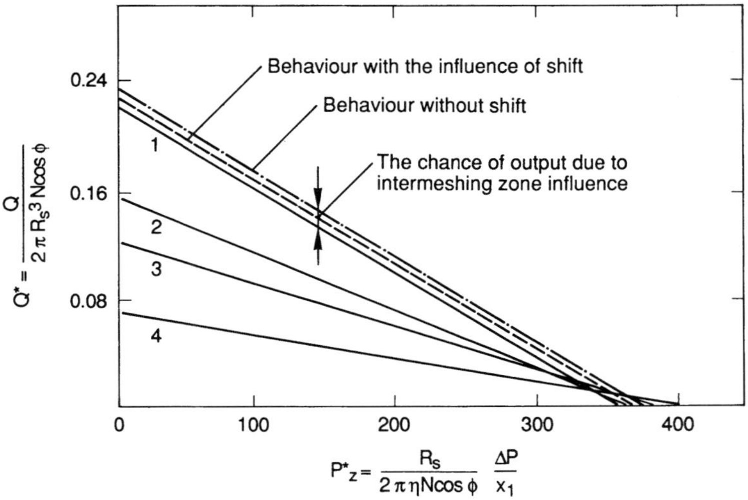

Figure 5.3: Screw characteristic curve calculations of Denson and Hwang [17] and Szydlowski and White [19] for selected screw designs

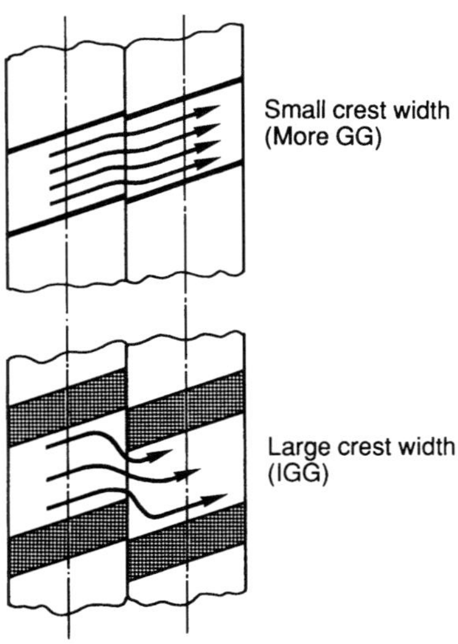

Figure 5.4: Flow in intermeshing co-rotating twin screw region [18]

These equations were solved simultaneously using a generalization of Tadmor et al. [20] FAN method. This in essence is a balance of fluxes based on Eq. (5.11), i.e.:

$$q_1(x_1+\Delta x_1,x_3)\,\Delta x_3+q_3(x_1,x_3+\Delta x_3)\,\Delta x_1-q_1(x_1,x_3)\,\Delta x_3-q_3(x_1,x_3)\,\Delta x_1=0 \quad (5.12a)$$

and is equivalent to a solution of:

$$\frac{\partial q_1}{\partial x_1}+\frac{\partial q_3}{\partial x_3}=0 \quad (5.12b)$$

which is a partial differential equation in $p(x_1,x_3)$.

In Figure 5.3, the results of Szydlowski and White [19] are compared with the computations of Denson and Hwang. Generally, the agreement is excellent for the screw designs considered. However, for large flight thicknesses, discrepancies arise with any modeling which neglects the intermeshing region. This is shown in Figure 5.5.

A more sophisticated numerical approach to the complex geometry of this problem is taken in a paper by Kalyon and his co-workers [21, 22], who attempted to take into consideration the full curvature of the machine, especially in the intermeshing region where the two screws come into contact. Crochet [23] reported similar finite element studies of flow in a screw channel of an intermeshing co-rotating twin screw extruder which includes curvature effects.

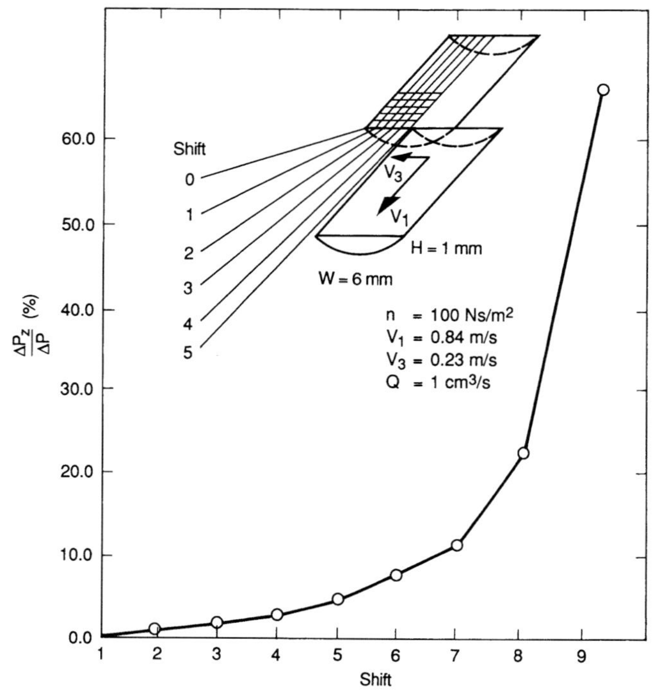

Figure 5.5: Effects of screw flight thickness on pressure development [18]

The analysis of Szydlowski and White described above has been generalized to non-Newtonian fluids by Wang and White [24]. They used the power fluid formulation for the non-Newtonian viscosity function. Specifically, they take η to be dependent upon the rate of deformation invariants as described in Chapter 2 (Eq. (2.45)), i.e.:

$$\eta = K\,(2\,t\,r\,\underline{d}^2)^{(n-1)/2} \tag{5.13a}$$

becomes

$$\eta = K\left[\left(\frac{\partial v_1}{\partial x_2}\right)^2 + \left(\frac{\partial v_3}{\partial x_2}\right)^2\right]^{(n-1)/2} \tag{5.13b}$$

This leads to in place of Eq. (5.3):

$$\begin{aligned} 0 &= -\frac{\partial p}{\partial x_1} + \frac{\partial \sigma_{12}}{\partial x_1} \\ &= -\frac{\partial p}{\partial x_1} + \frac{\partial}{\partial x_2} K\left[\left(\frac{\partial v_1}{\partial x_2}\right)^2 + \left(\frac{\partial v_3}{\partial x_2}\right)^2\right]^{(n-1)/2} \frac{\partial v_1}{\partial x_2} \end{aligned} \tag{5.14a}$$

$$\begin{aligned} 0 &= -\frac{\partial p}{\partial x_3} + \frac{\partial \sigma_{32}}{\partial x_1} \\ &= -\frac{\partial p}{\partial x_3} + \frac{\partial}{\partial x_2} K\left[\left(\frac{\partial v_1}{\partial x_2}\right)^2 + \left(\frac{\partial v_3}{\partial x_2}\right)^2\right]^{(n-1)/2} \frac{\partial v_3}{\partial x_2} \end{aligned} \tag{5.14b}$$

These equations may be integrated to yield:

$$\frac{\partial v_1}{\partial x_2} = \frac{1}{\eta}\frac{\partial p}{\partial x_1}(x_2 - C_1) \tag{5.15a}$$

$$\frac{\partial v_3}{\partial x_2} = \frac{1}{\eta}\frac{\partial p}{\partial x_3}(x_2 - C_3) \tag{5.15b}$$

suggesting (following Griffith [25]):

$$\eta = K\left[\frac{1}{\eta^2}\left(\frac{\partial p}{\partial x_1}\right)^2 (x_2 - C_1)^2 + \frac{1}{\eta^2}\left(\frac{\partial p}{\partial x_3}\right)^2 (x_2 - C_3)^2\right]^{(n-1)/2} \tag{5.16a}$$

$$\eta = K^{1/n}\left[\left(\frac{\partial p}{\partial x_1}\right)^2 (x_2 - C_1)^2 + \left(\frac{\partial p}{\partial x_3}\right)^2 (x_2 - C_3)^2\right]^{(n-1)/2n} \tag{5.16b}$$

Integrating Eqs. (5.15) to obtain v_1 and v_3 as well $\overline{v}_1$ and $\overline{v}_3$ yields:

$$q_1 = \overline{v}_1 H = \int_0^H \int_0^{x_2} K^{-1/n} \left[\left(x_2 \frac{\partial p}{\partial x_1} - C_1 \right)^2 + \left(x_2 \frac{\partial p}{\partial x_3} - C_3 \right)^2 \right]^{(1-n)/2n} \left(x_2 \frac{\partial p}{\partial x_1} - C_1 \right) \mathrm{d}x_2 \, \mathrm{d}x_2 \tag{5.17a}$$

$$q_3 = \overline{v}_3 H = \int_0^H \int_0^{x_2} K^{-1/n} \left[\left(x_2 \frac{\partial p}{\partial x_1} - C_1 \right)^2 + \left(x_2 \frac{\partial p}{\partial x_3} - C_3 \right)^2 \right]^{(1-n)/2n} \left(x_2 \frac{\partial p}{\partial x_3} - C_3 \right) \mathrm{d}x_2 \, \mathrm{d}x_2 \tag{5.17b}$$

Equations (5.17) are introduced into Eqs. (5.12a) which are then solved for the pressure field, which is in turn differentiated to yield the flow field q_1 and q_3.

Screw characteristic curves for non-Newtonian fluids are presented in Figure 5.6. These are presented in terms of the variables:

$$Q^* = \frac{Q}{2\,\pi\,R^3\,N\cos\phi} \qquad \frac{\mathrm{d}p^*}{\mathrm{d}x_1^*} = \frac{R}{K\,(2\,\pi\,R\,N\cos\phi)^n}\,\frac{\mathrm{d}p}{\mathrm{d}x_1} \tag{5.18}$$

These indicate a decrease in pumping characteristics as power law exponent decreases. The results are similar to the predictions of Griffith [25] and later investigations for single screw extrusion characteristics (see Section 3.3).

A more general view of this problem has also been taken by Wang and White [24]. They seek to simulate flow in a screw channel for a non-Newtonian fluid with a velocity field:

$$\underset{\sim}{v} = v_1(x_2, x_3)\,\underset{\sim}{e}_1 + v_2(x_2, x_3)\,\underset{\sim}{e}_2 + v_3(x_2, x_3)\,\underset{\sim}{e}_3 \tag{5.19}$$

They use as equations of motion:

$$0 = -\frac{\partial p}{\partial x_1} + \frac{\partial}{\partial x_2}\left(\eta \frac{\partial v_1}{\partial x_2} \right) + \frac{\partial}{\partial x_3}\left(\eta \frac{\partial v_1}{\partial x_3} \right) \tag{5.20a}$$

$$0 = -\frac{\partial p}{\partial x_2} + \frac{\partial}{\partial x_2}\left(2\,\eta \frac{\partial v_2}{\partial x_2} \right) + \frac{\partial}{\partial x_3}\left[\eta \left(\frac{\partial v_2}{\partial x_3} + \frac{\partial v_3}{\partial x_2} \right) \right] \tag{5.20b}$$

$$0 = -\frac{\partial p}{\partial x_3} + \frac{\partial}{\partial x_2}\left[\eta \left(\frac{\partial v_3}{\partial x_2} + \frac{\partial v_2}{\partial x_3} \right) \right] + \frac{\partial}{\partial x_3}\left(2\,\eta \frac{\partial v_3}{\partial x_3} \right) \tag{5.20c}$$

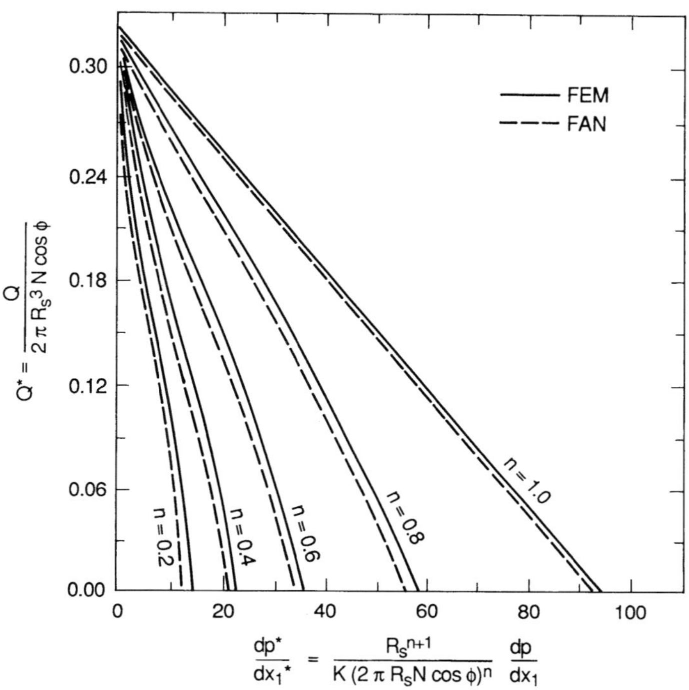

Figure 5.6: Screw characteristic curves for power law fluids as determined by Wang and White [24] comparing FAN and finite element (FEM) techniques

The shear viscosity was taken as:

$$\eta = K\left[\left(\frac{\partial v_1}{\partial x_2}\right)^2 + \left(\frac{\partial v_1}{\partial x_3}\right)^2 + \left(\frac{\partial v_3}{\partial x_2} + \frac{\partial v_2}{\partial x_3}\right)^2 + \left(\frac{\partial v_2}{\partial x_2}\right)^2 + \left(\frac{\partial v_3}{\partial x_3}\right)^2\right]^{(n-1)/2} \tag{5.21}$$

The problem was solved by a finite element method using the variational principle based on optimizing:

$$J = \int \frac{(2\,t\,r\,\underline{d}^2)^{(n+1)/2}}{(n+1)}\,\mathrm{d}V - \int p\left(\frac{\partial v_2}{\partial x_2} + \frac{\partial v_3}{\partial x_3}\right)\mathrm{d}V \tag{5.22}$$

where $2\,t\,r\,\underline{d}^2$ is as defined in Eq. (2.45).

A second finite element simulation of the flow of a power law fluid in a flattened out self-wiping screw channel has been given by Lai-Fook and co-workers [25, 26]. More recent finite element studies are by Goffart et al. [27] and by Bierdel [28].

The down channel velocity fields computed by Wang and White for open and closed discharge are contained in Figure 5.7 a and b. There are only positive components of v_1 for open discharge, but there are negative components as well for closed discharge. The magnitudes of the velocity fields are lower as expected for the non-Newtonian case as opposed to the Newtonian for closed discharge.

The above results were applied by these authors to calculations of screw characteristics. The results are contained in Figures 5.6. The differences are no more than 10% for the simple model of Eq. (5.13) to (5.17).

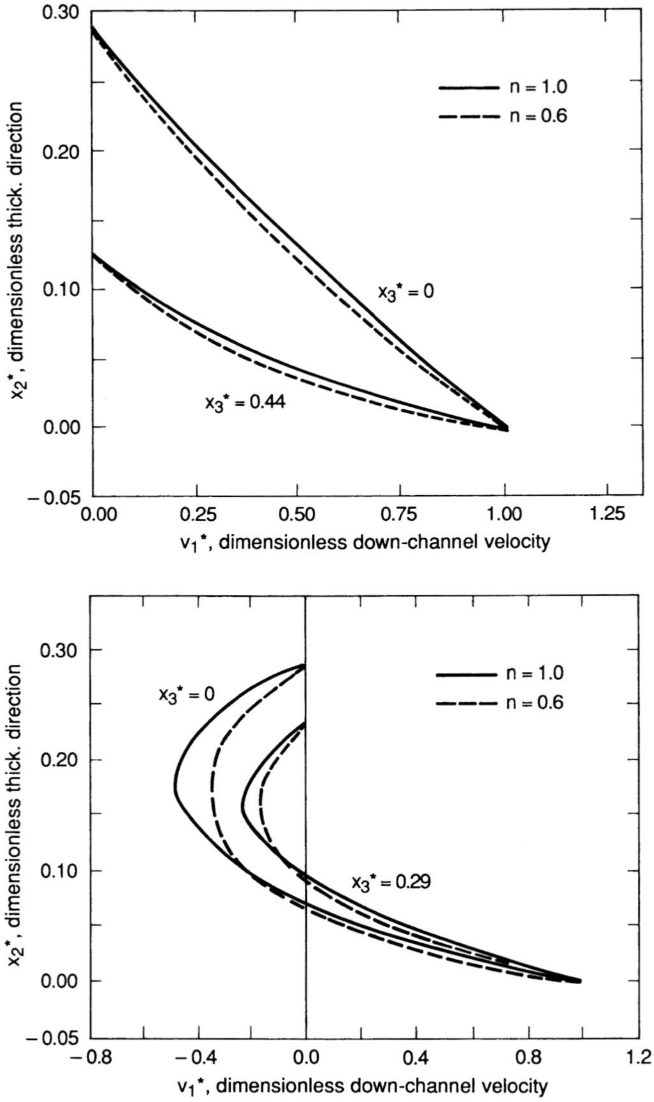

Figure 5.7 a, b: Down channel axial and transverse velocity fields at open and closed discharge of Wang and White [24]

5.4 Modeling Flow in Backward Pumping Screw Elements

The discussion of the previous section applies to forward pumping screw elements. The same modeling may be applied to backward pumping screw elements. The difference occurs in the boundary condition of Eq. (5.4). Instead of $v_1(H)$ being $U\cos\phi$, it is of form $(-U\cos\phi)$, i.e.:

$$\begin{aligned} v_1(0)&=0 \qquad & v_1(H)&=U_1=-U\cos\phi \\ v_3(0)&=0 \qquad & v_3(H)&=U_3=U\sin\phi \end{aligned} \tag{5.23}$$

The drag flow is reversed and tends to pull the fluid towards the hopper rather than the die. An analysis of this type has been carried for Newtonian fluids by White and Szydlowski [29]. This solution may be expressed as in Eq. (5.9), but the quantity U is negative, i.e.:

$$Q=-\frac{1}{2}U\,W\,H_0\,\cos\phi\,F_D-\frac{W\,H_0^3}{12\,\eta}\frac{dp}{dx_1}F_p \tag{5.24}$$

It should be seen from Eq. (5.24) that a positive throughput Q is only possible when dp/dx_1 is negative, i.e., the pressure decreases in the extrusion direction.

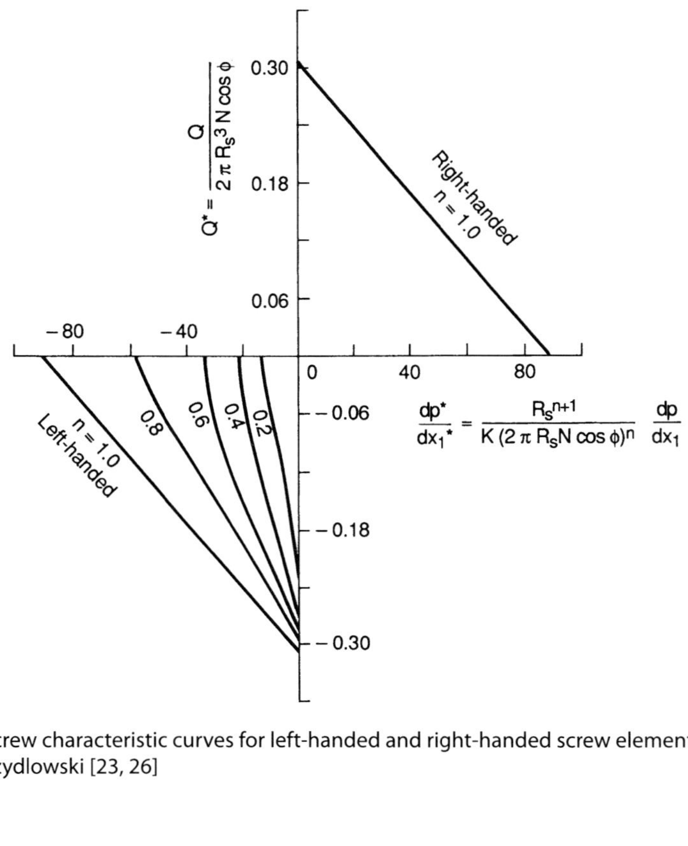

Figure 5.8: Screw characteristic curves for left-handed and right-handed screw elements of White and Szydlowski [23, 26]

Flow of Newtonian fluids in backward pumping screw elements was modeled by Tayeb et al. [30] in cylindrical coordinates.

Modeling the flow of power law fluids can be extended to left-handed screw elements. Wang and White [24] have presented computations of flow in left handed screw elements for power law fluids. Screw characteristic curves for left-handed screw elements are presented in Figure 5.8.

5.5 Flow Mechanisms and Modeling in Kneading Disc Block Elements

The heart of the intermeshing co-rotating twin screw extruder is the kneading disc block or paddle elements that melt the polymer and disperse ingredients. Kneading disc block/paddle elements, as we described in the previous chapter, develop from the work of Nelson [31] and Erdmenger [32–34]. Erdmenger [1] realized many of the basic mechanisms of flow in kneading disc block elements.

Unfortunately, much less attention has been given to flow in the kneading disc region than to screw elements. The first serious consideration of flow in the kneading disc region was given by Hans Werner [35], a Werner and Pfleiderer engineer, who did a doctoral dissertation at the University of Munich. This was also discussed by Werner and Eise [36] (see also Eise et al. [16]).

More recently, this subject has been developed in more detail by Szydlowski, Brzoskowski and White [37, 38]. Szydlowski et al. analyze flow in the kneading disc block region using hydrodynamic lubrication theory coupled with the FAN balance technique described in the previous section. This has been done for various kneading disc block geometries and staggering angles. They erect a coordinate system in which '1' represents the machine axis, '3' is the circumferential direction, and '2' the perpendicular distance between kneading disc and barrel (Figure 5.9). This leads to:

$$q_1 = -\frac{H^3}{12\,\eta}\frac{\partial p}{\partial x_3} \tag{5.25a}$$

$$q_3 = \frac{1}{2}U_3\,H - \frac{H^3}{12\,\eta}\frac{\partial p}{\partial x_3} \tag{5.25b}$$

where U_1 is the velocity in the direction of the screw axis is zero.

The coordinate system is embedded in the screw and the barrel is considered to move. The variation $H(x_3)$ is given by Booy's Eq. (5.1). The variation $H(x_1)$ is specified by kneading disc thickness and staggering angle. Szydlowski et al. [37] use Eq. (5.12a) as a balance of the fluxes.

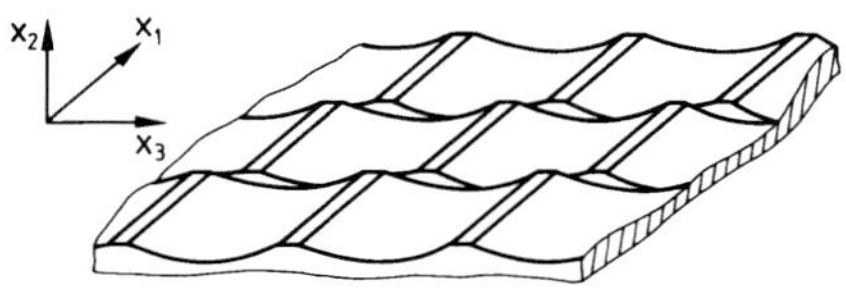

Figure 5.9: Flattened kneading discs in Szydlowski et al. model [37]

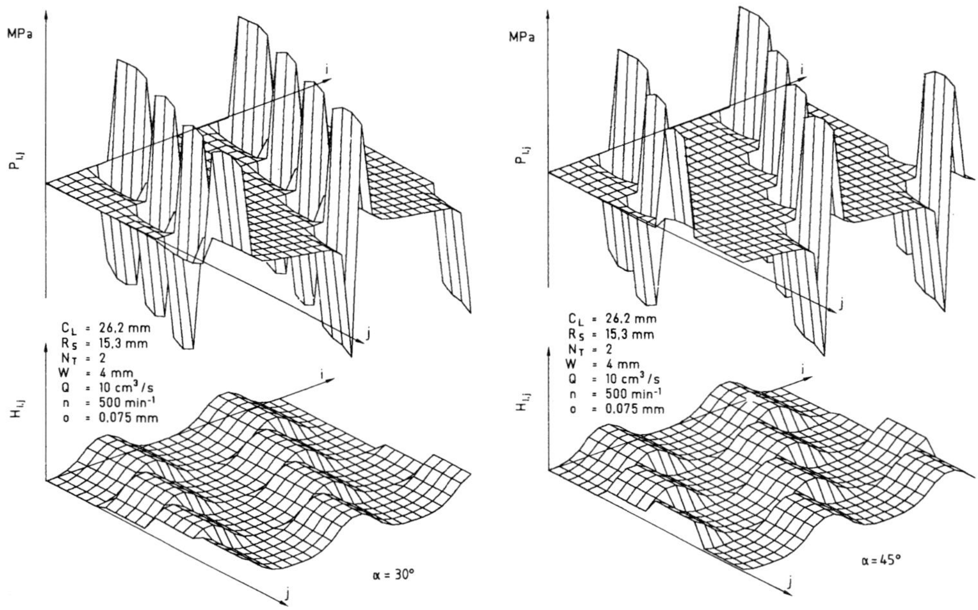

Figure 5.10: Pressure field in kneading disc block region [37]. Staggering angles are 30° and 45°

Szydlowski et al. [37] calculated pressure distributions among the kneading discs. This is shown for various kneading disc staggering angles in Figure 5.10. Using Eqs. (5.10) this may be related to the flux field q_1 and q_3.

The staggering of the kneading discs creates a helical channel resembling a screw dependent on the value of the staggering angle of the succeeding discs. These channels may pump fluid forward and backward along the screw. It is thus possible to compute screw characteristic curves for the pumping behavior of kneading disc assemblies if one also accounts for leakage flows. This is done in terms of the variables:

$$Q^* = \frac{Q}{2\,\pi\,R\,N\cos\phi} \tag{5.26a}$$

$$\frac{\mathrm{d}p^*}{\mathrm{d}x_1^*} = \frac{R}{2\,\pi\,\eta\,N\cos\phi}\,\frac{\mathrm{d}p}{\mathrm{d}x_1} \tag{5.26b}$$

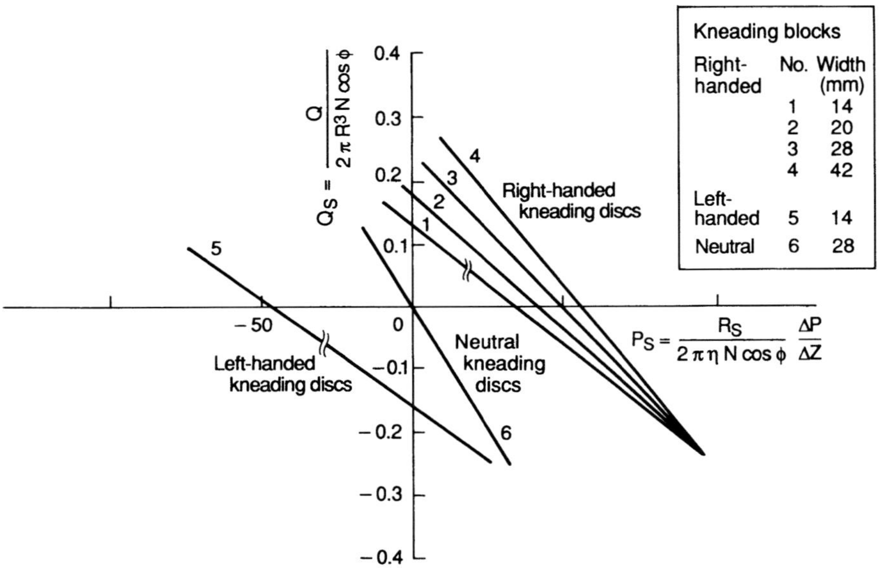

Figure 5.11: Screw characteristic curves for kneading disc blocks based on White and Szydlowski [29]

Typical $Q^* - \dfrac{dp^*}{dx_1^*}$ screw characteristic curves for kneading disc blocks are shown in Figure 5.11. These were computed by White and Szydlowski [29] and show a much steeper slope than the screw characteristic curves for screw elements shown in Figure 5.3. This indicates that a kneading disc block cannot pump against as large a pressure gradient as a screw. Figure 5.11 also shows screw characteristic curves for left handed kneading disc block elements, i.e., kneading disc block elements with staggering angles greater than 90° as well as right handed kneading disc block elements. As with the right-handed kneading disc blocks, these have higher slopes than screw elements.

Let us turn to flow patterns and distributive mixing. The flow field in kneading disc elements is shown in Figure 5.12. The flow in the kneading disc block basically resembles that in a poorly designed screw section. As we have noted, the staggering of the kneading disc at a constant angle defined a screw channel which may have a right handed or left handed configuration. Right handed kneading disc configurations built up pressure along the direction of flow which produces a backward motion in the interstices between the disc tips. These backward flows produce distributive mixing.

Szydlowski et al. [36] defined the axially forward flows Q_L, the circumferential flows Q_C, and the reverse flows Q_b caused by the staggering of the discs. The relative importance of the different flows is determined by the ratios:

$$f_L = \frac{Q_L}{Q_L + Q_C + Q_b} \tag{5.27a}$$

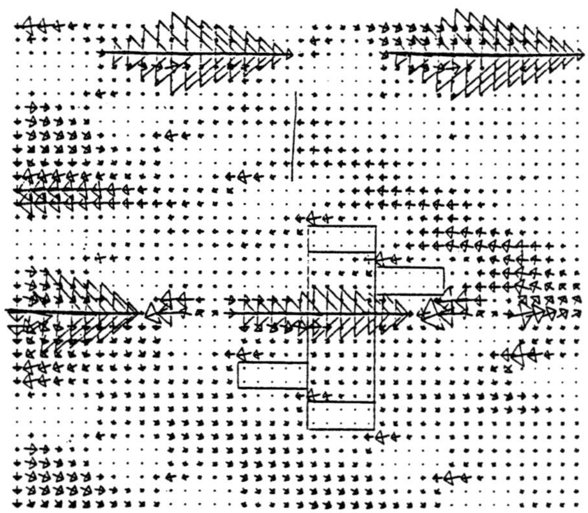

Figure 5.12: Flow fields in kneading disc region according to Szydlowski et al.

$$f_C = \frac{Q_C}{Q_L + Q_C + Q_b} \tag{5.27b}$$

$$f_b = 1 - f_L - f_C \tag{5.27c}$$

Mixing effects are produced by the Q_b and Q_C flows and may be evaluated by the magnitude of f_L and f_C; f_b, representing back flow, would seem to measure distributive mixing and f_C dispersive mixing as it represents flow over tips of kneading discs. Typical plots computed by Szydlowski et al. are shown in Figure 5.13. The staggering angle has a major effect on mixing. The greatest value of f_b is at about a staggering angle of 60°. f_b decreases with increasing machine throughput.

Werner [35] introduced a *G* factor to characterize the flow which has the form:

$$G = \frac{Q_b}{Q_L - Q_b} = \frac{Q_b}{Q} \tag{5.28}$$

Szydlowski et al. computed *G* as a function of processing conditions as shown in Figure 5.14.

The flow of non-Newtonian fluids in a kneading disc region was first analyzed in papers by Szydlowski and White [39] and Wang, White and Szydlowski [40]. They described four different methods of analyzing non-Newtonian flow in these machines. The most successful of these methods is the one that applies Eqs. (5.12) through (5.17). This has been applied to compute pressure fields, screw characteristic curves, and flow patterns, including f_L, f_C, f_b, and *G*. Pressure fields are much reduced in intensity with decreasing power law exponent. Screw characteristic curves for power law fluids in kneading disc regions are shown in Figure 5.15.

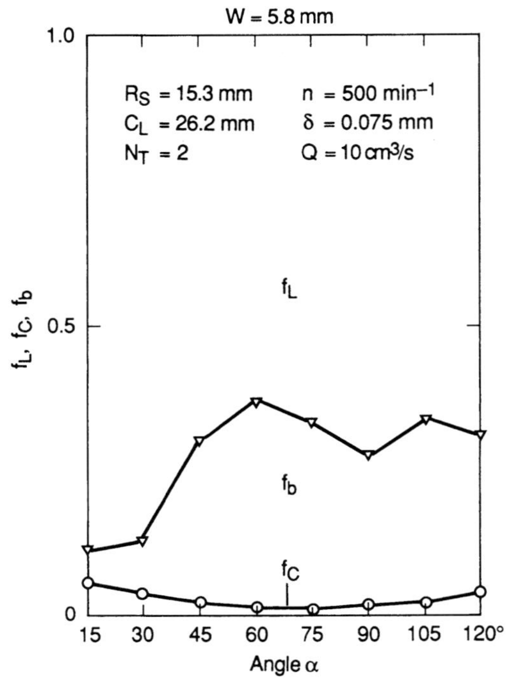

Figure 5.13: Szydlowski et al. [36] factors f_L, f_C, and f_b, as a function of process conditions in kneading disc

As with screw elements (Figure 5.6), increasing non-Newtonian character through decreasing power law exponents decreases pumping capability. Figure 5.16 shows f_L, f_C, and f_b for power law fluids. Generally, decreasing power law exponents increase f_b which in turn increases mixing. However, the effects of viscoelasticity are not included and the reader should be wary.

Szydlowski and White [38] have argued that the flow in the kneading disc block region is more complex than described above. There is an additional set of motions due to peristaltic pumping effects. This is due to the variation of the local value in the intermeshing regions between the screws. We illustrate this effect in Figure 5.17. Szydlowski and White analyzed this by introducing:

$$q_1(x_1+\Delta x_1, x_3)\,\Delta x_3 + q_3(x_1, x_3+\Delta x_1)\,\Delta x_1 - q_1(x_1,x_3)\,\Delta x_3 - q_3(x_1,x_3)\,\Delta x_1 = \frac{\mathrm{d}}{\mathrm{d}t}(H\,\Delta x_1\,\Delta x_3) \tag{5.29}$$

This was shown to introduce a cyclical motion during each screw revolution. For a right-handed arrangement of kneading discs, this gives a forward pumping motion; for a left-handed arrangement, it gives a backward pumping motion. This produces time varying pressures fluid fluxes Q_L and in addition Q_b and Q_C (Figure 5.18). As a result, f_L, f_C, and f_b vary in time during screw rotation.

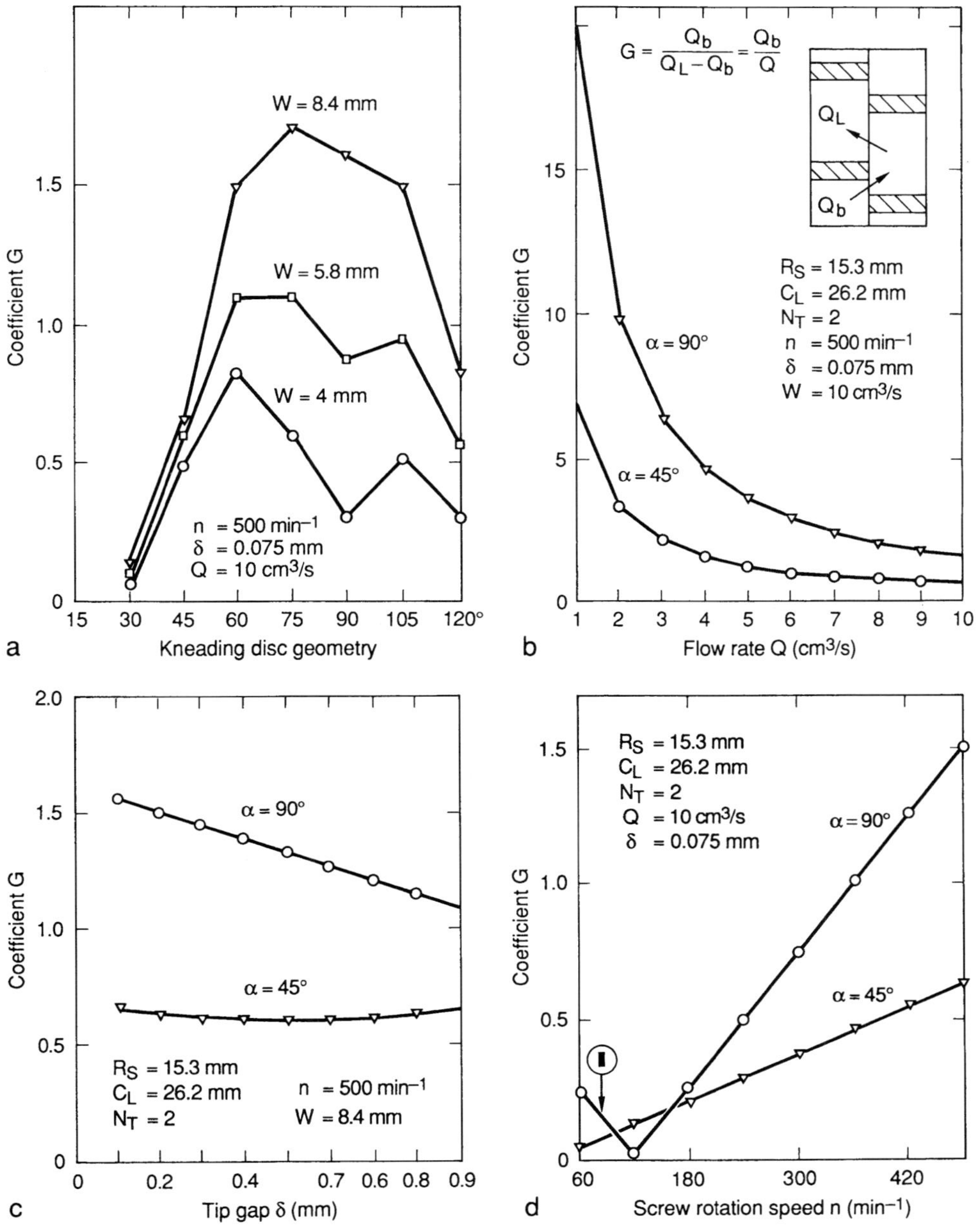

Figure 5.14: Werner's *G* Factor as a function of process conditions in kneading disc as computed by Szydlowski et al. [37]

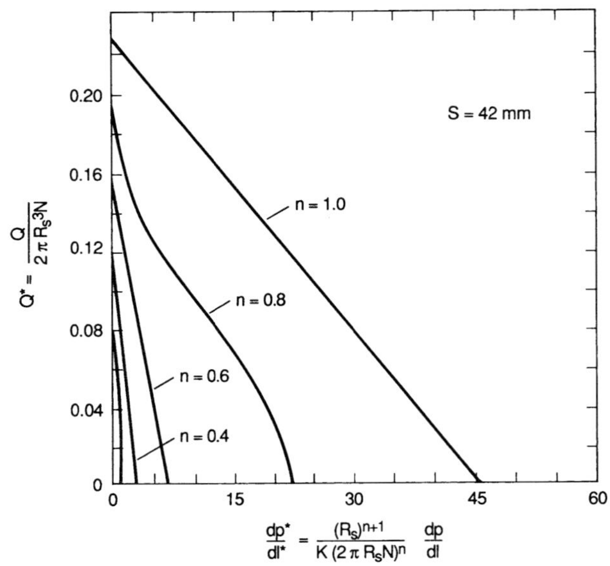

Figure 5.15: 'Screw characteristic curves' for kneading disc regions for power law fluids [40]

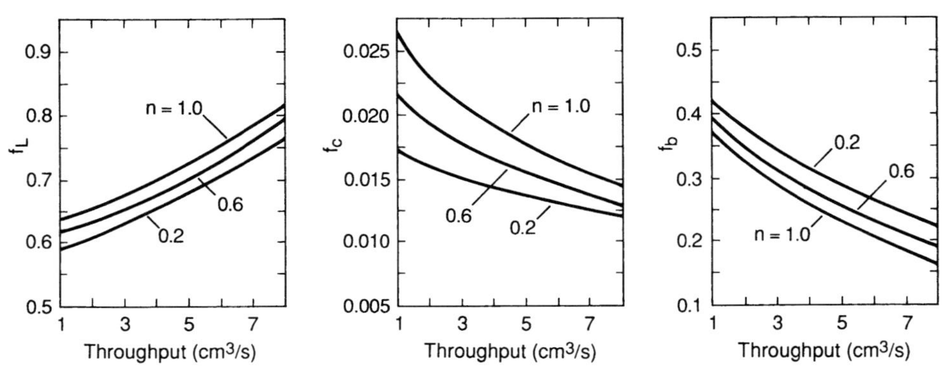

Figure 5.16: Factors f_L, f_C, and f_b for power law fluids in kneading disc regions [36]

Szydlowski and White made their calculations using a model in which the coordinate system is embedded in the kneading discs. Tadmor [41] pointed out that these results develop more naturally out of a formulation in which the coordinate system is embedded in the barrel and cylindrical coordinates are used. A formulation of this type has been given by Sebastian and Rakos [42].

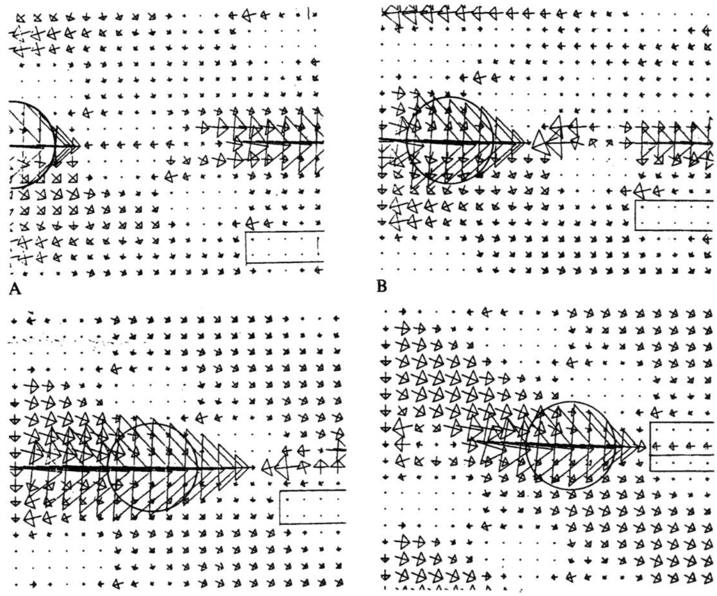

Figure 5.17: Peristaltic pumping motion in kneading disc region after Szydlowski and White [38]

Finite element simulation of flow in kneading disc blocks has been given variously by Gotsis and Kalyon [43], Yang and Manas Zloczower [44], Lawal, Kalyon and Ji [45], Kiani and Samann [46], van der Wal et al. [47] and Ishikawa et al. [48]. Gotsis and Kalyon [43] have calculated streamlines and velocity fields involving pairs of self-wiping co-rotating kneading discs for 3-dimensional flow of a Newtonian fluid. Yang and Manas-Zloczower [45] made calculations for a power law fluid and found similar velocity fields. They emphasized the probable importance of elongational flows, which were noted by earlier. They seek to represent this with a λ parameter defined by

$$\lambda = \frac{\dot{\gamma}}{\dot{\gamma} + \dot{\omega}} = \frac{\text{tr}\ \underline{d}^2}{\text{tr}\ \underline{d}^2 + \text{tr}\ \underline{\omega}^2} \tag{5.30}$$

where $\underline{d}$ and $\underline{\omega}$ are the deformation rate and vorticity tensors and tr( ) represents the sum of the diagonal components of the squares of $\underline{d}$ and $\underline{\omega}$. van der Wal et al. [47] presented calculated shear and elongation ratios at different disc staggering angles and screw speeds, extents of backflow, and pressure drops.

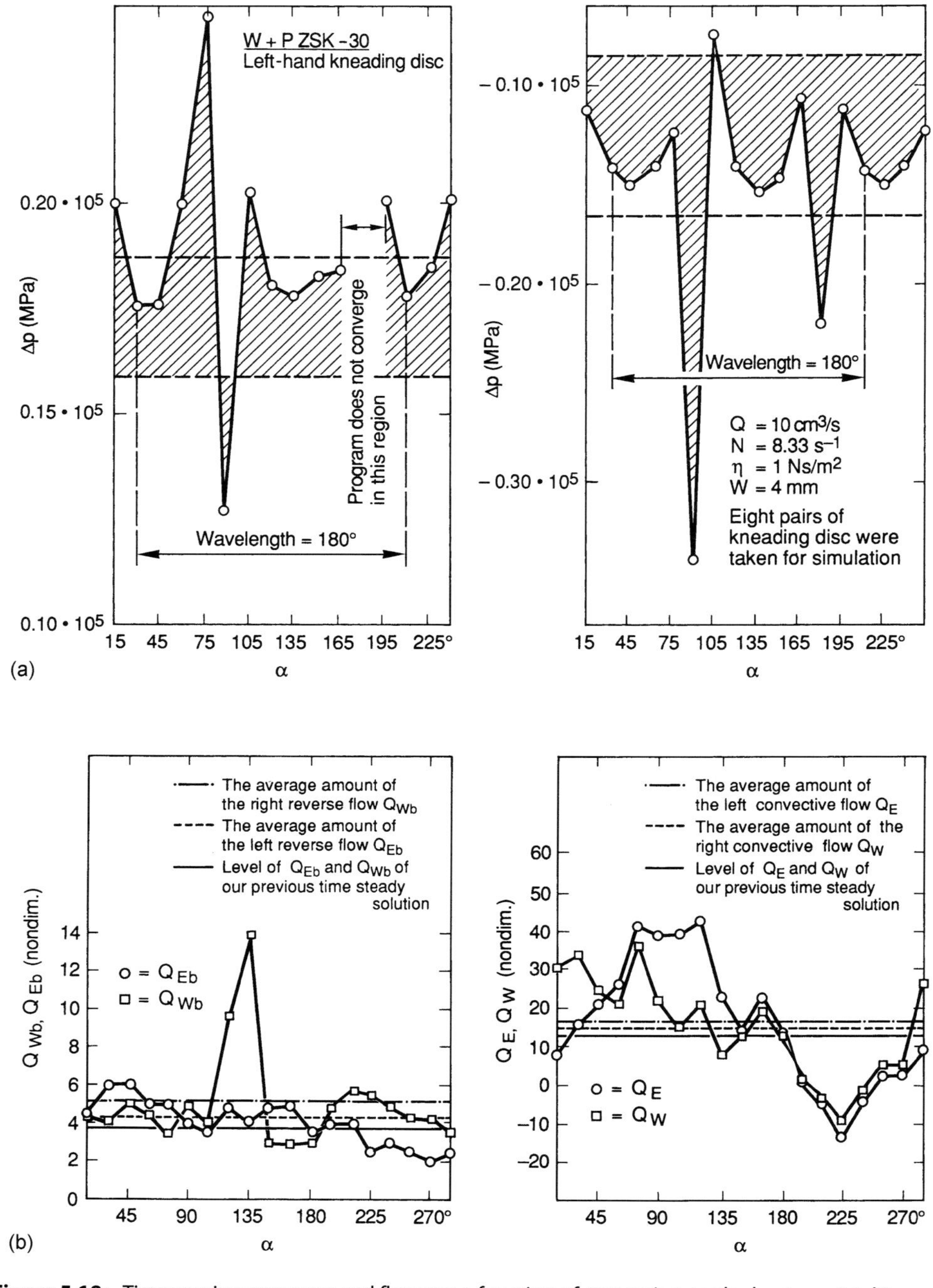

Figure 5.18: Time varying pressures and fluxes as a function of staggering angle due to peristaltic pumping effect; (a) pressure field, (b) time varying fluxes in right-handed kneading disc block [38]

5.6 Energy Balances

Energy balances in co-rotating twin screw extruders were first considered by Meijer and Elemans [49]. We need an energy equation compatible with the fluid mechanics developed in Sections 5.3 to 5.5. This was based on fluxes. This should be based on appropriate mean temperatures.

The energy equation has the form (Section 2.7):

$$\rho\, c\left[\frac{\partial T}{\partial t}+(\underset{\sim}{v}\cdot\nabla)\,T\right]=k\,\nabla^2 T+\sum_i\sum_j \sigma_{ij}\, d_{ij} \tag{5.31}$$

If we specialize this to a steady-state flow in a channel with direction '2' normal to the direction of flow and neglect heat conduction along the primary direction of flow '1', we have in Cartesian coordinates with

$$\underset{\sim}{v}=v_1(x_2)\,\underset{\sim}{e}_1+0\,\underset{\sim}{e}_2+v_3(x_2)\,\underset{\sim}{e}_3 \tag{5.32}$$

Then

$$\rho\, c\left(v_1\frac{\partial T}{\partial x_1}+v_3\frac{\partial T}{\partial x_3}\right)=k\left(\frac{\partial^2 T}{\partial x_2^2}+\frac{\partial^2 T}{\partial x_3^2}\right)+\sigma_{12}\frac{\partial v_1}{\partial x_2}+\sigma_{32}\frac{\partial v_3}{\partial x_2} \tag{5.33}$$

It is important to consider averaged temperatures through the cross-section. This is done by integrating through the depth x_2 and introducing the fluxes q_1 and q_3 which arise in Reynolds lubrication theory [50] (Section 2.8).

$$q_1=\int_0^H v_1\,\mathrm{d}x_2 \qquad q_3=\int_0^H v_3\,\mathrm{d}x_2 \tag{5.34}$$

and integrating Eq. (5.33)

$$\rho\, c\int_0^H\left(v_1\frac{\partial T}{\partial x_1}+v_3\frac{\partial T}{\partial x_3}\right)\mathrm{d}x_2=k\int_0^H\frac{\partial^2 T}{\partial x_2^2}\,\mathrm{d}x_2+\int_0^H\left[\sigma_{12}\frac{\partial v_1}{\partial x_2}+\sigma_{32}\frac{\partial v_3}{\partial x_2}\right]\mathrm{d}x_2 \tag{5.35}$$

to give

$$\rho\, c\left[q_1\frac{\partial \overline{T}}{\partial x_1}+q_3\frac{\partial \overline{T}}{\partial x_3}\right]=E(H)+E(0)+\sigma_{12}(H)\,U_1+\sigma_{32}(H)\,U_3-q_1\frac{\partial p}{\partial x_1}-q_3\frac{\partial p}{\partial x_3} \tag{5.36}$$

where

$$E(H)=k\frac{\partial T}{\partial x_2}(H) \qquad E(0)=-k\frac{\partial T}{\partial x_2}(0) \tag{5.37a}$$

$$\frac{\partial \overline{T}}{\partial x_j} = \frac{1}{q_j} \int_0^H v_j \frac{\partial T}{\partial x_j} \, dx_2 \tag{5.37b}$$

Equation (5.36) was given by White and Chen [51].

If the coordinate axis is along the screw channel, we will have:

$$q_3 = 0 \tag{5.38}$$

and Eq. (5.36) becomes

$$\rho \, c \, q_1 \frac{\partial \overline{T}}{\partial x_1} = E(H) + E(0) + \sigma_{12}(H) \, U_1 + \sigma_{32}(H) \, U_3 - q_1 \frac{\partial p}{\partial x_1} \tag{5.39}$$

If we now integrate with respect to x_3 through the channel width, we obtain

$$\rho \, c \, Q \frac{\partial \overline{\overline{T}}}{\partial x_1} = E(H) \, W_b + E(0) \, W_s + \pi \, D \, N \int_0^W [\sigma_{12}(H) U_1 + \sigma_{32}(H) U_3] \, dx_3 - Q \frac{\partial p}{\partial x_1} \tag{5.40}$$

where

$$\begin{aligned} Q &= \int_0^H \int_0^W v_1 \, dx_2 \, dx_3 \quad \overline{\overline{T}}(x_1) = \frac{1}{Q} \int_0^W q_1 \, \overline{T}(x_1, x_3) \, dx_3 \\ &= \frac{1}{Q} \int_0^W \int_0^H v_1 \, T(x_1, x_2, x_3) \, dx_2 \, dx_3 \end{aligned} \tag{5.41}$$

is known as the cup mixing temperature.

The heat fluxes $E(H)$ and $E(0)$ are usually represented in terms of heat transfer coefficients:

$$\begin{aligned} E(H) &= h_b (T_b - \overline{\overline{T}}) \\ E(0) &= h_s (T_s - \overline{\overline{T}}) \end{aligned} \tag{5.42}$$

5.7 Melting

The investigation of melting in modular co-rotating twin screw extruders begins with a 1992 study by Todd [52], who concluded by inspecting carcasses that melting was induced by interparticle friction. Subsequent studies were published by Bawiskar and White [53], Curry [54], and Potente and Melisch [55] in 1995/1996. The first efforts at modeling melting were made by Potente and Melisch [55]. They presumed melting begins at the surfaces of pellets which are packed in a crystalline lattice. The heat fluxes in the pellets are driven by high temperatures at the pellet surfaces. In 1997/1998, Bawiskar and White [56, 57], who had studied both carcasses and cross-sectioned carcasses, found evidence of melt layers at hot ($T_b > T_m^0$) barrel surfaces. They envisaged melting being initiated at the barrel and forming a melt layer at the barrel surface. In their view, the Potente-Melisch study followed the later stages of the melting process and not the initiation.

The early work of Todd [52] was continued with Gogos, Esseghir, Yu, and Tadmor [58], leading to quantitative modeling of melting by Gogos et al. [59,60] based on inter-pellet dissipation heating.

The modeling of Potente and Melisch [55] and Bawiskar and White [56, 57] used appropriate energy balances in each phase and considered boundary conditions at phase boundaries. Gogos and Kim [59] and Qian and Gogos [60] wrote their energy equation, Eq. (5.31), which represents energy conservation at a point in a continuum as

$$\rho c \frac{\mathrm{D}T}{\mathrm{D}t} = k \nabla^2 T + \sum_i \sum_j \sigma_{ij} d_{ij} + P\,E\,D + F\,E\,D \tag{5.43}$$

where the dissipation function term has been expanded three fold. The first term represents viscous dissipation, the second term plastic pellet deformation, and the third term inter-particle friction. Equation (5.43) is not really correct for a melting process where there are both pellets and phase boundaries. The melting of a thermoplastic pellet bed is not a continuum.

In a subsequent study by Vergnes et al. [61] the authors see both a melt layer at the barrel and subsequently individual pellets melting. They propose a model which contains aspects of those of both Bawiskar and White [56, 57] and Potente and Melisch [55].

More recently, Jung and White [62] made an experimental study over a very wide range of conditions of screw speeds, barrel temperatures, and throughputs. They found, depending on operating conditions, that melting initiated by the Bawiskar-White hot barrel or Todd-Gogos pellet friction/deformation mechanisms as well as a screw initiated melting process may occur (see Section 6.7). At low screw speeds, the beds held together during melting and at higher screw speeds, the bed breaks up. Under the conditions where the bed breaks up, melting may proceed by the Potente-Melisch mechanism.

Jung and White [63] later argued that one may model and make predictions for each melting mechanism and apply the model which indicates that melting occurs first. They describe specific models for each region.

5.8 Global Composite Modular Twin Screw Extruder Modeling

The presentation thus far in this chapter has concerned flow and heat transfer in individual machine elements. We now turn to the machine as a whole, which is a composite structure of different machine modules.

For single screw extruders, the global calculations generally involve simultaneously solving the fluid mechanical force balance and an energy flux balance forward from the hopper toward the die. The material properties are known at the hopper, as is the screw rotation speed, N. The mass flow rate, $G = \rho\ Q$, and the exit temperature, $\bar{\bar{T}}$, are not known.

The situation for a modular co-rotating twin screw extruder is different. It is generally used for compounding or reactive extrusion. The mass fluxes of all of the ingredients entering the machine are known and they are specified by the nature of the product and the customer requirements. The machine is operated in a metered starved fed manner. The mass flow rates are known throughout the machine. This method of feeding also results in having both fully filled and starved filled regions along the axis of the twin screw extruder during its operation. This has been observed by various experimental investigations dating to Erdmenger and will be discussed in the next chapter.

If we know the mass flow rate, G, through the machine, and the pressure at the die exit, $p(L)$ (atmospheric), we may calculate the pressure profile backward from the die to the beginning of the screw using the die dimensions and the melt rheological properties. The temperature of the compound exiting the die, $\bar{\bar{T}}(L)$, is however unknown. If we presume a cup mixing temperature profile $\bar{\bar{T}}(x)$, the resulting calculation will have to be corrected later using an energy balance. This approach is contained in the 1987 paper of White and Szydlowski [29] and in more detail in the paper by Wang et al. [40] in 1989. Using the screw characteristics of the various machine elements (such as those in Figure 5.13 and 5.15) we can calculate pressure profiles backward along the modular screw.

It is invariably found in such calculations that the pressure $p(x)$ will fall to atmospheric as we proceed. This indicates that starved flow develops. The equation relating flow and pressure gradient along the a screw element is roughly

$$Q = \alpha(n)\, N - \frac{\beta(n)}{K} \left(\frac{\mathrm{d}p}{\mathrm{d}x} \right)^{n} \tag{5.44}$$

The mass flow rate is less than it would be for drag flow. When the pressure drops below atmospheric, starvation develops and Eq. (5.44), which was derived for a filled cross-section, must replaced by

$$Q = \phi\ \alpha'\, N \tag{5.45}$$

where ϕ is the fraction of the cross-section that is filled. Thus we can have two expressions for the mass flow rate along the screw axis

$$G = \rho\, Q = \alpha(n)\, N - \frac{\beta(n)}{K}\left(\frac{\mathrm{d}p}{\mathrm{d}x}\right)^{n} = \phi\, \alpha'\, N \tag{5.46}$$

Calculations such as those described in the above paragraphs can be continued along the backward length of the extruder. They produce pressure profiles such as those shown in Figure 5.19.

The analysis of this machine proceeds by iterative backward hydrodynamic analyses and forward calculated energy balances which include melting. This approach is outlined in papers by Chen and White [64] in 1994 and with a melting model by Bawiskar and White [65] in 1997.

The energy balances necessary for forward marching involve energy balances of Section 5.6, in particular Eqs. (5.40) and a suitable melting model adapted from Section 5.7. The energy balance would calculate $\bar{\bar{T}}$ at the end of the screw and die. This would be compared to that initially presumed. The hydrodynamic model would be corrected.

In Figure 5.20 a, b, c, d, e we show pressure, fill factor, cup mixing temperature, torque, and power profiles along the axis of a modular co-rotating twin screw extruder for one set of conditions calculated by the methods of the above paragraph.

More ambiguously stated similar developments are given by Potente and his coworkers [66, 67] in 1996 and 1999 and by Canedo [68] in 1999. A better and more readable presentation of such a development was given by Vergnes et al. [69, 70] in 1996/1998. They proposed and described forward fluid mechanical and energy balances (see Table 5.1).

Table 5.1: Global Software for Modular Co-Rotating Twin Screw

Software name	Features	References
Akro Co Twin Screw		
Version 1 (1989)	Isothermal Non-Newtonian melt flow	Wang et al. [40] White et al. [42]
Version 2 (1993)	Non-isothermal Non-Newtonian melt flow	Chen and White [64]
Version 3 (1997)	Non-isothermal Non-Newtonian melt flow with melting	Bawiskar and White [65]
Version 4 (2003)	Add reactive extrusion and devolatilization to Version 3	White et al. [70]
Sigma (1996–1999)	Non-isothermal Non-Newtonian melt flow with melting	Potente et al. [66] Potente et al. [67]
Ludovic (1996–1998)	Non-isothermal Non-Newtonian melt flow	Vergnes et al. [69]
TXS (1999)	Non-isothermal Non-Newtonian melt flow with melting Oversimplified screw characteristic curve calculations	Canedo [68]

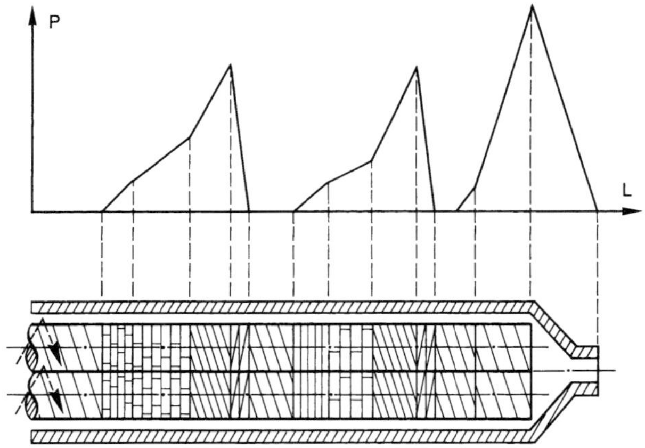

Figure 5.19:
Pressure profiles and starvation in an intermeshing co-rotating twin screw extruder of White and Szydlowski [29]

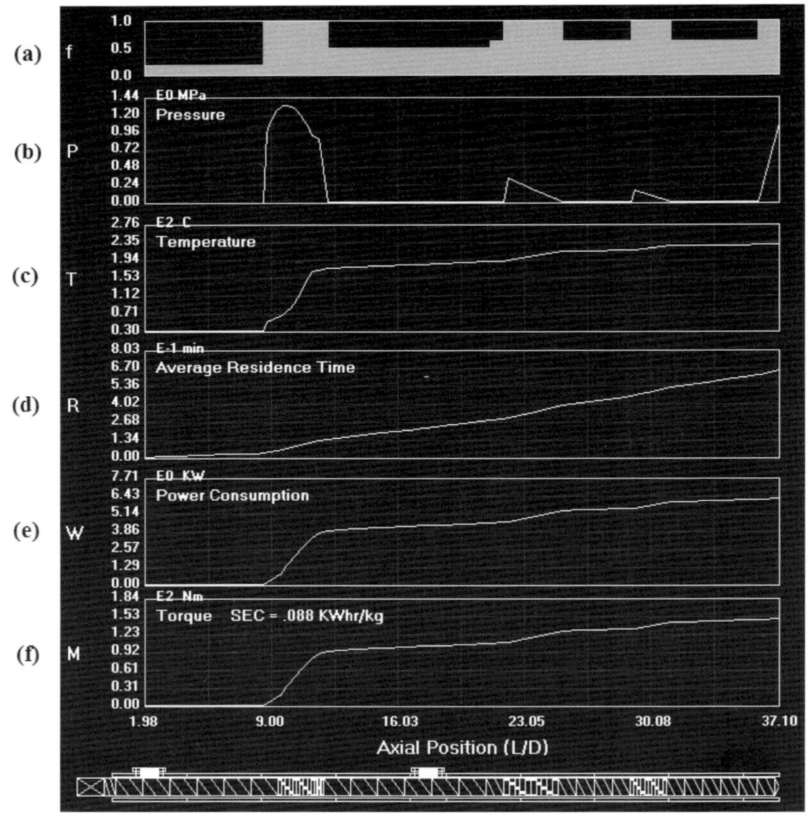

Figure 5.20: Simulation of (a) fill factor, (b) pressure, (c) cup mixing temperature, (d) mean residence time, (e) power consumption, and (f) torque profiles in a 40 mm diameter co-rotating twin screw extruder for a polypropylene at 300 lb/h and 300 rpm

It is possible to include additional features in this type of software. In a 2006 paper, White et al. [71] described a software which allows the user to do all the calculations described by Bawiskar and White [65] but also to model reactive extrusion and devolatilization. There have also been efforts at modeling mixing (see Section 5.9).

5.9 Modeling Mixing

There have been various efforts to model mixing in flow through modular co-rotating twin screw extruders [45, 69, 72–74]. The earliest of these papers by Schmidt et al. [72] described an effort to model the development of interfacial area through the calculation of the shear strain. They write the total shear strain as

$$\gamma_{\mathrm{T}} = \gamma_{\mathrm{s}} \cdot \gamma_{\mathrm{k}}^{\mathrm{T}} \tag{5.47}$$

where γ_{s} is the shear strain calculated for the screw elements and $\gamma_{\mathrm{k}}^{\mathrm{T}}$ the transverse strain imposed by the kneading disc blocks. Montes and White [73] are also concerned with interfacial area development through increasing shear strain. They write

$$\gamma_{\mathrm{T}} = \gamma_{\mathrm{s}} \cdot \gamma_{\mathrm{k}} \tag{5.48}$$

and seek to use the model of Wang et al. [40].

Delamare and Vergnes [69] model droplet morphology development in blends by considering the mechanisms of break-up and coalescence. This modeling is integrated into their Ludovic software [70].

Chen and Manas-Zloczower [74] describe finite element calculations of flow patterns in kneading disc blocks. They use the λ parameter of Eq. (5.30) to characterize the mixing.

5.10 Transient Analyses

Transient problems in co-rotating twin screw extruders were first discussed by Curry et al. [75] in 1991. They were concerned with the influence of the extent of feeder error.

There seems little additional effort until 2000, when papers by E. K. Kim and White [76, 77] model isothermal and non-isothermal transient startup flow in modular co-rotating twin screw extruders. They consider the influence of modular screw configurations.

In a subsequent paper, Kim and White [78] model superposed hydrodynamic disturbances from feeders in a starved flow modular twin screw extruder.

References

1. R. Erdmenger, *Chem. Ing. Tech.*, **36**, 175 (1964).
2. A. Wunsche, German Patent 131,392 (1901).
3. R. W. Easton, U.S. Patent 1,468,37 (1923).
4. F. F. Pease, U.S. Patent 2,048,286 (1936).
5. R. Colombo, Italian Patent 370,578 (1939).
6. W. Meskat and R. Erdmenger, German Patent (filed July 7, 1944) 862,668 (1953).
7. M. Ullrich in Co-Rotating Twin Screw Extruders, edited by K. Kohlgrüber, Hanser, Munich (2007).
8. M. L. Booy, *Polym. Eng. Sci.*, **18**, 978 (1978).
9. R. Erdmenger, U.S. Patent 3,254,367 (1966).
10. B. A. Loomans, U.S. Patent 3,900,187 (1975).
11. F. Martelli, *SPE Journal* **27**, (1) 26 (1971).
12. O. Armstroff and H. D. Zettler, *Kunststofftechnik*, **12**, 240 (1973).
13. H. Herrmann and U. Burkhardt, in 5th Leobener Kunststoffkolloquium Doppelschnecken – Extruder, p. 11, Lorenz Verlag, Vienna (1978).
    U. Burkhardt, H. Herrmann and S. Jakopin, *Plastics Compounding*, (Nov/Dec) 73 (1978)
14. W. A. Mack and K. Eise in Science and Technology of Polymer Science and Technology of Polymer Processing, edited by N. P. Suh and N.H, Sung, MIT, Cambridge (1979).
15. K. Eise, H. Herrmann, H. Werner and U. Burkhardt, *Adv. Polym. Technol.*, **1**, (2) (1981).
16. K. Eise, J. E. Curry and J. F. Nangeroni, *Polym. Eng. Sci.*, **23**, 642 (1983).
17. C. D. Denson and B. K. Hwang, *Polym. Eng. Sci.*, **20**, 965 (1980).
18. M. L. Booy, *Polym. Eng. Sci.*, **20**, 1220 (1980).
19. W. Szydlowski and J. L. White, *Adv. Polym. Tech.*, **7**, 177 (1987).
20. Z. Tadmor, E. Broyer, C. Gutfinger, *Polym. Eng. Sci.*, **14**, 660 (1974).
21. D. M. Kalyon, A. D. Gotsis, C. G. Gogos and C. Tsenoglou, *SPE ANTEC Tech Papers*, **34**, 64 (1988).
22. D. M. Kalyon, A. D. Gotsis, U. Y. Imazer, C. G. Gogos, H. N. Sangani, B. Aral, and C. Tsenoglou, *Adv. Polym. Technol.*, **8**, 337 (1988).
23. M. J. Crochet, Personal Communication, (1988).
24. Y. Wang and J. L. White, *J. Non-Newt. Fluid Mech.*, **32**, 19 (1989).
25. R. M. Griffith, *IEC Fund*, **1**, 180 (1962).
26. R. A. Lai-Fook, A. Senouci, A. C. Smith, and D. P. Isherwood, *Polym. Eng. Sci.*, **29**, 433 (1989).
    R. A. Lai-Fook, Y. Li, and A. C. Smith, *Polym. Eng. Sci.*, **31**, 1157 (1991).
27. D. Goffart, D. J. van der Wal, E. M. Klomp, H. W. Hoogstraten, L. P. B. M. Janssen, L. Breysse and Y. Trolez, *Polym. Eng. Sci.*, **36**, 901 (1996).
28. M. Bierdel, in Co-Rotating Twin Screw Extruders, edited by K. Kohlgrüber, Hanser, Munich (2007).
29. J. L. White and W. Szydlowski, *Adv. Polym. Technol.*, **7**, 419 (1987).
30. J. Tayeb, B. Vergnes, and G. Della Valle, *J. Food Sci.*, **53**, 616 (1988).
31. W. K. Nelson, U.S. Patent 1,868,671 (1932).

32. R. Erdmenger, German Patent 813,154 (1951).
33. R. Erdmenger, U.S. Patent 2,670,188 (1954).
34. R. Erdmenger, German Patent 940,109 (1956).
35. H. Werner, Dr. Ing. Dissertation, University of Munich (1976).
36. H. Werner and K. Eise, *SPE ANTEC Tech. Papers*, **25**, 181 (1979).
37. W. Szydlowski, R. Brzoskowski, and J. L. White, *Int. Polym. Process.*, **1**, 207 (1987).
38. W. Szydlowski and J. L. White, *Int. Polym. Process.*, **2**, 142 (1988).
39. W. Szydlowski and J. L. White, *J. Non-Newt. Fluid Mech.*, **28**, 29 (1988).
40. Y. Wang, J. L. White and W. Szydlowski, *Int. Polym. Process.*, **4**, 262 (1989).
41. Z. Tadmor, Personal Communication, Nice, April (1990).
42. D. H. Sebastian and R. Rakos, *SPE ANTEC Tech. Papers*, **36**, 135 (1990).
43. A. D. Gotsis and D. M. Kalyon, *SPE ANTEC Tech. Papers*, **35**, 44 (1989) and with Z. Ji, *ibid*, **36**, 139 (1990).
44. H. H. Yang and I. Manas-Zloczower, *Polym. Eng. Sci.*, **32**, 1411 (1992).
45. A. Lawal, D. M. Kalyon, and Z. Ji, *Polym. Eng. Sci.*, **33**, 140 (1993); *ibid*, **35**, 1325 (1995).
46. A. Kiani and H. J. Samann, *SPE ANTEC Tech. Papers*, **39**, 2758 (1998).
47. D. J. van der Wal, D. Goffart, E. M. Klomp, H. W. Hoggstraten and L. P. B. M. Janssen, *Polym. Eng. Sci.*, **36**, 912 (1996).
48. T. Ishikawa, S. Kihara and K. Funatsu, *Polym. Eng. Sci.*, **40**, 357 (2000).
49. H. E. H. Meijer and P. H. M. Elemans, *Polym, Eng, Sci.*, **28**, 275 (1988).
50. O. Reynolds, *Phil. Trans. Roy. Soc.*, **4177**, 157 (1886).
51. J. L. White and Z. Chen, *Polym. Eng. Sci.*, **34**, 229 (1994).
52. D. B. Todd, *SPE ANTEC Tech. Papers*, **38**, 2528 (1992).
53. S. Bawiskar and J. L. White, *Int. Polym. Process.*, **10**, 105 (1995).
54. J. E. Curry, *SPE ANTEC Tech. Papers*, **41**, 92 (1995).
55. H. Potente and U. Melisch, Paper presented at Polymer Processing Society Meeting, Seoul, Korea, March (1995): *Int. Polym. Process.*, **11**, 101 (1996).
56. S. Bawiskar and J. L. White, *Int. Polym. Process.*, **12**, 331 (1997).
57. S. Bawiskar and J. L. White, *Polym. Eng. Sci.*, **38**, 727 (1998).
58. M. Esseghir, D. W. Yu, C. G. Gogos, D. B. Todd and Z. Tadmor, *SPE ANTEC Tech. Papers*, **43**, 3684 (1997).
59. C. G. Gogos and M. H. Kim, *SPE ANTEC Tech. Papers*, **46**, 134 (2000).
60. B. Qian and C. G. Gogos, *SPE ANTEC Tech. Papers*, **47**, 124 (2001).
61. B. Vergnes, G. Souveton, H. L. Delaceua and A. Ainson, *Int. Polym. Process.*, **16**, 351 (2001).
62. H. Jung and J. L. White, *Int. Polym. Process.*, **18**, 127 (2003).
63. H. Jung and J. L. White, *Int. Polym. Process.*, **23**, 242 (2008).
64. Z. Chen and J. L. White, *Int. Polym. Process.*, **9**, 310 (1994).
65. S. Bawiskar, J. L. White, *Int. Polym. Process.*, **12**, 331 (1997).
66. H. Potente, U. Melisch and J. Flecke, *SPE ANTEC Tech. Papers*, **42**, 334 (1996).
67. H. Potente, M. Bastian and J. Flecke, *Adv. Polym. Technol.*, **18**, 147 (1999).
68. E. L. Canedo, *SPE ANTEC Tech. Papers*, **45**, 310 (1999).

69. L. Delamare and B. Vergnes, *Polym. Eng. Sci.*, **36**, 1685 (1996).
70. B. Vergnes, G. Dellavalle and L. Delamare, *Polym. Eng. Sci.*, **38**, 1781 (1998).
71. J. L. White, J. M. Keum, H. Jung, K. Ban and S. H. Bumm, *Polym. Plastics Technol. Eng.*, **45**, 539 (2006).
72. L. R. Schmidt, E. M. Lovgren and P. C. Meissner, *Int. Polym. Process.*, **4**, 270 (1989).
73. S. Montes and J. L. White, *Int. Polym. Process.*, **6**,156 (1991).
74. H. Cheng and I. Manas-Zloczower, *Polym. Eng. Sci.*, **37**, 1082 (1997).
75. J. E. Curry, A. Kiani, and A. Dreiblatt, *Int. Polym. Process.*, **6**, 148 (1991).
76. E. K. Kim and J. L. White, *Polym. Eng. Sci.*, **40**, 543 (2000).
77. E. K. Kim and J. L. White, *Int. Polym. Process.*, **15**, 233 (2000).
78. E. K. Kim and J. L. White, *Polym. Eng. Sci.*, **41**, 232 (2001).

6 Experimental Studies of Intermeshing Co-Rotating Twin Screw Extruders

6.1 Introduction

We now turn to a discussion of experimental investigations of flow mechanism in intermeshing co-rotating twin screw extruders. Such studies are relatively limited in number and do not form a coherent literature. Certain basic characteristics of the flow have, however, become clear. We will describe these results in the present chapter.

In Section 6.2, we summarize efforts aimed at flow visualization of fluid motions. We turn in Section 6.3 to investigations of fluid pumping characteristics. In Section 6.4, we consider investigations of residence time distributions. Mixing is considered in Section 6.5. We then turn to considerations of heat transfer in Section 6.6. Research on melting is summarized in Section 6.7.

6.2 Flow Visualization

The earliest reported studies of mechanisms of fluid motions in an intermeshing co-rotating twin screw extruder are by Erdmenger [1] of Bayer AG in 1964, who reported experiments in which screws were removed and inspected. These studies indicated the fluid moved in 'figure-8' motions around the periphery of the screw (Figure 6.1). Erdmenger [1] also notes that the flow characteristics in the kneading disc blocks are different from those in the screw regions (Figure 6.2).

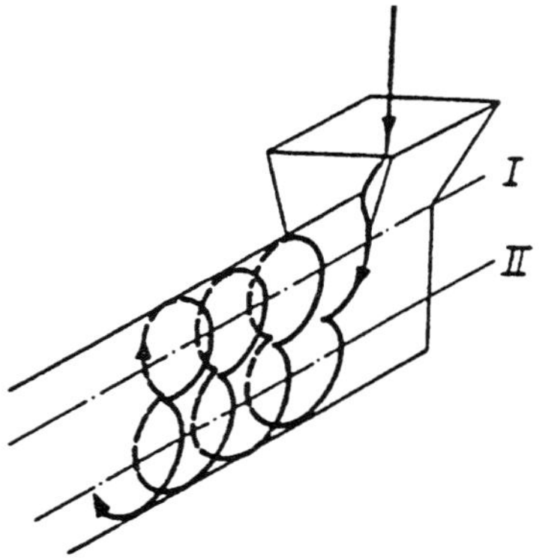

Figure 6.1: Flow in screw regions of intermeshing co-rotating twin screw extruder (Erdmenger [1]) (reprinted with permission of Chemie Ingenieur Technik)

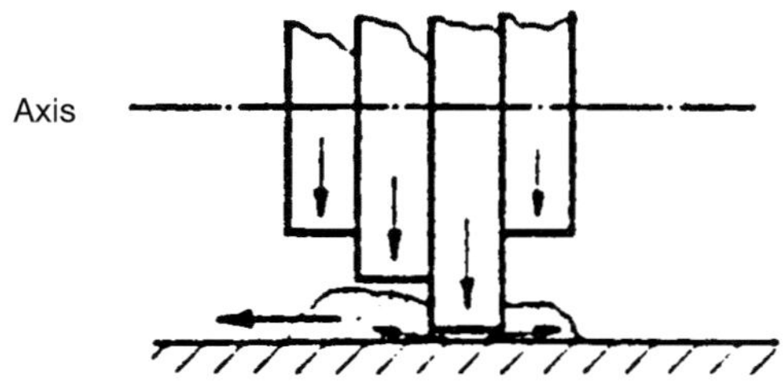

Figure 6.2: Flow in kneading disc block section (Erdmenger [1]) (reprinted with permission of Chemie Ingenieur Technik)

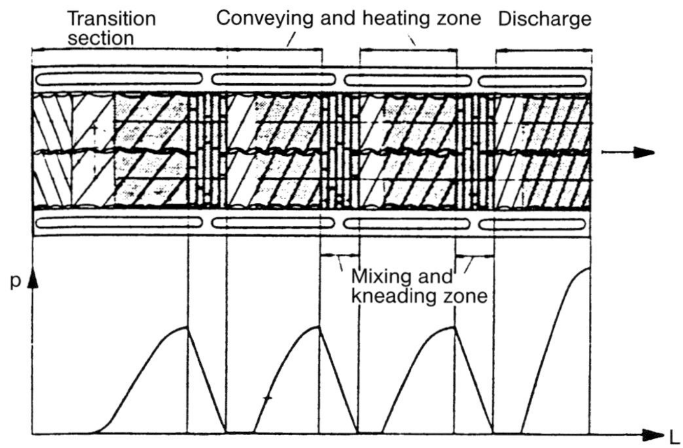

Figure 6.3: Starvation flow observations of Armstorff and Zettler [2]

A second flow visualization study of fluid motions was published by Armstroff and Zettler [2] of BASF in 1973. These authors put modular screws of a Werner and Pfleiderer ZSK into a transparent plastic barrel and studied the motions of the fluid. The modular screws had right-handed and left-handed screw elements as well as kneading discs along the shafts. Armstroff and Zettler observe alternate regions of filled and starved channels. This observation is summarized in Figure 6.3 taken from their paper. Kneading disc elements are observed to be fully filled, while right-handed screw elements tend to be starved.

A videotape showing flow of a viscous oil in a transparent plastic barrel containing Werner and Pfleiderer modular screws was made by Werner and Pfleiderer. This was studied by Szydlowski and his co-workers [3, 4]. They note that it shows peristaltic pumping effects in the kneading disc regions and starvation effects in the modular machine.

Sakai [5] of Japan Steel Works has reported flow visualization studies on intermeshing co-rotating machines compared to other types of twin screw machines. He uses a Newtonian fluid in a transparent plastic barrel. He reports the 'figure-8' motions found by Erdmenger.

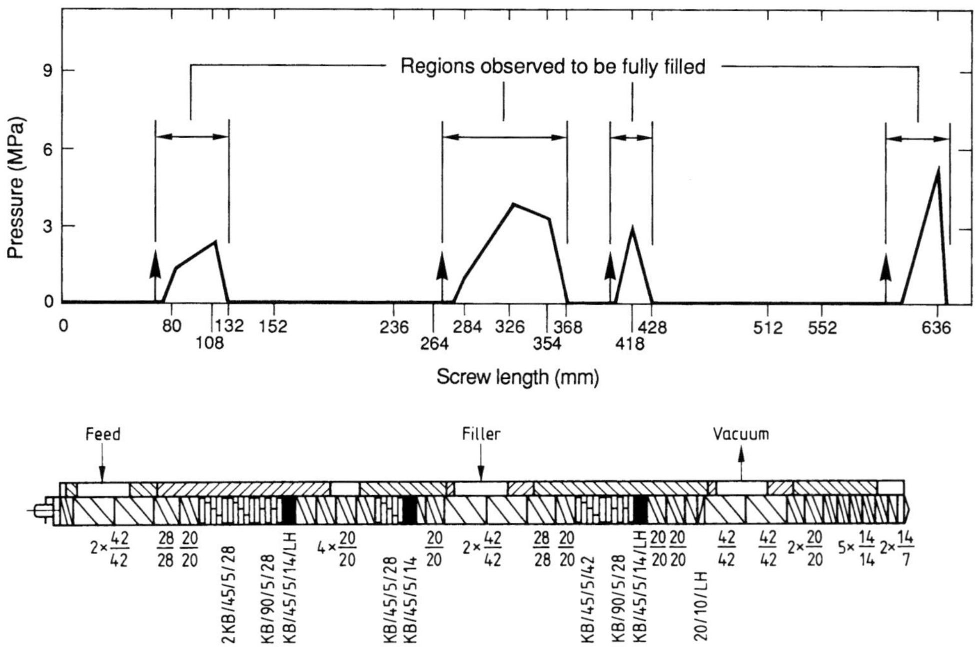

Figure 6.4: Observations of screw filling and starvation and comparison with experiment after Wang et al. [8]

Meijer and Elemans [6] have made flow visualization studies in a modular intermeshing co-rotating twin screw extruder in a transparent polymethyl methacrylate barrel. Glycerin was used as a process fluid. The positions of fill and starvation along the length of the screw were determined. These were compared to predictions of the same authors of extent of fill in different sections along the screw. Their work bares similarities to that of Armstroff and Zettler [2].

Observations of the distribution of polymer on the modular screws of a Baker Perkins twin screw extruder in a clam shell barrel have been reported by K. L. Nichols, Jayaraman, and Grulke [7]. These authors also observed a sequence of fully-filled and starved regions along the length of the modular screw.

Wang et al. [8] and White et al. [9] have studied the distribution of polymer along the length of screws of modular intermeshing co-rotating twin screw extruders using polyethylenes and polypropylenes. They compared their results with models of the type developed by White and Szydlowski [10] and themselves. Power law fluid model screw characteristic curves were used for the different machine elements. Such calculations were described in Section 5.6. Comparisons of experiment and predictions are shown in Figure 6.4 and it can be seen that agreement is quite good.

A subsequent study of fill factor distribution along the axis in a co-rotating twin screw machine was given by Kalyon et al. [11].

6.3 Pumping Characteristics

Certain points are to be borne in mind for global pumping characteristics. First, commercial modular machines are generally operated in a starved manner with the feed(s) controlled at the hoppers. Under such circumstances the output is determined by a programmed input. If, however, the machine is flood fed, as with single screw extruders, the output will be determined by the screw speed. Increase of screw speed would increase the rate of output in the feed section. This greater throughout would exist in the different elements and at the output. Thus, the extrusion rate would increase with screw speed.

If an intermeshing co-rotating twin screw extruder existed as continuous right-handed screws from hopper to the die, its performance would resemble that of a single screw pump and give similar screw characteristics, i.e., if flood-fed with a Newtonian fluid, it would be expected to give an output Q response representable as:

$$Q = \alpha\, N - \frac{\beta}{\eta}\,\Delta p \tag{6.1}$$

where N is screw speed, η viscosity, and Δp pressure rise.

The first study of fluid pumping characteristics in a modular intermeshing co-rotating twin screw extruder was due to Armstroff and Zettler [2] in the paper described in the previous section. They presented dimensionless pressure drops for flow through left-handed screw channels and kneading disc blocks as a function of screw speed for various polymer melts. The authors presented their results in a formulation reminiscent of low viscosity, high inertia flows. They use the variables:

$$\frac{\Delta p}{\rho\, d_H^2\, N^2} \qquad \frac{d_H^2\, N\, \rho}{\eta}$$

where d_H is a hydraulic radius of the flow channel, $\Delta p / \rho\, d_H^2\, N^2$ is a pressure drop divided by momentum, and $d_H^2\, N\, \rho / \eta$ is a Reynolds number. They correlate their data in equations of form:

$$\frac{\Delta p}{\rho\, d_H^2\, N^2} = K \left(\frac{L}{d_H}\right)^a \left(\frac{d_H^2\, N\, \rho}{\rho}\right)^{-b} \tag{6.2}$$

for each of the screw elements. Here, a and b are positive exponents.

Experimental studies of the pumping characteristics were subsequently presented by Todd [12] of APV Chemical Machinery (Baker Perkins) for screw and paddle/kneading disc elements. He reports work of himself and H. G. Karian representing pumping characteristics through:

$$Q = \alpha\, N - \frac{\beta'}{\eta\, L}\,\Delta p \tag{6.3}$$

where L is the length of the barrel section so that β' is β L. $Q - \Delta p$ data for stationary kneading disc assemblies are presented which allow direct determination of β'. Studies of drag flow without pressure yielded α. Todd and Karian's results are summarized in Table 6.1. This shows most clearly the superior pumping characteristics of screws as opposed to kneading disc blocks.

Most experimental studies of screw element pumping in recent years have involved comparisons with flow simulations of the type described in Chapter 5 [8, 9, 13, 14].

An exception to this is a 2002 paper by Brouwer, Todd and Janssen [15] which contrasts the pumping characteristics of screw elements and various mixing elements (SME, ZME and Turbine Mixing (Gear) Elements (TME)) (see Figures 4.27 and 4.29) with viscous Newtonian liquids. Equation. (6.3) was used here again. The results generally indicated for the α term in Eq. (6.3).

$$\text{Screw} > \text{SME} > \text{ZME} \tag{6.4}$$

with left handed elements having negative values.

Table 6.1: Todd and Karian's Drag and Pressure Flow Parameters (50 mm Twin Screw Extruder): (a) Paddle/Kneading Disc Configurations

Staggering angle	Width / Diameter	a (cm^3) (Karian)	β' (cm^4) (Karian)	β' (cm^4) (Todd)
30	0.25	51.1	0.508	0.533
45	0.125	18.7	0.198	0.194
45	0.25	31.1	0.348	0.338
45	0.5	36.4	0.603	–
60	0.125	5.7	0.228	–
60	0.25	17.9	0.366	0.355
60	0.5	22.9	0.487	–
90	0.25	0	–	0.429

(b) Screws

Helix Angle	a (cm^3) (Karian)	a (cm^3) (Todd)	β' (cm^4) (Karian)	β' (cm^4) (Todd)
18	40.8	42.0	0.118	0.112
6.1	12.4	14.0	0.021	0.011

6.4 Residence Time Distributions

There have been no totally satisfactory studies of residence time distributions in intermeshing co-rotating twin screw extruders. A number of results do exist. These may actually be misleading in part because of the narrowness of variables investigated. The first studies of residence time distribution in intermeshing co-rotating twin screw extruders were reported in 1966 by Herrmann [16] of Werner and Pfleiderer. These results date earlier as they are quoted in Erdmenger's 1964 paper [1].

Residence time distribution measurements have been presented by Sakai [5], who contrasted the behavior of the various types of twin screw extruders. Sakai finds intermeshing co-rotating machines to produce a narrower distribution of residence time than a single screw extruder. He argues that this is associated with the self-wiping character of the screws which prevents the retention of stagnant layers on the screw surface. At the same time, the intermeshing counter-rotating is found to have a narrower residence time distribution, which he associates with the C-chamber mechanism (Chapter 8].

Residence time distributions in modular intermeshing co-rotating twin screw extruders were investigated between 1969 and 1975 by Todd and Irving [17] and Todd [18] of Baker Perkins. Todd and Irving [17] describe a study in which sodium or potassium nitrate tracer was added to glucose which was processed through a Baker Perkins intermeshing co-rotating twin screw extruder with both complete screws and various combinations of screw and paddle elements. Electrical conductivity was used to probe concentrations in the output stream. Todd [18] describes the alternative use of methylene blue dyes or MnO_2 radioactive tracer techniques. A plot of relative concentration of tracer as a function of residence time, taken from Todd [18], is shown in Figure 6.5. This contrasts the behavior of various elements with a two-tipped paddle/kneading disc element without pumping capability (i.e., 90° staggering angle).

Todd and Irving interpret their tracers studies using an axial diffusion model, i.e., they write:

$$\frac{\partial c}{\partial t} + v_1 \frac{\partial c}{\partial x_1} = E \frac{\partial^2 c}{\partial x_1^2} \tag{6.5}$$

for a pulse input. They introduce N_{pe} the Peclet number, and a reduced time t_r defined by:

$$N_{pe} = \frac{LU}{E} \tag{6.6a}$$

$$t_r = \frac{t}{\theta} \tag{6.6b}$$

where θ is the mean residence time. The larger E and the smaller N_{pe}, the greater the dispersion and the broader the residence time distribution. Generally, the neutral kneading disc block (paddle) element introduced less band broadening and axial dispersion than the screw elements.

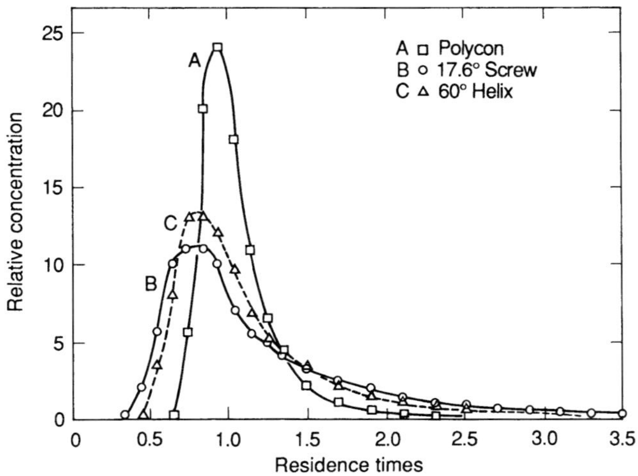

Figure 6.5: Residence time distribution breadth of a pulse produced in flow through different elements after Todd [17, 18] (reprinted with permission of the Society of Plastics Engineers)

A more recent study of residence times in intermeshing co-rotating twin screw extruders was presented by Hornsby [19] using the radioactive decay of $^{56}MnO_2$. He used a machine with trapezoid-shaped rather than self-wiping screw flights.

Subsequent studies of residence time distributions in modular co-rotating twin screw extruders have been published by Potente and Ansahl [20], Cassagnau et al. [21], Bur and Gallant [22], P. J. Kim and White [23], Kye and White [24], and Shon et al. [25]. Most of these studies used tracers [20, 23–25]. Cassagnau et al. [21] substituted ultra-violet absorbing groups into the polymer being studied, while Bur and Gallant [22] used fluorescence.

Certain general conclusions are found in these papers which are general among all the research groups. Generally the mean residence times increase with decreasing screw speed and reduced throughput rates. The residence time distribution is broadened by introducing mixing elements such as kneading disc blocks. This clearly means that increased productivity results from higher screw speeds, but extents of mixing or chemical reaction may be less.

6.5 Mixing

Bigio and Erwin [26] have made a basic study of mixing in a modular intermeshing co-rotating twin screw extruder using the apparatus design shown in Figure 6.6. They studied the reduction in striation thickness along the length of the twin screw extruder. In the screw sections the number of striations and interfacial area increased linearly with position along the length of the screw (Figure 6.7). In the kneading disc block elements there was rapid increase in number of striations and graying as shown in Figure 6.8.

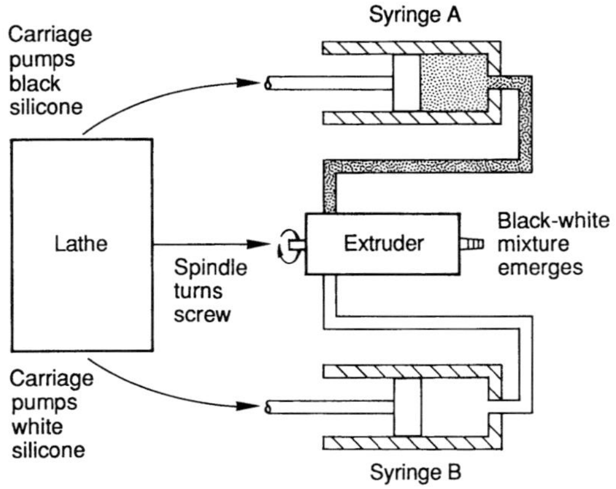

Figure 6.6: Apparatus of Bigio and Erwin [26] for studying flow in an intermeshing co-rotating twin screw extruder (reprinted with permission of the Society of Plastics Engineers)

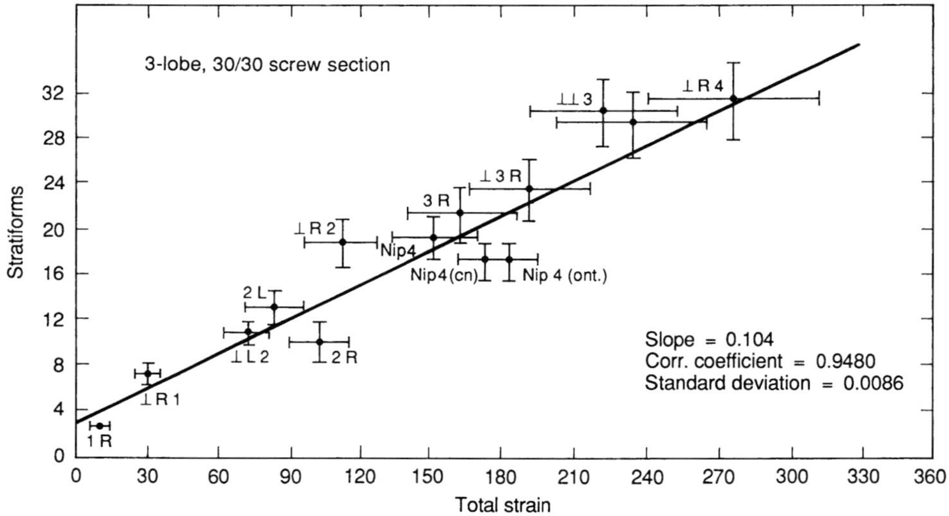

Figure 6.7: Number of striations versus total average strain in screw channel after Bigio and Erwin [26] (reprinted with permission of the Society of Plastics Engineers)

In a subsequent study, Cassidy and Bigio [27] have studied the influence of starved flow. Distributive mixing in similar twin screw extruders has also been studied by Ess and Hornsby [28] using automated image analysis. No specific conclusions relating to extruder performance are discussed.

Kalyon and Sangani [29] have studied optical microscope cross-sections of solidified polymers containing pigments which were compounded in a modular intermeshing co-rotating twin screw extruder. They have used image analysis to follow the mixing process.

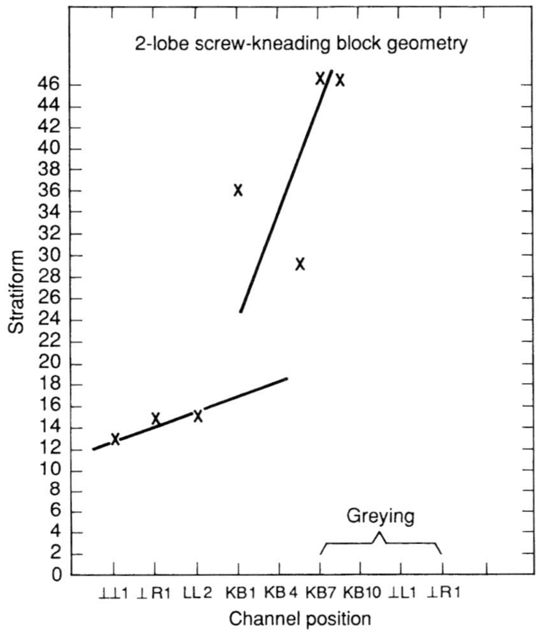

Figure 6.8: Number of striations versus position in channel for screw-kneading disc combination after Bigio and Erwin [17] (reprinted with permission of the Society of Plastics Engineers)

In a more recent study, Ess and Hornsby [30], using their trapezoid-shaped screw channel twin screw machine, investigated the development of dispersion of particle agglomerates. Generally, it is found that increasing intense shearing breaks down agglomerates.

Sinton, Kalyon, and their co-workers [31] have used Nuclear Magnetic Resonance (NMR) imaging to characterize compounds produced in a co-rotating twin screw extruder.

Shon et al. [32] have studied the damage to glass fibers and dispersive mixing of small particle agglomerates such as calcium carbonate by optical and scanning electron microscopy in various twin screw extruders and continuous mixers. They tried to model the damage and agglomerate break up with the expression

$$\frac{\mathrm{d}'d'}{\mathrm{d}t} = -k\,d + k'\left(\frac{1}{d}\right) \tag{6.7}$$

where d is the agglomerate size, k and k' breakdown and agglomeration rate constants.

Various investigators have investigated the blending of polymer melts [33–41] in co-rotating twin screw extruders. These essentially show that severe screw configurations with long series of kneading disc blocks were most effective in producing fine phase morphologies.

Shon et al. [41] have sought to apply Eq. (6.7) to the polymer blend dispersion process. The parameter k is found to be 50 to 150 times smaller for a polypropylene-polyamide 6 blend but k' is 300 to 1200% greater as compared dispersive nixing of smaller particles such as calcium carbonate.

6.6 Heat Transfer

An experimental study of heat transfer in intermeshing co-rotating twin screw extruders has been presented by Todd [42] of APV Chemical Machinery (Baker Perkins). The heat transfer rates to fluid moving through a two-tipped kneading disc (paddle) configuration with a 90° helix angle in an intermeshing co-rotating twin screw extruder with 100 mm diameter screws were determined (see Figure 6.9). The process fluids were glucose, polyvinyl chloride, polyethylene terephthalate, and nylon. Various heat transfer media were used in the screws and the barrel.

Todd [42] correlates his data for the local heat transfer coefficient h (Figure 6.10) with a formulation for the Nusselt number $h\,D/k$, as a function of Reynolds number $D^2\,N\rho/\eta$ and Prandtl number $c\,\eta/k$. Specifically, he finds:

$$\frac{h\,D}{k} = 0.94\left(\frac{D^2\,N\,\rho}{\eta}\right)^{0.28}\left(\frac{c\,\eta}{k}\right)^{0.33}\left[\frac{\eta}{\eta_w}\right]^{0.14} \tag{6.8}$$

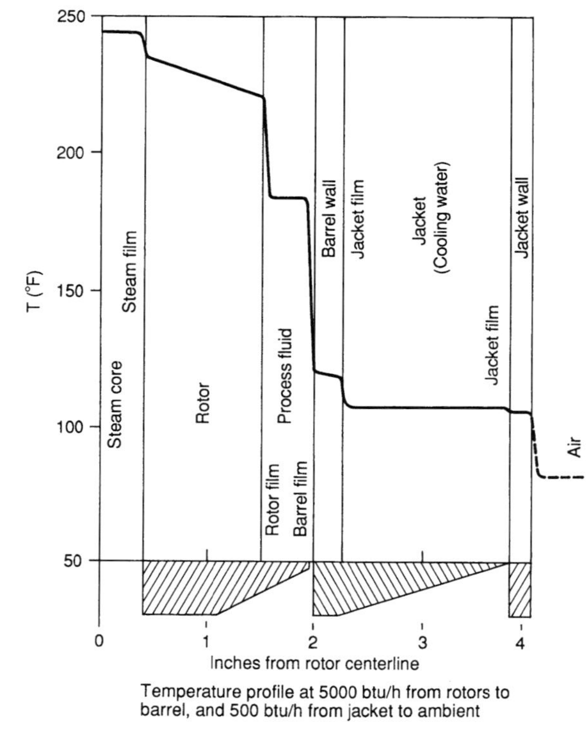

Figure 6.9: Temperature profile in twin screw extruder after Todd [42] (reprinted with permission of the Society of Plastics Engineers)

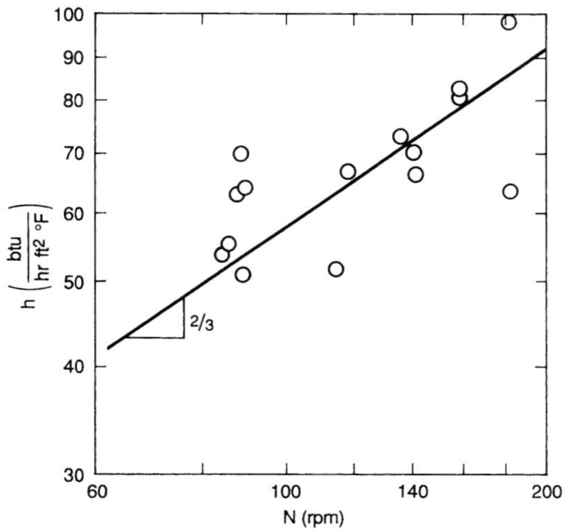

Figure 6.10: Internal heat transfer coefficient according to Todd [42] in intermeshing co-rotating twin screw extruder (reprinted with permission of the Society of Plastics Engineers)

6.7 Melting

Until the last decade there has been little study of the melting mechanisms in modular twin screw extruders. An early investigation was reported in 1985 by Karian [43] of Baker Perkins, who studied the 'melting profile' for polyvinyl chloride in the paddle/kneading disc block section of a modular twin screw machine. The melt appeared to form very rapidly compared to single screw machines.

A more thorough study of melting in a co-rotating twin screw extruder was subsequently reported by Todd [44] in 1992, using polyvinyl chloride, styrene acrylonitrile copolymer, polycarbonate and polyethylene carcasses formed during the melting/fusion process.

There have been many subsequent experimental investigations of melting [45–52]. However, there is much disagreement in the interpretation of observations. Potente and Melisch [48] observe pellets dispersed and melting in a melt matrix. Bawiskar and White [49] observe a melt layer formed along the barrel which propagated towards the screw with increasing residence time. Vergnes et al. [51] observe both the dispersed melting pellets and the melt layer along the barrel. The most extensive observations seem to be those of Jung and White [52, 53], who describe experiments on various polymers over a range of conditions of screw speed and feed rate. They published a series of diagrams that show melting initiated at the barrel surface, in the bulk of bed and at the screw surface, depending on temperature and screw speed. At high barrel temperatures, melting was initiated at the barrel. At the lower temperatures, melting was initiated in the bulk. At still lower temperatures, it was sometimes initiated at the screw.

The propagation of melting thus can involve a moving front or occur uniformly in the bulk. At moderate and higher screw speeds, the pellet bed becomes unstable leading to the Potente-Melisch mechanism. The behavior is similar to that found by Vergnes et al. [51].

References

1. R. Erdmenger, *Chem. Ing. Tech.*, **36**, 175 (1964).
2. O. Armstroff and H. D. Zettler, *Kunststofftechnik*, **12**, 240 (1973).
3. W. Szydlowski and J. L. White, *Int. Polym. Process.*, **2**, 142 (1988).
4. W. Szydlowski, R. Brzoskowski, and J. L. White, *Int. Polym. Process.*, **1**, 207 (1987).
5. T. Sakai, *Gosé Jushi*, **24**, 7 (1978).
6. H. E. H. Meijer and P. H. M. Elemans, *Polym. Eng. Sci.*, **28**, 275 (1988).
7. K. L. Nichols, K. Jayaraman, and E. G. Grulke, Paper presented at the 4th Annual Polymer Processing Society Meeting, Orlando, May (1988).
8. Y. Wang, J. L. White, and W. Szydlowski, *Int. Polym. Process.*, **4**, 262 (1989).
9. J. L. White, S. Montes, and J. K. Kim, *Kautschuk Gummi Kunststoffe*, **43**, 20 (1990).
10. J. L. White and W. Szydlowski, *Adv. Polym. Technol.*, **7**, 419 (1987).
11. D. M. Kalyon, C. Jacob, and P. Yaras, *Plastics Rubber Comp. Proc. Appl.*, **16**, 193 (1991).
12. D. B. Todd, *SPE ANTEC Tech. Papers*, **35**, 168 (1989); *Int. Polym. Process.*, **6**, 143 (1991).
13. H. Potente, J. Ansahl and R. Witteneier, *Int. Polym. Process.*, **5**, 28 (1990).
14. R. A. Lai-Fook, Y. Ci, and A. C. Smith, *Polym. Eng. Sci.*, **31**, 1857 (1991).
15. T. Brouwer, D. B. Todd, and L. P. B. M. Janssen, *Int. Polym. Process.*, **17**, 26 (2002).
16. H. Herrmann, *Chem. Ing. Tech.*, **38**, 25 (1966).
17. D. B. Todd and H. F. Irving, *Chem. Eng. Prog.*, **65** (9), 85 (1969).
18. D. B. Todd, *Polym. Eng. Sci.*, **15**, 437 (1975).
19. P. R. Hornsby, *Plastics Compounding*, (Sept/Oct) 65 (1983).
20. H. Potente and J. Ansahl, *Kunststoffe*, **80**, 926 (1990).
21. P. Cassagnau, C. M. Jangos and A. Michel, *Polym. Eng. Sci.*, **31**, 772 (1991).
22. A. J. Bur and F. M. Gallant, *Polym. Eng. Sci.*, **31**, 1365 (1991).
23. P. J. Kim and J. L. White, *Int. Polym. Process.*, **9**, 108 (1994).
24. H. Kye and J. L. White, *J. Appl. Polym. Sci.*, **52**, 1249 (1994).
25. K. Shon, D. H. Chang, and J. L. White, *Int. Polym. Process.*, **14**, 44 (1999).
26. D. I. Bigio and L. Erwin, *SPE ANTEC Tech. Papers*, **31**, 45 (1985).
27. K. Cassidy and D. I. Bigio, *SPE ANTEC Tech. Papers*, **37**, 92 (1991).
28. J. W. Ess and P. R. Hornsby, *Polymer Testing*, **6**, 205 (1986).
29. D. M. Kalyon and H. N. Sangani, *Polym. Eng. Sci.*, **29**, 1018 (1989).
30. J. W. Ess and P. R. Hornsby, *Plastics Rubber Proc.*, **8**, 147 (1987).
31. S. W. Sinton, J. C. Crowley, G. A. Lo, D. M. Kalyon, and C. Jacob, *SPE ANTEC Tech. Papers*, **36**, 116 (1990).

32. K. Shon, D. Liu and J. L. White, *Int. Polym. Process.*, **20**, 322 (2005).
33. A. P. Plochocki, S. S. Dagli, and M. H. Mack, *Kunststoffe*, **78**, 254 (1988).
34. U. Sundararaj, C. W. Macosko, R. J. Rolando, and H. T. Chen, *Polym. Eng. Sci.*, **32**, 1814 (1992).
35. V. Bordereau, Z. H. Shi, L. A. Utracki, P. Samhut and M. Carrega, *Polym. Eng. Sci.*, **32**, 1846 (1992).
36. S. Lim and J. L. White, *Int. Polym. Process.*, **8**, 119 (1993).
37. S. Lim and J. L. White, *Polym. Eng. Sci.*, **34**, 221 (1994).
38. C. G. Gogos, M. Esseghir, D. W. Yu, D. B. Todd and J. E. Curry, *SPE ANTEC Tech. Papers*, **40**, 270 (1994).
39. T. P. Vaiwio, A. Harlin and U. V. Sepphla, *Polym. Eng. Sci.*, **35**, 225 (1995).
40. M. A. Huneault, M. F. Champagne and A. Luciani, *Polym. Eng. Sci.*, **36**, 1694 (1996).
41. K. Shon, S. H. Bumm and J. L. White, *Polym. Eng. Sci.*, **48**, 756 (2008).
42. D. B. Todd, *SPE ANTEC Tech. Papers*, **34**, 54 (1989).
43. H. G. Karian, *J. Vinyl Technol.*, **7** (1985).
44. D. B. Todd, *SPE ANTEC Tech. Papers*, **38**, 2528 (1992).
45. S. Bawiskar and J. L. White, *Int. Polym. Process.*, **10**, 105 (1995).
46. H. Potente and U. Melisch, Paper presented at 7th Annual Meeting, Polymer Processing Society (1995).
47. J. E. Curry, *SPE ANTEC Tech. Papers*, **41**, 92 (1995).
48. H. Potente and U. Melisch, *Int. Polym. Process.*, **11**, 101 (1996).
49. S. Bawiskar and J. L. White, *Int. Polym. Process.*, **12**, 331 (1997); *Polym. Eng. Sci.*, **38**, 727 (1998).
50. M. Esseghir, D. W. Yu, C. G. Gogos and D. B. Todd, *SPE ANTEC Tech. Papers*, **43**, 3684 (1997).
51. B. Vergnes, G. Souveton, M. L. Delacour and A. Ainser, *Int. Polym. Process.*, **16**, 351 (2001).
52. H. Jung and J. L. White, *Int. Polym. Process.*, **18**, 127 (2003).
53. H. Jung and J. L. White, *J. Appl. Polym. Sci.*, **102**, 1990 (2006).

7 Technology of Intermeshing Counter-Rotating Twin Screw Extruders

7.1 Introduction

In this chapter we will describe the development of intermeshing counter-rotating twin screw extrusion technology. These machines derive from positive displacement twin screw pumps, a development of the early 20th century. Intermeshing counter-rotating twin screw extruders were, during World War II and in the post-war period, important processing machines for extruding rigid polyvinyl chloride. They retain this application today, while finding broader uses in a range of technologies, including compounding, pharmaceutical packaging, devolatilization, and reactive extrusion.

We shall proceed as in Chapter 4, using a roughly historical perspective to describe the development of intermeshing counter-rotating twin screw machines. Early technological developments are described in Section 7.2. In Section 7.3, we discuss the joint program on kneading pumps of the I.G. Farbenindustrie (Hoechst) and Leistritz in the mid 1930s and early 1940s. Section 7.4 describes the origins of thermoplastics extrusion in intermeshing counter-rotating twin screw machines. Conical twin screw extruders are considered in Section 7.5. The various post-war new machine designs and machinery companies developed in Europe are considered in Section 7.6 to 7.10. The Japanese machinery manufacturers are considered in Sections 7.11, 7.13, and 7.14. Leistritz development of modular machines for compounding is considered in Sections 7.12 and 7.15.

7.2 Origins

Counter-rotating intermeshing mixing/pumping machines date to the 19th century. An 1874 U.S. patent application by S. Lloyd Wiegand [1] of Philadelphia, PA, explicitly describes a fully intermeshing counter-rotating twin screw extruder intended for extruding and sheeting dough (Figure 7.1). Wiegand, interestingly, recommends conical twin screws.

Fully intermeshing twin screw pumps are clearly discussed in a 1912 German patent application by Karl Werner [2] and a 1915 U.S. patent application by Holdaway [3] (Figure 7.2). In succeeding years, one saw the development of many fully intermeshing counter-rotating screw pumps intended for handling viscous Newtonian liquids. This was part of a much broader ranging technological development, involving positive displacement counter-rotating rotary pumps (Rollkolbenpumpen) of varying designs [3–17], including not only screw but cam and various vane designs.

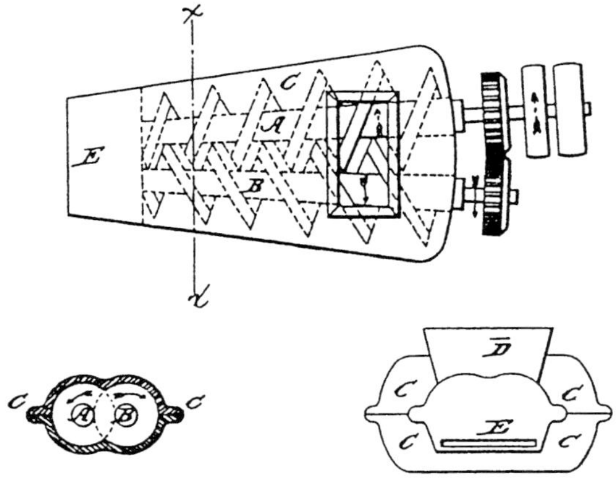

Figure 7.1: Wiegand's 1874 intermeshing counter-rotating twin screw machine for sheeting dough [1]

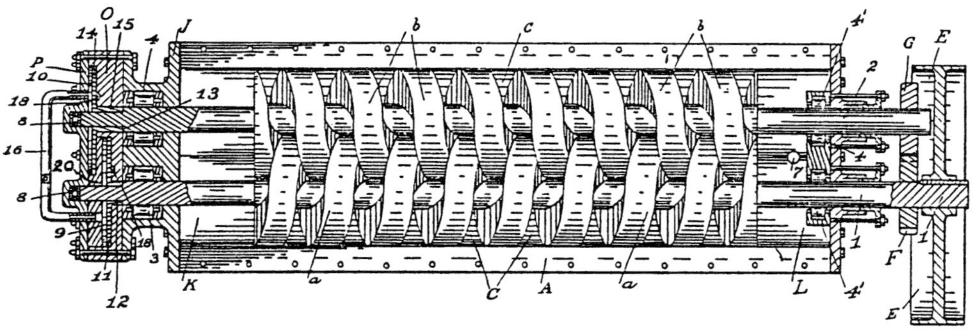

Figure 7.2: Holdaway 1915 intermeshing counter-rotating twin screw pump [3]

A 1922 patent to Societe Anonyme des Establissements A. Olier of Clermont-Ferrand, France [8] (inventor not named) developed a partially intermeshing counter-rotating twin screw machine intended for continuous extrusion of profiles of rubber compounds (Figure 7.3). It was called a 'die molding machine'. This most remarkable early patent claims that the new machine is superior to single screw extruders because it produces a more uniform output.

The twin screw pumps involve the displacement of a closed constant volume C-shaped chamber filled with fluid through the twin screw system. In a 1925 U.S. patent application the Swedish engineer C. O. J. Montelius [11] argued that the most effective twin screw pump is obtained when the two screws possess different numbers of threads per screw. A single threaded screw matched with a double threaded screw is better than two single threaded or two double threaded screws. Formulae were presented for both twin screw and multiple screw machines. We describe this in more detail in Section 8.2.

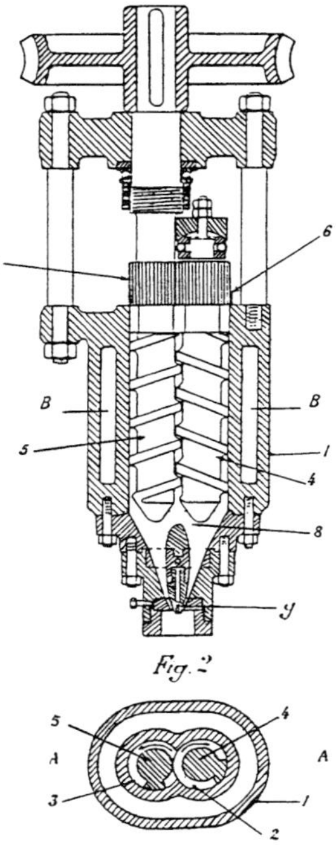

Figure 7.3: Olier 1922 intermeshing counter-rotating twin screw machine [8]

7.3 I.G. Farbenindustrie – Leistritz Knetpumpe Development Program

A most important development took place in Germany in the 1930s. It brought together the Frankfurt-Hoechst part of the I.G. Farbenindustrie and Maschinenfabrik Paul Leistritz of Nuremberg to develop a new generation of twin screw extrusion machines.

A brief discussion of each of these firms is worthwhile. The I.G. Farbenindustrie A.G. (Interessen Gemeinschaft der Farbenindustrie Aktiengesellschaft) was founded in 1925, combining all the major chemical firms of Germany. It was a result of the deteriorated economic situation in Germany following World War I. The firms involved included Friederich Bayer and Company, BASF, and Farbwerke Hoechst. The new firm played the major role in the development of synthetic rubber and thermoplastics in the period of its existence from 1925 to 1945. The I.G. Farbenindustrie was initiated by Carl Duisberg from Bayer, and Carl Bosch from BASF. Duisberg was the first head of the I.G. Farbenindustrie and was later succeeded by Bosch. It was headquartered in Frankfurt in the previous headquarters of Farbwerke Hoechst (see also Section 4.3). The I.G. Farbenindustrie was broken up by the victorious allies at the end of World War II to become the later Bayer AG, BASF AG, Hoechst AG, and Chemische Werke

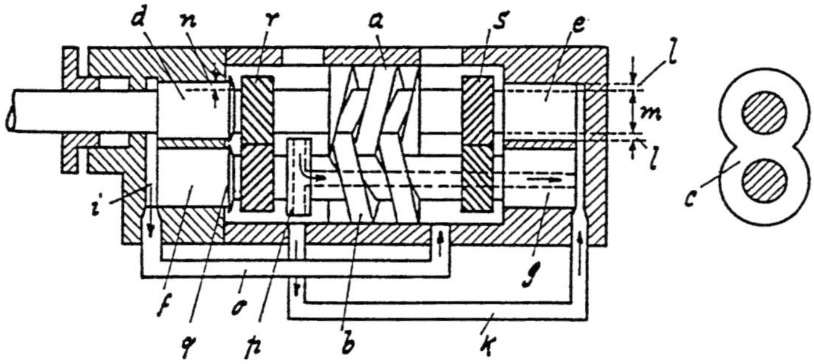

Figure 7.4: 1926 twin screw pump design of Leistritz and Burghauser [10]

Hüls (Hüls was in the 1990s absorbed by Degussa AG). In 2000, Hoechst AG was divided into several separate companies of which Celanese AG and Ticona are still involved in the chemical and polymer industries. The companies comprising the I.G. Farbenindustrie are today competitors and among the very largest chemical companies in the world. They have today annual sales in excess of 100 billion dollars.

Maschinenfabrik Paul Leistritz had been founded in Nuremberg (Nürnberg) in the kingdom of Bavaria in the German empire in 1905 by Paul Leistritz to manufacture blades/buckets and profiled strip stock for steam turbines [18]. After World War I, in the 1920's, Paul Leistritz and Franz Burghauser [10, 18, 19] invented, and the Leistritz firm began to manufacture, counter-rotating intermeshing twin and multiple screw pumps ("Schraubenpumpen mit zwei oder mehr Schraubenspindeln"). These pumps act like gear pumps and produce a throughput determined by the intermeshing geometry and screw speed. The special design of Leistritz and Burghauser [10] is shown in Figure 7.4. Maschinenfabrik Paul Leistritz also manufactured mufflers for internal combustion engines.

In the 1930s, the efforts of the I.G. Farbenindustrie and Leistritz were combined to apply the twin screw pump principle to difficult viscous industrial fluids, such as coal-oil masses, ceramic masses, and rubber compounds. The I.G. Farbenindustrie effort, which was based in Frankfurt-Hoechst, was led by Siegfried Kiesskalt, the Leistritz effort in Nuremberg by Franz Burghauser and Paul Leistritz.

Before joining the I.G. Farbenindustrie, Siegfried Kiesskalt had been with the Maschinenlaboratorium of the Technische Hochschule of Karlsruhe. Kiesskalt was involved in this period (ca. 1926/1927) with vane pumps and screw pumps for oils [20]. Kiesskalt's early activities with the I.G. Farbenindustrie (ca 1928/1929) concerned studies of lubricating oils [21]. This background made him familiar with Leistritz pumps.

In transporting coal, ceramic, and polymer systems there was also a necessity to knead and mix. The first patent I.G. Farbenindustrie was filed by Kiesskalt with H. Tampke, F. Winnacker, and E. Weingaertner [22] in 1935. It describes a counter-rotating machine for this purpose with the part of the screw near the discharge fully closed and the feed port region only partially closed (Figure 7.5). The exit region of the I.G. Farbenindustrie machine was redesigned by Kiesskalt [23].

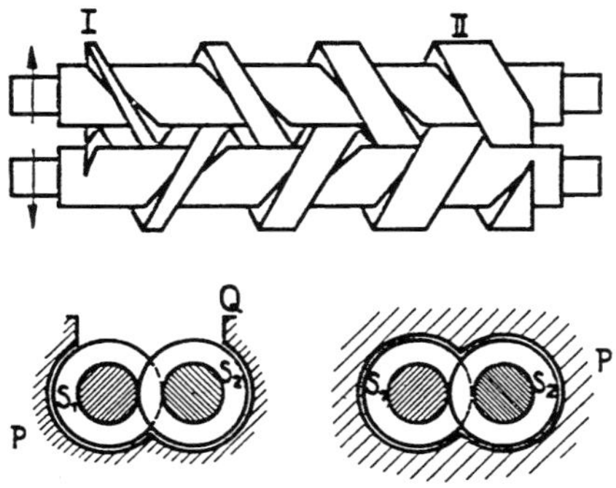

Figure 7.5: Kiesskalt et al. 1935 Knetpumpe [22]

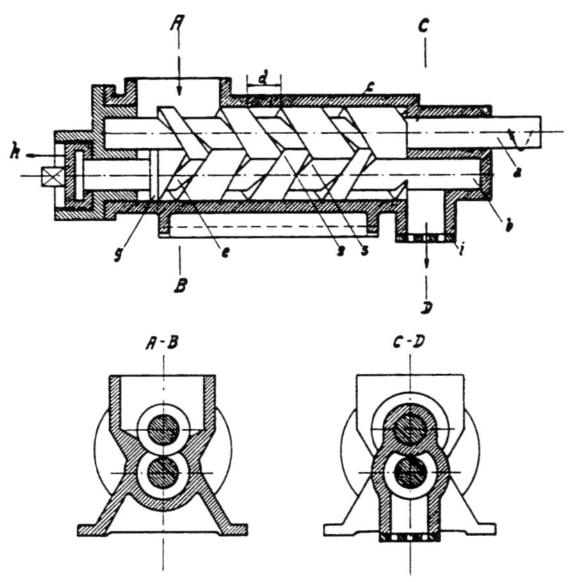

Figure 7.6: Leistritz-Burghauser 1935 Knetpumpe [24]

A patent filed shortly after that of Kiesskalt et al. [22] by Maschinenfabrik Paul Leistritz involves Franz Burghauser with Paul Leistritz [24,25] and treats a machine with tight fitting fully closed and intermeshing screws where the distance between flights steadily decreases in the pumping direction (Figure 7.6). They describe their new machine as a 'Knetpumpe' or kneading pump. Improved designs for feeding viscous fluids into the machine were subsequently described by Burghauser and Erb [26].

The logic for the design of the Burghauser-Leistritz machine is explained succinctly as follows. It was known that the mechanism of the machine was to displace C-shaped closed chambers forward to the die [11] (see Chapter 8, Section 8.2). If the pitch is decreased in the machine direction, so will be the volume of the C-shaped chamber. This will force the material out

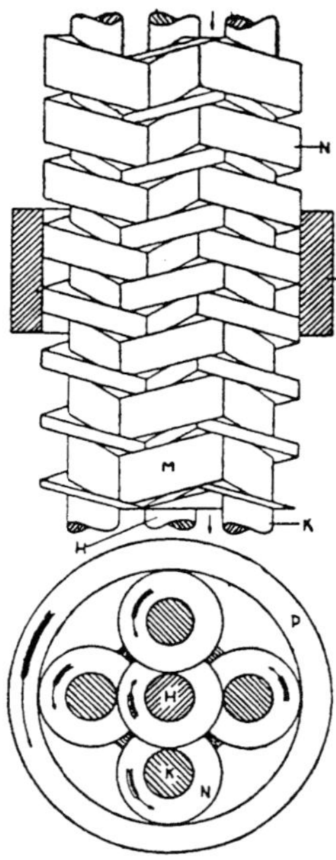

Figure 7.7: I.G. Farbenindustrie 1938 five-screw Knetpumpe [27]

of the C-chamber into the region over the flights, where the high shearing will result in the desired kneading and homogenization.

A 1939 Italian patent [27], apparently from the same I.G. Farbenindustrie group, describes 3-, 5-, and 7-screw intermeshing counter-rotating screw extruders. The purpose of those machines was again to pump and masticate rubber as well as oil, water, coal, and ceramic masses (Figure 7.7).

In patents filed from 1935, Kiesskalt with Tampke, Winnacker and Weingartner [28] describe a process for producing ash-free coal oil dispersions that require the use of kneading machines, probably involving those described above. In a patent filed in 1938, Kiesskalt and Borgwardt [29] describe a continuous intermeshing counter-rotating machine for homogenization of mixtures to produce emulsions.

The first publication in the open literature on an intermeshing counter-rotating twin screw 'knetpumpe' was by Kiesskalt [30] in 1942. Kiesskalt returned to the subject in papers in 1950 [31] and 1951 [32]. Riess and Meskat [33] and Riess and Erdmenger [34] also reviewed the I.G. Farbenindustrie-Leistritz developments in papers in the early 1950s. Kiesskalt seems to have left the I.G. Farbenindustrie/Hoechst following World War II and became a professor and head of the Institut für Verfahrenstechnik at the Technische Hochschule of Aachen [35].

7.4 Beginnings of Commercial Twin Screw Extruders for Polyvinyl Chloride (PVC)

By 1935, polyvinyl chloride pipes were being laid as pressure pipes for water supplies in Germany [36]. These were largely produced by intermeshing counter-rotating twin screw extruders. Maschinenfabrik Paul Leistritz was one of the major manufacturers of these twin screw extruders [18]. During World War II, they were manufacturing twin screw extruders for PVC with screw diameters of 200 mm. In the 1980s, H. Ocker [37] told one of us (J. L. White) about visiting manufacturing plants of I.G. Farbenindustrie plants in Germany during the early 1940s and observing many of these counter-rotating twin screw extruders producing pipes and profiles.

The final months of World War II saw extensive destruction of Germany, including their cities and infrastructure. The Nuremberg facility of Maschinenfabrik Paul Leistritz was destroyed.

In the post World War II period, Leistritz emphasized traditional smaller screw pumps for lubricating oils and other applications. They considered intermeshing counter-rotating twin screw extruders as a specialty item, made only on request. This was to change in time.

7.5 Conical Twin Screw Extruders

Conical twin screw extruders were first discussed in the early 1940s. In a 1943 patent application, Steinmann and Heyne [38] of Dynamit Nobel in Troisdorf, Germany proposed the use of intermeshing counter-rotating twin screw extruders with conical screws. Their patent drawing is shown in Figure 7.8. This design allows high production with lower torques. These machines would become of importance in the post-war period.

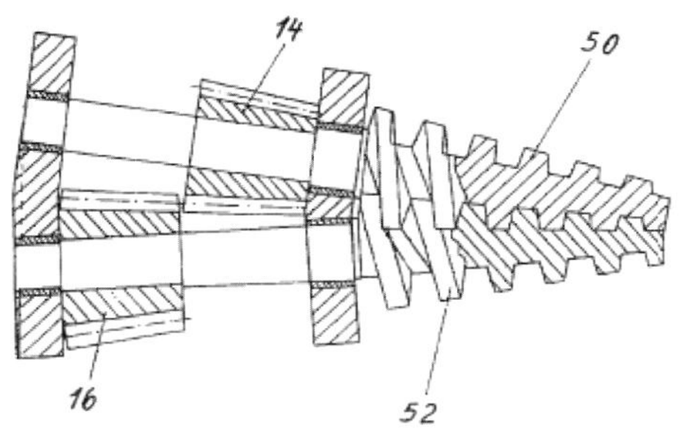

Figure 7.8: Steinmann-Heyne 1944 conical intermeshing counter-rotating twin screw extruder [38]

7.6 Pasquetti Bitruder

With much of Europe destroyed during World War II, there was a great need for new infrastructure, including polyvinyl chloride pipes which led to intense activity in this field.

In 1950, Carlo Pasquetti [39] of Vanese-Masnago, Italy, filed for patents on an intermeshing counter-rotating twin extrusion screw, which was divided into segments, each of which with the number and pitch of screw threads decreasing in the flow direction. Pasquetti, as an afterthought in his patent, speaks of the possibility of using his design with conical screws. The screw design from Pasquetti's patent is shown in Figure 7.9.

The Pasquetti design was licensed and manufactured by Schloemann AG of Düsseldorf and marketed as the Bitruder. The machine seems to have been primarily applied for the extrusion of rigid polyvinyl chloride. The Pasquetti-Schloemann Bitruder and its application as an extruder for polyvinyl chloride is discussed in papers of the 1950s and 60s by Schaerer [40], Baigent [41], Schutz [42], Prause [43], Zielonowski [44], and Schneider [36].

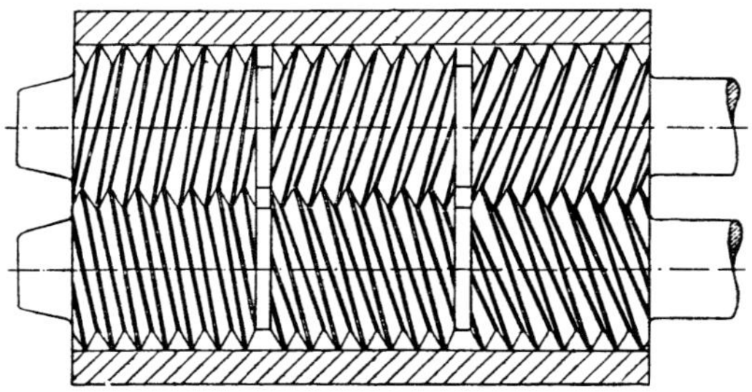

Figure 7.9: Pasquetti 1950 intermeshing counter-rotating twin screw extruder screw design [39]

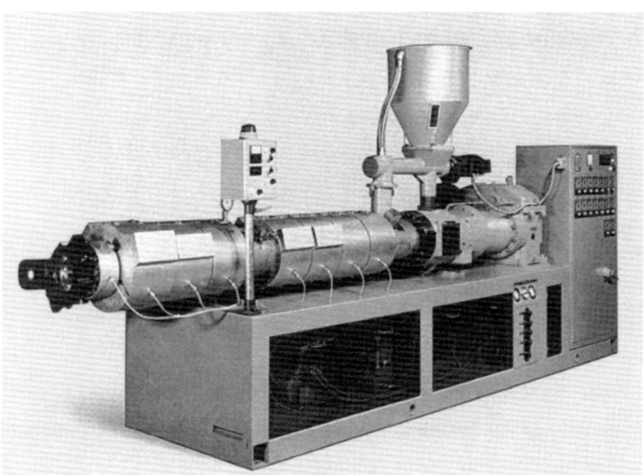

Figure 7.10: Photograph of modern Reifenhauser Bitruder (courtesy of Reifenhauser)

The manufacturer of the Bitruder was taken over by Reifenhauser KG of Troisdorf in the 1970s. Predohl [45] of Reifenhauser describes applications to polyolefins and polystyrene, including foam extrusion as well as the traditional application to rigid polyvinyl chloride.

Today, the Reifenhauser Bitruder is manufactured with both machined and modular screws. It is still primarily used as an extrusion machine. A photograph of the Reifenhauser Bitruder is shown in Figure 7.10.

7.7 Mapre Extruder

Other companies announced the manufacture of twin screw machines in the 1950s. This included Nouvelle Mapre SA [41–44, 46, 47] of Diekirch, Luxemburg. The Mapre machine involved screws with a rectangular profile with a progressively increasing flight thickness, along the axis of the screw (Figure 7.11).

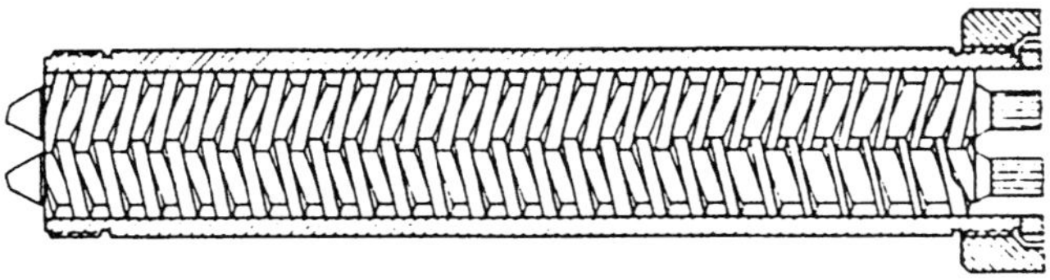

Figure 7.11: Mapre intermeshing counter-rotating twin screw extruder screw design

It was used as a compounding machine and as a profile extruder primarily for rigid polyvinyl chloride but also for polypropylene and polyamides.

7.8 Kestermann Extruder

The firm Gerhard Kestermann Zahnräder und Maschinenfabrik (later Rolf Kestermann), Bad Oeynhausen, West Germany also entered the manufacture of intermeshing counter-rotating twin screw extruders [36, 43, 44, 46–48]. This machine seems to have been introduced after the Pasquetti/Schloemann and Mapre machines. It is not mentioned by Schaerer [40] or Baigent [41] in their reviews published in the 1950s. The Kestermann extruder was described in more detail in papers by Schneider [36], Prause [43], Zielonowski [44], and notably by Selbach [48–50] of Kestermann in the 1960s. This machine was also primarily used for the extrusion of rigid polyvinyl chloride. During the 1960s, H. W. Selbach [51] developed a new complex screw design similar to a kneading pump of the general Burghauser-Leistritz type, except that a new flight grows along the length which penetrates into the thick flights existing at the feed end of the extruder (Figure 7.12).

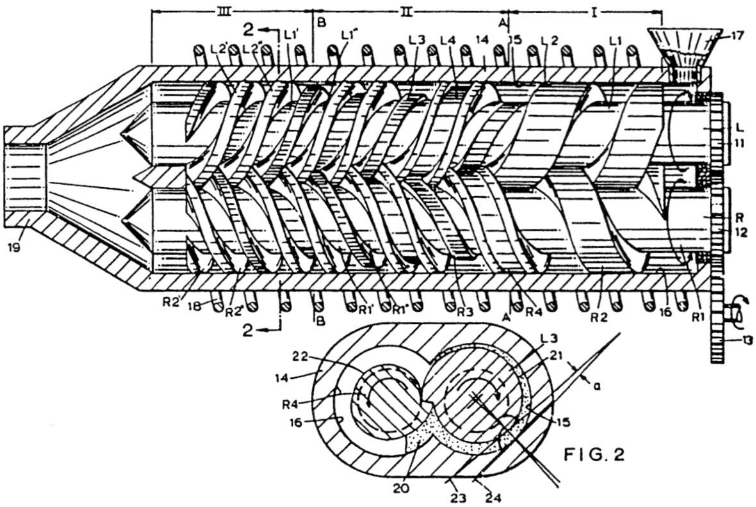

Figure 7.12: Selbach's 1962 Kestermann intermeshing counter-rotating twin screw extruder screw design [48]

Selbach [52] also introduced a perforated disc that preceded a major widening of the screw channel. This was to reduce pressure and allow for devolatilization.

Rheinstahl acquired Kestermann's activities in 1970 [36]. Thyssen and Rheinstahl merged in 1972 and Kestermann was the basis of the Thyssen PlastikMaschinen founded in 1975.

The Kestermann firm was taken over by the Battenfeld machinery combine in 1980. Battenfeld Extrusionstechnik today maintains a Bad Oeynhausen manufacturing plant and manufactures intermeshing counter-rotating twin screw extruders for polyvinyl chloride profile extrusion.

7.9 Anger Extruders

The Anger brothers Anton and Wilhelm entered the counter-rotating twin screw extruder business in the 1950s and by 1955 were producing intermeshing counter-rotating twin screw extruders. They founded the company Kunststoffwerk Gebrüder Anger GmbH and Co. [36]. Its headquarters were in Munich and the production shop was in Bogen on the Danube [36]. The first extruders were marketed in 1959. In 1962, Rheinstahl acquired a majority of the firm [36].

In 1964, Anton Anger founded Anton Anger Allgemeine Maschinenbau Gmbh (AGM) in Linz, Austria [36, 41–43, 53, 54]. AGM built both conical and parallel twin screw extruders. Specifically, AGM introduced the manufacture of tapered or conical twin screw machines [44, 54] (Figure 7.13). These were used for profile extrusion of rigid polyvinyl chloride.

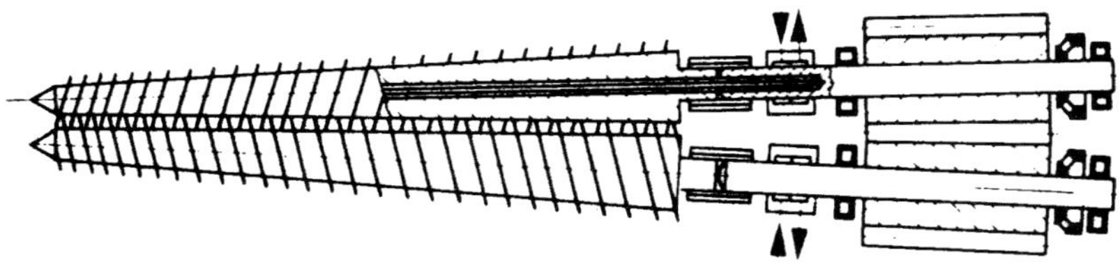

Figure 7.13: AGM conical intermeshing counter-rotating twin screw extruder [44]

His brother, Wilhelm Anger, founded Anger Plastic Verarbeitungsmaschinen GmbH (APM) located in Vienna. APM concentrated on the production of one-stage and two-stage parallel extruders.

APM manufactured machined screws where the pitch and flight thickness varied along the length of the screw [44]. This was applied to devolatilization. Slices in screw flights were also introduced to develop mixing.

Cincinnati Milacron absorbed the Austrian AGM firm and continued the manufacture of their conical twin screw machines in Vienna [36]. It also absorbed APM. Today it has become Cincinnati Extrusion based in Vienna.

7.10 Overview of European Post World War II Intermeshing Counter-Rotating Twin Screw Extruders

A brief summary of machinery companies manufacturing intermeshing counter-rotating twin screw extruders seems appropriate at this point. The 1954 article of Schaerer [40] and the 1956 article by Baigent [41] summarize manufacturers of intermeshing counter-rotating twin screw extruders. They mention Pasquetti (Schloemann), Mapre, and Trudex.

The 1962 article of Schutz [42] on the same topic mentions Wilhelm Anger, Gerhard Kestermann, Mapre, and Schloemann (Pasquetti).

The 1968 article of Prause [43] mentions AGM, Anger Plastic Verarbeitungsmaschinen (APM), Bausano (Turin, Italy), Rolf Kestermann (formerly Gerhard Kestermann), Mapre, and Schloemann (Pasquetti).

The 1968 article of Zielonowski [44] mentions AGM, Anger-APM, Rolf Kestermann, Paul Leistritz, Mapre, Schloemann (Pasquetti), and Hans Weber KG.

A 1981 article by Barth [57] mentions Battenfeld (former Kestermann), Cincinnati Milacron, Krauss-Maffei, Paul Leistritz, Maplan, Mapre, Oswag (formerly Anton Anger), Reifenhauser (formerly Schloemann-Pasquetti), and Weber. These machines were primarily for the extrusion of rigid polyvinyl chloride.

A 2005 paper of Schneider [36] describes the technology of Krauss-Maffei and mentions as other manufacturers Cincinnati Extrusion GmbH of Vienna, Austria; Battenfeld Extrusions Technik, and Maplan.

7.11 Japanese Intermeshing Counter-Rotating Twin Screw Extruders Manufacturers

European and American authors have generally not discussed Japanese intermeshing counter-rotating twin screw extruder manufacturers. The fully intermeshing counter-rotating twin screw extruder was apparently first manufactured in Japan by the Toshiba Machine Company Ltd. of Tokyo, Japan, in 1955. It was intended for extrusion of rigid polyvinyl chloride. Toshiba Machine has manufactured intermeshing counter-rotating twin screw extruders since that time; they have applied these machines to both profile extrusion and compounding (TED series). They also manufacture conical twin screw machines for profile extrusion (TEC series).

A second major manufacturer of intermeshing counter-rotating twin screw extruders is the Japan Steel Works of Hiroshima, Japan [58, 59]. A third manufacturer of counter-rotating twin screw machines is Mitsubishi Heavy Industries of Tokyo, Japan. Mitsubishi produces conical twin screw machines for extrusion profiles (VTE series). A fourth manufacturer of intermeshing counter-rotating twin screw extruders was the Ikegai Corp. of Tokyo, Japan. Their machine was intended for profile extrusion of rigid polyvinyl chloride (GTC series). Ikegai was the first manufacturer of twin screw extruders in Japan in 1951. These first machines were intermeshing co-rotating twin screw extruders (see Chapter 4).

7.12 Leistritz Modular Machines-1

Maschinenfabrik Paul Leistritz, in the late 1960s, developed a modular intermeshing counter-rotating twin screw extruder for compounding. There is an important pre-history that deserves discussion.

In a 1949 German patent application, E. Hack [60, 61] of Maschinenfabrik Paul Leistritz described an intermeshing twin screw pump with two modules. One module was for pumping, the other module was for mixing.

A similar idea is mentioned in a 1949 American patent application by Sennet [62] of the De Laval Steam Turbine Company, which contains a single screw forward pumping device followed by a multiple (two or more) screw device which pumps in the opposite direction.

While Maschinenfabrik Paul Leistritz (now Leistritz AG) did not aggressively market intermeshing counter-rotating twin screw extruders in the 1950s and 60s, they did maintain a significant level of activity. They continued to manufacture positive displacement screw pumps for lubricating oils. In a 1951 paper, Kiesskalt [32] describes a Leistritz intermeshing counter-rotating twin screw electrically heated gelatinizing machine. A kneading pump, reminiscent of the Burghauser-Leistritz kneading pump, is described in a U.S. patent filed in 1951 by Kleinlein and Bernhardt [64]. In a 1956 paper, Tanner [64] presented the first study of rigid polyvinyl chloride extrusion in an intermeshing counter-rotating Leistritz twin screw extruder.

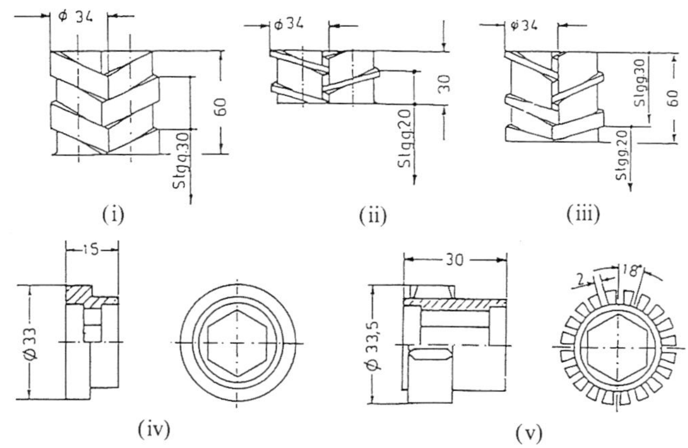

Figure 7.14: Modular elements of the Leistritz GG modular intermeshing counter-rotating twin screw extruder: (i) fully intermeshing thick flighted pumping screw, (ii) fully intermeshing thin flighted pumping screw, (iii) Kiesskalt kneading pump element, (iv) non-pumping "shearing" element, and (v) non-pumping "slit stowing" element [72]

In the mid 1960s, Doboczky [65, 66] of Leistritz published the first fundamental studies of leakage flows in an intermeshing counter-rotating twin screw extruder (see Section 8.4).

From the late 1960s, Leistritz gave increased attention to the commercial development of intermeshing counter-rotating twin screw extruders. Modular intermeshing counter-rotating twin screw machines were developed, notably by Hellmut Tenner [67] and manufactured as ZSE machines from then on. Special mixing element modules were introduced. The machines were marketed both in Germany and in the United States, where American Leistritz headed by William Thiele was established. The first paper describing the machine was published by Tenner [67] in 1976. Subsequent papers describing the machine were published in succeeding years by Thiele and his coworkers [68–70], Tenner [71], and others [72].

The modular elements of the Leistritz GG modular intermeshing counter-rotating twin screw extruder are shown in Figure 7.14. The elements consist of (i) tight fitting screws, (ii) screws with thin flights, (iii) screw elements similar to Kiesskalt et al.'s knetpumpe [22] (iv) special mixing elements including shearing discs.

7.13 Japan Steel Works Activities

During the 1980s, Japan Steel Works made a strong case for the usefulness of modular intermeshing counter-rotating twin screw extruders for various applications. The machines included screw elements that are not completely intermeshing. Sakai and Hashimoto [58] have described the use of Japan Steel Works TEX intermeshing counter-rotating twin screw

Figure 7.15: Photograph of modern JSW machine (courtesy of Japan Steel Works)

extruders for devolatilization. Sakai [59] also described the application of TEX intermeshing counter-rotating twin screw extruders to compounding. They argue that these machines do a better job dispersing small particles than do intermeshing co-rotating machines.

Japan Steel Works introduced TEX laboratory twin screw extruders, which may also be intermeshing counter-rotating or co-rotating.

The machine contained screw elements and kneading disc blocks similar to modular co-rotating twin screw extruders of Erdmenger design [73] (see Chapter 4).

A photograph of a modern JSW TEX machine is shown in Figure 7.15.

7.14 Polyplastics Counter-Rotating Paddle Machine

A 1981 patent application by Komazawa, Mori, Ikenaga, Hotta, and Nakashima [74], of Polyplastics of Osaka, Japan (a joint venture company of Celanese, and later Hoechst AG through Hoechst-Celanese), describes a continuous reactor involving shafts with intermeshing counter-rotating paddles (kneading disc blocks). The specific application was polymerization of trioxane to polyformaldehyde. It appears that this machine was the modular Japan Steel Works twin screw extruder described in the previous section. The machine of Komazawa et al. [74] is summarized in Figure 7.16.

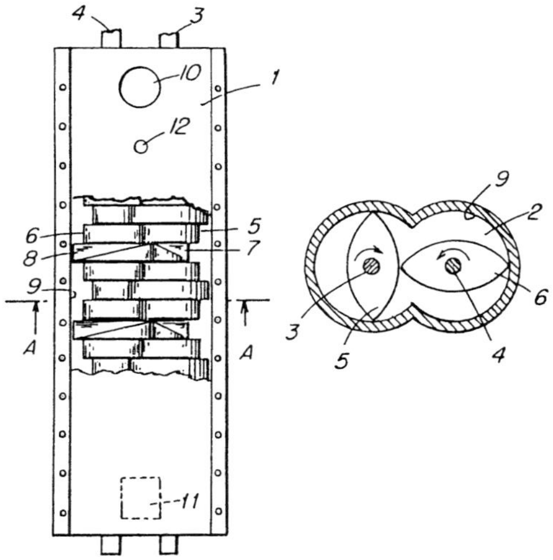

Figure 7.16: Polyplastic's intermeshing counter-rotating paddle machine [74]

7.15 Leistritz Modular Machines-2

In the 1990s, W. C. Thiele [75, 76] of American Leistritz developed a second generation of modular intermeshing counter-rotating twin screw extruders. The machine was called "counterflight" compared to earlier Leistritz modular counter-rotating GG machine. The level of intermesh is less than in the earlier Leistritz modular machine, the screws are longer (larger L/D), and the achievable screw speeds are much higher. The machine elements are different including hexilobal mixing screws. We show the Leistritz modular counterflight extruder in Figure 7.17.

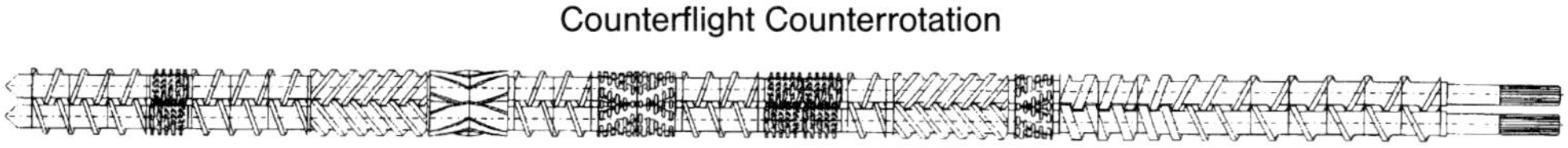

Figure 7.17: Leistritz modular counterflight extruder [75]

7.16 Detailed Design and Manufacture of Polyvinyl Chloride Twin Screw Extruders

There has been little discussion in the literature regarding the detailed design and manufacture of flood-fed counter-rotating twin screw extruders for polyvinyl chloride pipes and profiles. In recent years, Stasiek and his coworkers [77, 78] have discussed this in detail, including the detailed machine tools used to fabricate the screws and the problems involved.

References

1. S. L. Wiegand, U.S. Patent (filed April 28, 1874) 155,602 (1874).
2. K. Werner, German Patent (filed July 12, 1912) 281,104 (1914).
3. W. S. Holdaway, U.S. Patent (filed June 2, 1915) 1,218,602 (1917).
4. Anonymus (Gasapparate und Maschinen-Fabrik Gebrudern Pintsch) German Patent 75,506 (1892).
5. O. Erfurth, German Patent (filed Oct. 29, 1904) 162,565 (1905).
6. O. Erfurth, German Patent (filed Jan. 22, 1905) 163,440 (1905).
7. F. Wissiak, Austrian Patent (filed Aug. 20, 1907) 35,106 (1908).
8. Anoymous (Societe Anonyme Des Establissements A. Olier, British Patent (filed April 8, 1922) 180,638 (1923).
9. Anonymous, *Shipbuilding and Shipping Record*, p. 73 (1922).
10. P. Leistritz and F. Burghauser, German Patent (filed April 24, 1926) 453,727 (1927).
11. C. O. J. Montelius, U.S. Patent (filed March 20, 1925) 1,698,802 (1929).
12. C. O. J. Montelius, U.S. Patent (filed March 1, 1929) 1,965,557 (1934).
13. L. R. Schmidt, U.S. Patent (filed June 6, 1930) 1,846,692 (1932).
14. G. A. Ungar, U.S. Patent (filed June 11, 1930) 1,846,700 (1932).
15. E. Blau, *Chem. Zeit*, **54**, 801 (1930).
16. C. O. J. Montelius, *Teknisk Tidskrift*, **6**, 61 (1933).
17. H. A. Hartner, U.S. Patent (filed March 2, 1936) 2,119,162 (1938).
18. J. L. White, *Int. Polym. Process.*, **8**, 286 (1993).
19. H. Plachenka, Leistritz Schraubenspindelpumpen aus Nüremberg, Leistritz, Nuremberg (1965); also various anonymous published Leistritz brochures.
20. S. Kiesskalt, *VDI Zeitschr.*, **71**, 453 (1927).
21. S. Kiesskalt, *VDI Zeitschr.*, **73**, 1502 (1929).
22. S. Kiesskalt, H. Tampke, F. Winnacker, and E. Weingaertner, German Patent (filed July 26, 1935) 652,990 (1937).
23. S. Kiesskalt, U.S. Patent filed July 23, 1937) 2,148,205 (1939).
24. P. Leistritz and F. Burghauser, German Patent (filed Dec. 1, 1935) 682,787 (1939) and German Patent (filed July 31, 1936) 699,757 (1940).

25. F. Burghauser, U.S. Patent (filed Nov. 24, 1936) 2,115,006 (1938).
26. F. Burghauser and K. Erb, German Patent (filed Feb. 5, 1938) 690,990 (1939) and U.S. Patent (filed Feb. 1, 1939) 2,231,357 (1941).
27. I.G. Farbenindustrie (Frankfurt am Main), Italian Patent (filed April 9, 1938) 373,183 (1939).
28. S. Kiesskalt, H. Tampke, K. Winnacker, and E. Weingaertner, German Patent (filed May 25, 1935) 676,045 (1939) and German Patent (filed May 8, 1940) 690,831 (1940).
29. S. Kiesskalt and E. Borgwardt, German Patent (filed July 10, 1938) 739,278 (1943).
30. S. Kiesskalt, *VDI Zeitschr.*, **86**, 752 (1942).
31. S. Kiesskalt, *VDI Zeitschr.*, **92**, 551 (1950).
32. S. Kiesskalt, *Kunststoffe*, **41**, 414 (1951).
33. K. Riess and W. Meskat, *Chem. Ing. Tech.*, **23**, 205 (1951).
34. K. Riess and R. Erdmenger, *VDI Zeitschr.*, **93**, 633 (1951).
35. S. Kiesskalt, *Chem. Ing. Tech.*, **38**, 573 (1966).
36. H. P. Schneider, *Kunststoffe*, **95**, (5) 44 (2005).
37. H. Ocker, Personal Communication, Werner and Pfleiderer Stuttgart, Germany (ca 1987).
38. H. Steinmann and F. Heyne, German Patent (filed May 23, 1943) 834,900 (1951); German Patent (filed April 4, 1944) 846,012 (1952).
39. C. Pasquetti, British Patent (filed June 7, 1950) 677,945 (1952).
40. A. J. Schaerer, *Kunststoffe*, **44**, 105 (1954).
41. K. Baigent, *Trans. Plastics Inst.*, (Apr), 134 (1956).
42. F. C. Schutz, *SPE Journal*, **18**, (Sept.) 1147 (1962).
43. J. J. Prause, *Plastics Technol.*, (Nov.) 41 (1967), (Feb.) 30 (1968), and (Mar.) 52 (1968).
44. W. Zielonowski, *Kunststoffe*, **58**, 394 (1968).
445. W. Predohl in 5th Leobener Kunststoffkolloquium Doppelschnecken-Extruder, p. 66, Lorenz Verlag, Vienna (1978).
46. F. C. Schutz, *SPE Journal*, **18**, (Sept.) 1147 (1962).
47. Anonymous, *Plastics (Plastics Institute)*, **5**, 80 (1961).
48. H. W. Selbach, *Plastverarbeiter*, **13**, 595 (1962).
49. H. W. Selbach, *Kunststoffe*, **52**, 232 (1962).
50. H. W. Selbach, *Kunststoffe*, **57**, 784 (1967).
51. H. W. Selbach, U.S. Patent (filed Dec. 5, 1961) 3,104,420 (1963).
52. H. W. Selbach, U.S. Patent (filed June 27, 1966) 3,407,438 91968).
53. F. Burger, W. Hanslick, P. Heilmeyr, and A. Kerschbaumer, *Kunststoffe*, **56**, 278 (1966).
54. Anonymous, *Kunststoffe*, **57**, 691 (1967).
55. F. Alber in 5th Leobener Kunststoffkolloquium Doppelschnecken-Extruder, p. 31, Lorenz Verlag, Vienna (1978).
56. Anonymous, *Kunststoffe*, **67**, 326 (1977).
57. H. Barth, *Kunststoffe*, **71**, 636 (1981).
58. T. Sakai and N. Hashimoto, *SPE ANTEC Tech. Papers*, **32**, 360 (1986).
59. T. Sakai, *SPE ANTEC Tech. Papers*, **33**, 146 (1987).
60. E. Hack, German Patent (filed Sept. 10, 1949) 815,103 (1959).
61. E. Hack, German Patent (filed Oct 2, 1948) 830,167 (1952).

62. M. B. Sennet, U.S. Patent (filed Sept. 14, 1949) 2,548,451 (1952).
63. F. Kleinlein and M. Bernhardt, U.S. Patent (filed Nov. 23, 1951) 2,686,336 (1954).
64. K. Tanner, *Kunststoffe*, **46**, 431 (1956).
65. Z. Doboczky, *Plastverarbeiter*, **16**, 57 (1965).
66. Z. Doboczky, *Plastverarbeiter*, **16**, 395 (1965).
67. H. Tenner, *Kunststoffberater*, **6**, 1 (1976).
68. W. C. Thiele, *SPE ANTEC Tech. Papers*, **29**, 127 (1983).
69. W. C. Thiele, W. Petrozelli, and D. Lorene, *SPE ANTEC Tech. Papers*, **36**, 120 (1990).
70. W. C. Thiele, W. Petrozelli, and C. Martin, *SPE ANTEC Tech. Papers*, **37**, 1849 (1991).
71. H. Tenner, *Kunststoff Journal*, (Dec.) 102 (1987).
72. S. Lim and J. L. White, *Int. Polym. Process.*, **9**, 33 (1994).
73. J. W. Cho and J. L. White, *Int. Polym. Process.*, **11**, 21 (1996).
74. H. Komazawa, T. Mori, Y. Ikenaga, H. Hotta, and T. Nakashima, U.S. Patent (filed Aug. 12, 1981) 4,390,684 (1983).
75. W. C. Thiele in Plastics Compounding, edited by D. B. Todd, Hanser, Munich (1998).
76. W. C. Thiele in Pharmaceutical Extrusion Technology, edited by I. Ghebre-Sellasie and C. Martin, Dekker, New York (2003).
77. J. Stasiek, *Plasty Kaucuk*, **35**, 70 (1998).
78. J. Stasiek and T. Nieszporek, *Int. Polym. Sci. Tech.*, **30**, 769 (2003); *Polimery*, **47**, 441 (2002).

8 Flow Mechanisms and Modeling of Intermeshing Counter-Rotating Twin Screw Extruders

8.1 Introduction

We are concerned in this chapter with developing a fundamental understanding of the fluid mechanics of intermeshing counter-rotating twin screw extruders. The basic understanding of the mechanism for this type of machine predates its applications to polymer processing technology, and appears to be contained in patents before open publications. The engineering analysis of this machine for plastics processing begins with the monographs of G. Schenkel [1] on extrusion around 1960. Schenkel [1] argued, following early investigators, that this machine is basically a positive displacement pump in which "C-chambers" along the screw flights are moved forward. Critical contributions were made by various researchers to describe the flow mechanisms in this machine in the period from 1965 to 1976. Many of these activities have been summarized in a monograph published by Janssen [2] in 1978, which is dedicated to intermeshing counter-rotating twin screw machines.

We will begin in Section 8.2 with a discussion of screw geometry of the intermeshing counter-rotating machine and the positive displacement pump principle. The fluid mechanics of flow in C chambers is developed in Section 8.3. In Section 8.4, we describe the various leakage flows that occur in intermeshing counter-rotating twin screw machines. In Section 8.5 we describe FAN models of intermeshing counter-rotating twin screw elements. Modeling of the melting process is considered in Section 8.6. We then turn (Section 8.7) to models for composite modular counter-rotating twin screw elements. Screw-screw-barrel interaction is considered in Section 8.8.

8.2 Screw Geometry and Positive Displacement Pump Principle

Intermeshing twin screw extruders involve the fitting together of two different screws. Obviously, this involves screws of the same pitch. If the screws are fully intermeshing, we are obviously dealing with screws of the same flight depth. If two screws are to rotate in the same direction, they must both have the same hand, i.e., both be right-handed screws or both be left-handed screws. On the other hand, if the screws rotate in opposite directions, the screws must have opposite hands, i.e., one is a right-handed screw and the other is a left-handed screw.

The pumping action of two counter-rotating fully intermeshed screws has long been a subject of serious attention among engineers. The forward pumping of filled C-chambers is implicit and appears to have been well understood in patents of the 19th century and in the pre-World War I period. This mechanism is described in the 1874 patent application of Wiegand [3], which concerns extrusion of baking dough.

> *"The dough enters the spiral or helical spaces between the screw threads and were only a single screw used would rotate with the screw excepting in so for it is detained by the friction of the portion in contact with the case but using a screw made right and left, with the threads of each screw inter-locking in the spaces of the other screw, the rotation of the dough is prevented and the dough is forced lengthwise by the actions of the screws towards the slot E in the end of the case."*

The concept is clear in a 1915 patent application by Holdaway [4]:

> *"The perimeters of said screws a and b contact with the interior of said intersecting cylinders thus forming a plurality of fluid chambers by which when the screws are operated a given quantity fluid is advanced from the inlet chamber to the outlet chamber and from the outlet chamber is forced through the outlet pipe."*

Further critical discussions about machines of this general type date to the 1920s [5–7]. A clear mathematical discussion is given in a 1929 patent application of Montelius [6] indicating how such a machine operates as a positive displacement pump. We quote Montelius:

> *"U.S. Patent No. 1,698,802 describes an arrangement intended to be used as a pump or motor consisting of two cooperating screws of certain shapes, said screws being of opposite pitch and rotating in opposite directions, and being tightly bound by a casing. These two screws act together as a piston which on the rotation of the screws advances the fluid liquid on gas present within the casing, in the direction along the axis of the screw.*
>
> *It is obvious that when the screws A and B (see Figure 8.1) are in any position given, their threads form a closure, shutting off the S-shaped bore of the casing like a piston. When the screws are rotated, the closure is axially displaced by a distance corresponding to the thread pitch of screw B for each revolution of screw B. The volume then displaced corresponds to a cylinder having a bottom surface equal to the cross-section area of the S-shaped bore, the height equaling the pitch of B, which is, however, to be reduced by the volume of the same length of the screws. On the other hand, the screw volume in a pitch length L is equal to the volume of a cylinder having a base area equal to the total cross-sectional area of the screw and having a height equal to the thread pitch.*
>
> *Mathematically expressed, the volumetric displacement, i.e., the volume delivered, by a single revolution of screw B is:*
>
> $$V = S\,(A_S - G_1 - G_2) \tag{8.1}$$
>
> *where A_S designates the S-shaped bore of the screw casing, G_1, the cross-sectional shaded area of the single thread screw A, G_2, the cross-sectional area of the double thread screw B, all of these cross-sections being taken at right angles to the axis, and S designates the pitch of screw B."*

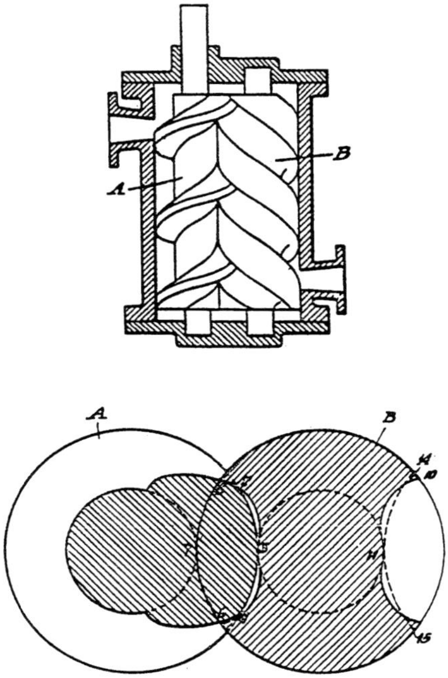

Figure 8.1: Montelius [6] patent drawing of an intermeshing counter-rotating twin screw pump to explain the mechanism of operation

Kiesskalt, in a 1927 paper [7], describes the forward pumping capacity of an intermeshing twin screw pump as:

$$Q = N\,V_C - Q_{leak} \tag{8.2}$$

where V_C represents the total C-chamber volume and Q_{leak} is a backward leakage flow.

The concepts contained in the above paragraphs were well known to Kiesskalt and his co-workers with the I.G. Farbenindustrie and to Burghauser and Leistritz of Maschinenfabrik Paul Leistritz in their 1935–1945 joint program on 'kneading pumps' described in Chapter 7 (Section 7.3). This may be seen in the following quote from a 1936 U.S. patent application by Burghauser [8]:

> *"The kneading pump to which the present invention relates has the known features of loosely enmeshing screw or worm rotors. Known rotary pumps, however, are constructed with a constant pitch and a spacing between adjacent flanks of the threads on the respective rotors which is constant throughout the length of the thread."*
>
> *"The kneading pump of the present invention (Figure 7.5) differs from the known pumps in one of the most important features of the known apparatus inasmuch as worms or screws having a constantly changing chamber volume along the length of the thread as used."*

This principle is discussed in other patents of the period. It is again described by Maschinenfabrik Paul Leistritz researchers in post World War II patents [9] and papers [10, 11].

The first discussion of the intermeshing counter-rotating multiple screw extruders for thermoplastics in the open literature is in the treatise by Schenkel [1], who writes:

$$Q = m\, N\, V_C \tag{8.3a}$$

where Q is the throughput, m the number of screws, N is the screw rotation rate, and V_C the C-chamber volume per screw. The volume of the chamber volume was computed by Schenkel [1]. Doboczky [10, 11] of Leistritz wrote for twin screw machines:

$$Q = 2\, i\, N\, V_C \tag{8.3b}$$

where i is the number of thread starts.

One may calculate the volume V_C in the following manner [1]. This is:

$$V_C = V_1 - V_2 - V_3 \tag{8.4}$$

Here, V_1 is the cross-section of one barrel-half times the pitch S.

$$V_1 = S\left[\left(\pi - \frac{\alpha}{2}\right)R^2 + \sqrt{\left(R\,H - \frac{H^2}{4}\right)\left(R - \frac{H}{2}\right)}\right] \tag{8.5a}$$

R is the outer radius, H is the depth of the chamber, and α the overlap angle in radians. The angle α may be expressed:

$$\alpha = 2\tan^{-1}\frac{\sqrt{2\,D\,H - H^2}}{D - H} \tag{8.5b}$$

V_2 is the volume of the screw root:

$$V_2 = \pi\,(R - H)^2\, S \tag{8.5c}$$

V_3 is the volume of the screw flight:

$$V_3 = \int_{R-H}^{R} b(r)\, 2\,\pi\, r\, \mathrm{d}r \tag{8.5d}$$

The formulation of the above paragraph does allow one to calculate the volume of the C-chamber, but the result is not stated in a physically meaningful manner. In terms of a flattened-out model of a twin screw machine, we may write:

$$V_C = H\, W\, L_C \tag{8.6}$$

where L_C is the length of the chamber. The value of L_C is clearly related to the diameter of the screws and the pitch or helix angle. The length of travel around a screw channel between points of intersection with the other screw will be of form:

$$L_C = \frac{\pi D}{\cos\phi} \tag{8.7a}$$

which suggests that V_C may be expressed approximately:

$$V_C = \frac{\pi D H W}{\cos\phi} \tag{8.7b}$$

A 1924 Swedish patent and a 1925 U.S. patent application of Montelius [5] described the detailed design of screws needed for a fully intermeshing counter-rotating machine, including detailed flight designs and the number of operating thread starts which is highly desired for tight fitting screws. For the best pumping characteristics, i.e., to minimize leakage, Montelius argues that, if one has two co-operating screws, the number of thread starts, v_1, on one screw should not be equal to the number of thread starts on the second screw, v_2. He considers it best if one screw be single-flighted; the other should best be double-flighted, i.e.:

$$v_2 = v_1 - 1 \tag{8.8}$$

In such a case, screw 2 which is single-threaded, would have to rotate twice as fast as screw 1, which is double-threaded, if both screws have the same diameter.

Montelius [5] also considers the case where n screws, with v_1 threads per screw cooperate with one central screw with v_2 threads per screw. Here, he argues that the best pump design can be obtained when:

$$v_2 = n\,(v_1 - 1) \tag{8.9}$$

In a three-screw pump, Montelius argues it is best, if all the screws are double-flighted and thus rotate with the same speed.

For a series of n screws of type 1, with N screws of type 2, Montelius gives:

$$N\,v_2 = n\,(v_1 - N) \tag{8.10}$$

8.3 Flow in C-Chambers

The fluid within the C-chamber should not be considered stagnant. It undergoes a circulating flow driven by the relative motion of the screw and barrel. The fluid in the C-chamber is also bounded by the moving walls at the back and front of the chamber. The flow patterns in C-chambers have been modeled by Kim, Skatschkow, and Jewmenow [12], Janssen et al. [13], and Wyman [14]. The model of Janssen et al. [13] and Wyman [14] is simpler and will be presented here. It is based on the consideration of the velocity component along the screw channel alone. The equation of motion specifying the velocity field involves consideration of shearing between the screw and barrel. If we neglect the contribution of the screw flights to the shearing, the equation of motion becomes:

$$0 = -\frac{\partial p}{\partial x_1} + \frac{\partial \sigma_{12}}{\partial x_2} \tag{8.11a}$$

$$0 = -\frac{\partial p}{\partial x_3} + \frac{\partial \sigma_{32}}{\partial x_2} \tag{8.11b}$$

With boundary conditions:

$$v_1(O) = -V_N \qquad v_1(H) = U_1 - V_N \qquad 0 = \int_0^H v_1 \, dx_2 \tag{8.12a}$$

$$v_3(O) = 0 \qquad v_3(H) = U_3 \qquad 0 = \int_0^H v_3 \, dx_2 \tag{8.12b}$$

where V_N is the velocity of the coordinate system which advances along the screw channel in such a way that the relative position in a C-chamber is unchanged. This is:

$$V_N = \frac{\pi D N}{\cos\phi} \tag{8.13}$$

For a Newtonian fluid, Eq. (8.11) becomes:

$$0 = -\frac{\partial p}{\partial x_1} + \eta \frac{\partial^2 v_1}{\partial x_2^2} \tag{8.14a}$$

$$0 = -\frac{\partial p}{\partial x_3} + \eta \frac{\partial^2 v_3}{\partial x_2^2} \tag{8.14b}$$

The solution of Eq. (8.14a, b), subject to the boundary conditions of Eq. (8.12a, b) yields:

$$v_1 = -V_N + U_1 \left(\frac{x_2}{H}\right) - \frac{1}{2\eta} \frac{\partial p}{\partial x_1} \left[\left(\frac{x_2}{H}\right) - \left(\frac{x_2}{H}\right)^2\right] \tag{8.15a}$$

$$v_3 = U_3 \left(\frac{x_2}{H}\right) - \frac{1}{2\eta} \frac{\partial p}{\partial x_3} \left[\left(\frac{x_2}{H}\right) - \left(\frac{x_2}{H}\right)^2\right] \tag{8.15b}$$

It follows that:

$$q_1 = 0 = \int_0^H v_1 \, \mathrm{d}x_2 = -V_N \, H + \frac{U_1 \, H}{2} - \frac{H^3}{12 \, \eta} \frac{\partial p}{\partial x_1} \tag{8.16a}$$

$$q_3 = 0 = \int_0^H v_3 \, \mathrm{d}x_2 = \frac{U_3 \, H}{2} - \frac{H^3}{12 \, \eta} \frac{\partial p}{\partial x_3} \tag{8.16b}$$

The pressure gradient in the C-chamber is:

$$\frac{\partial p}{\partial x_1} = -6 \, \eta \frac{\pi \, D \, N}{H^2} \left(\frac{2}{\cos \phi} - \cos \phi \right) \tag{8.17a}$$

$$\frac{\partial p}{\partial x_3} = -6 \, \eta \frac{\pi \, D \, N \sin \phi}{H^2} \tag{8.17b}$$

which means that the highest pressure is at the hopper end of the C-chamber.

8.4 Leakage Flows

8.4.1 General

Leakage in intermeshing twin screw pumps and its implications has long been realized. The effect is described in the previously cited Eq. (8. 5) of Kiesskalt [7]. It was later clearly more pointed out in 1965 by Doboczky [10, 11] of Leistritz with application to modern twin screw extruders. Doboczky classifies these as:

$$Q = Q_C - Q_{CL} - Q_{PL} - Q_{FL} \tag{8.18}$$

where Q_{CL}, Q_{PL}, and Q_{FL} represent different leakage flows. These leakage flows are described by Doboczky as follows. Q_{CL} is the 'Flankenschleppstrom' (*calendering leakage* flow between the screws). Q_{PL} is the 'Druckstrom zwischen den Flanken' (*pressure leakage*). Q_{FL} is the 'Leckstrom über die Schneckenstege' (*flight leakage*). These are illustrated in Figure 8.2. The problem of leakage flows in intermeshing counter-rotating twin screw extruders was subsequently considered by Klenk [15–17] and Janssen et al. [2, 13, 18–20].

We now turn to the problem of calculating the magnitudes of the individual leakage flows. In each of the next three successive sections we shall analyze and predict the values of these leakage flows. Generally, it is considered that:

$$Q_{CL} > Q_{PL} > Q_{FL} \tag{8.19}$$

i.e., the calendering leakage flow is the largest and the screw flight leakage the least.

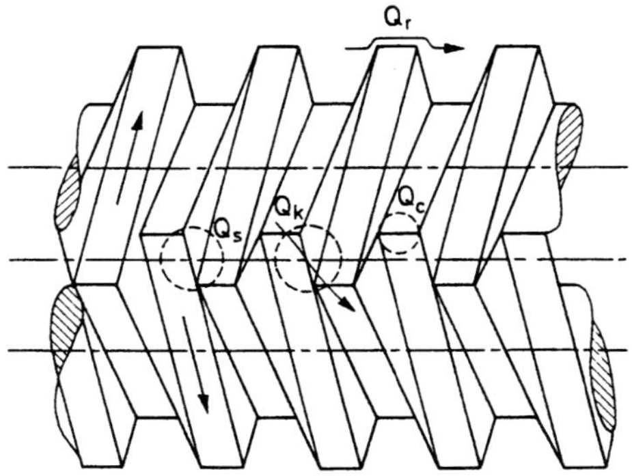

Figure 8.2: Leakage flows

8.4.2 Calendering Leakage (Q_{CL})

The calendering leakage flow represents flow in the gap between the screws. This effect is discussed by Doboczky [10] and by Janssen et al. [2, 13, 20]. The calendering leakage flow effect is sketched in Figure 8.3.

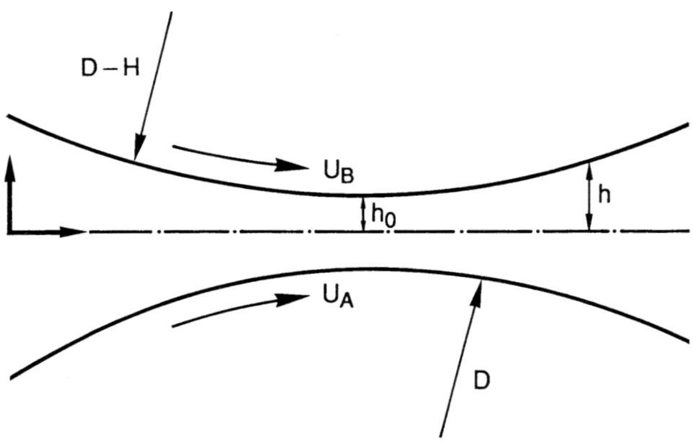

Figure 8.3: Calendering leakage flow

The flow in the calendering region may be represented by the force balance:

$$0 = -\frac{\partial p}{\partial x_1} + \frac{\partial \sigma_{12}}{\partial x_2} \tag{8.20}$$

with boundary conditions:

$$v_1(-h) = U_A \qquad v_1(h) = U_B \tag{8.21}$$

where 2 h is the gap between the screws which is much smaller than the channel depth:

$$2\,h \ll H \tag{8.22}$$

and U_A and U_B are the linear velocities of the two screws. These are:

$$U_A = \pi\, D\, N \qquad U_B = \pi\,(D - H)\, N \tag{8.23}$$

For a Newtonian fluid, Eq. (8.20) becomes:

$$0 = -\frac{\partial p}{\partial x_1} + \eta\,\frac{\partial^2 v_1}{\partial x_2^2} \tag{8.24}$$

Solving for the velocity field we obtain:

$$v_1(x_2) = \frac{U_A + U_B}{2} + \frac{U_B - U_A}{2}\left(\frac{x_2}{h}\right) - \frac{h^2}{2\,\eta}\,\frac{\partial p}{\partial x_1}\left[1 - \left(\frac{x_2}{h}\right)^2\right] \tag{8.25}$$

The total flow is:

$$Q_{CL} = W \int_{-h}^{h} v_1(x_2)\,\mathrm{d}x_2 = (U_A + U_B)\,W\,h + \left(\frac{U_B - U_A}{2}\right) W\,h - \frac{2\,h^3}{3\,\eta}\,\frac{\partial p}{\partial x_1} \tag{8.26}$$

To calculate Q_{CL}, we must integrate Eq. (8.26) to obtain Δp. To carry this out, we must know h as a function of x_1. If we estimate this to be (following Janssen et al. [13]):

$$h = h_0 + \frac{x_1^2}{2\,R} \tag{8.27}$$

where h_0 is the minimum nip and R the screw radius, we may integrate Eq. (8.26). Janssen et al. give the form:

$$\Delta p = \frac{3}{4}\,\pi\,\eta\,\frac{\sqrt{2\,R\,h_0}}{h_0^3}\left[\frac{3\,Q_C}{4}\left(\frac{S}{2} - H\,\tan\psi\right) - (U_1 + U_2)\,h_0\right] \tag{8.28}$$

where S is the screw pitch and ψ the flight wall angle. Janssen [2] later wrote this as:

$$\Delta p = 6\,\eta\,\frac{\sqrt{(2\,R - H)\,h_0/2}}{h_0^3}\left[-\frac{3\,i\,Q_C}{4\,(S - Bi)} + N\,\pi\,(2\,R - H)\,h_0\right] \tag{8.29}$$

N is the screw speed and i is the number of thread starts.

Subsequently, Speur, Mavridis, Vlachopoulos, and Janssen [20] reported a numerical simulation of flow in the calender gap between the screws. The simulation is for Newtonian fluids and applies to finite element techniques. A vortex is predicted at the entry to the calender gap. Calculations showed that the vortex intensity is determined uniquely by the value of the relative throughput and disappears when it exceeds 24% for the geometries examined by the authors.

Speur et al. [20] also made calculations for a power law fluid model and found that shear thinning suppressed the vortices.

8.4.3 Pressure Flow Leakage (Q_{PL})

Doboczky [10] and Janssen et al. [2, 12] have considered the pressure flow backward through the tetrahedron gap between the flanks of the screws. Janssen et al. represent this leakage through the expression:

$$Q_{\mathrm{PL}} = 0.0026 \frac{1}{\eta} \left(\frac{H}{D} \right)^{1.8} \psi^2 D^3 \Delta p \tag{8.30}$$

which seems to have been developed experimentally. ψ is the slope in radians of the sidewall of the screw flight. Here, D is the screw diameter and Δp the pressure difference between opposite C-chambers.

8.4.4 Flight Leakage Effects (Q_{FL})

In this section, we turn to leakage flows over the tips of the flights in the C-chambers. As in leakage flows in single screw extruders (Section 3.4), it is due to the combined effects of drag and pressure flow over the tips of screw flights. Analyses of leakage over screw flights date at least to Mohr and Mallouk [21] for single screw machines and for intermeshing counter-rotating twin screw extruders, to Doboczky [10] and Janssen et al. [13]. The velocity field normal to the screw flight is the v_3 component. The form of this v_3 velocity field is given in Eq. (8.15b).

The flight leakage flow Q_{FL} over a screw flight tip of length e is:

$$Q_{\mathrm{FL}} = \int_0^{L_{\mathrm{C}}} \int_0^{\delta_{\mathrm{F}}} v_3 \, \mathrm{d}x_2 \, \mathrm{d}x_1 = \frac{1}{2} U \, L_{\mathrm{C}} \, \delta_{\mathrm{F}} \sin\phi + \frac{L_{\mathrm{C}} \, \delta_{\mathrm{F}}^3}{12\,\eta} \frac{\Delta p_{\mathrm{f}}}{e} \tag{8.31}$$

Here, Δp_f is the pressure drop over a single flight caused by the pressure distribution across the screw channel and the pressure build up along the screw axis to the die. This may be estimated from Eq. (8.17b).

8.5 FAN Models of Flow in Modular Elements

Flow Analysis Network (FAN) modeling of individual elements of an intermeshing counter-rotating twin screw extruder was published in 1998/1999 by Hong and White [23, 24]. The same basic method described in Section 5.3 was applied to the flow in an intermeshing counter-rotating machine, but using cylindrical coordinates because of the geometry. Because of the positive displacement effect described in Section 8.2, the method was modified to include its contributions. Specifically, they wrote:

Between Screw and Barrel (Region I):

$$\begin{aligned} &q_\theta^{\mathrm{I}}(\theta-\Delta\theta,z)\,\Delta z+q_z^{\mathrm{I}}(\theta,z-\Delta z)\,R_{\mathrm{CH}}\,\Delta\theta \\ &=q_\theta^{\mathrm{I}}(\theta+\Delta\theta,z)\,\Delta z+q_z^{\mathrm{I}}(\theta,z+\Delta z)\,R_{\mathrm{CH}}\,\Delta\theta \end{aligned} \tag{8.32}$$

where

$$q_\theta=\int_{R_i}^{R_0} v_\theta\,\mathrm{d}r \qquad q_z=\frac{1}{R_{\mathrm{ch}}}\int_{R_i}^{R_0} r\,v_z\,\mathrm{d}r \tag{8.33 a, b}$$

This formulation had been given earlier by Bang and White [25] for a tangential counter-rotating twin screw extruder (see Section 11.4).

Entrance Between Screws (Region I – Region II):

$$\begin{aligned} &q_\theta^{\mathrm{IR}}(\theta-\Delta\theta,z)\,\Delta z+q^{\mathrm{IL}}(\theta-\Delta\theta,z)\,\Delta z+q_z^{\mathrm{I}}(\theta,z-\Delta z)\,R_{\mathrm{CH}}\,\Delta\theta \\ &=q_\theta^{\mathrm{II}}(\theta+\Delta\theta,z)\,\Delta z+q_z^{\mathrm{II}}(\theta,z+\Delta z)\,R_{\mathrm{CH}}\,\Delta\theta \end{aligned} \tag{8.34}$$

where superscript *I* refers to fluxes in Region I and superscript *II* refers to fluxes in Region II. *IL* means the left screw and *IR* the right screw. R_{CH} is a characteristic radius such as the screw radius.

In the Interscrew Region:

$$\begin{aligned} &q_\theta^{\mathrm{II}}(\theta-\Delta\theta,z)\,\Delta z+U_\theta(R_0-R_i)\,(\theta-\Delta\theta,z)+q_z^{\mathrm{II}}(\theta,z-\Delta z)\,R_{\mathrm{CH}}\,\Delta\theta \\ &=q_\theta^{\mathrm{II}}(\theta+\Delta\theta,z)\,\Delta z+U_\theta(R_0-R_i)\,(\theta+\Delta\theta,z)+q_z^{\mathrm{II}}(\theta,z+\Delta z)\,R_{\mathrm{CH}}\,\Delta\theta \end{aligned} \tag{8.35}$$

where the term $U_\theta(R_0 - R_i)$ represents positive displacement effects. An equation similar to Eq. (8.34) is used to represent the exit flows from the intermeshing region.

The fluxes q_j are related to the screw velocities and local pressure gradients through as Eq. (8.33 a, b) and the equations of motion:

$$0 = -\frac{1}{r}\frac{\partial p}{\partial \theta} + \frac{1}{r^2}\frac{\partial}{\partial r}(r^2\,\sigma_{\theta r})$$
$$0 = -\frac{\partial p}{\partial z} + \frac{1}{r}\frac{\partial}{\partial r}(r\,\sigma_{zr}) \qquad (8.36\text{ a, b, c})$$
$$0 = -\frac{\partial p}{\partial r}$$

with

$$\sigma_{\theta r} = \eta\, r\,\frac{\partial}{\partial r}\left(\frac{v_\theta}{r}\right)$$
$$\sigma_{zr} = \eta\,\frac{\partial v_z}{\partial r} \qquad (8.37\text{ a, b})$$

Equations (8.32) to (8.35) are thus finite difference equations for the pressure field in the barrel. They may be used to compute screw characteristic curves of

$$Q^* = \frac{Q}{2\,W\,H} \quad \text{vs.} \quad \Delta p^* = \frac{H^2\cos\theta}{12\,K\,\pi\,D\,N}\frac{\Delta p}{L} \qquad (8.38)$$

The Leistritz intermeshing counter-rotating GG elements they modeled are shown in Figure 8.4. Figure 8.5 shows such screw characteristic curves. It is found that the pumping characteristics are much better than for tangential counter-rotating machines or for co-rotating twin screw extruders. The thick-flighted elements have the best pumping characteristics.

It is also possible to carry out calculations for non-Newtonian fluids. Hong and White [24] describe calculations for power law fluids, where 'n' is the power law exponent. Screw characteristic curves in the form

$$Q^* = \frac{Q}{2\,W\,H} \quad \text{vs.} \quad \Delta p^* = \frac{H^{n+1}\cos\theta}{2\,K\,(\pi\,D\,N)^n}\frac{\Delta p}{L} \qquad (8.39)$$

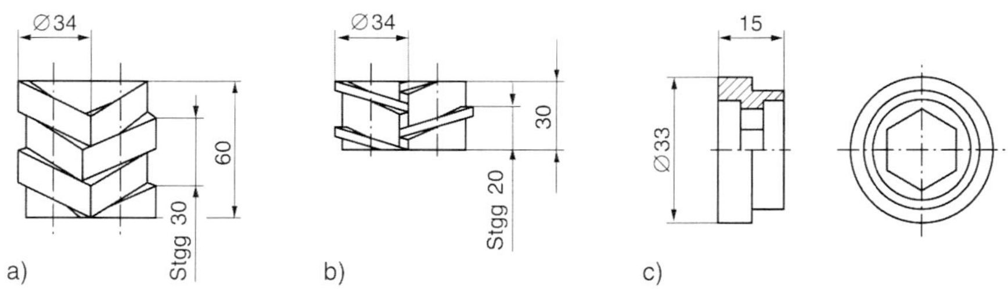

Figure 8.4: Leistritz intermeshing counter-rotating modules modeled by Hong and White [23, 24]; (a) thick-flighted screw; (b) thin-flighted screw; (c) shearing disc

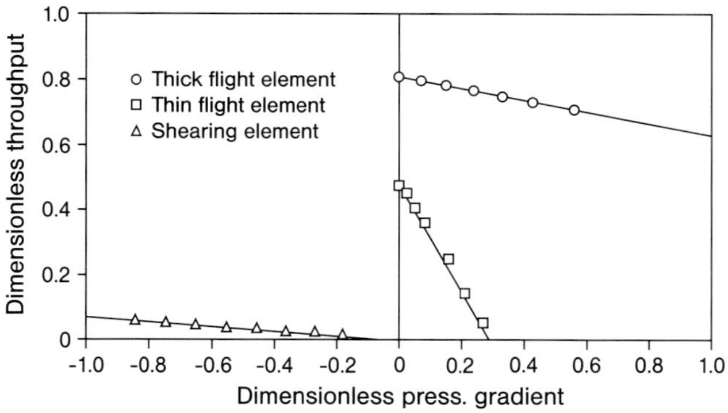

Figure 8.5: Hong and White's [23] computed screw characteristic curves for Newtonian fluids

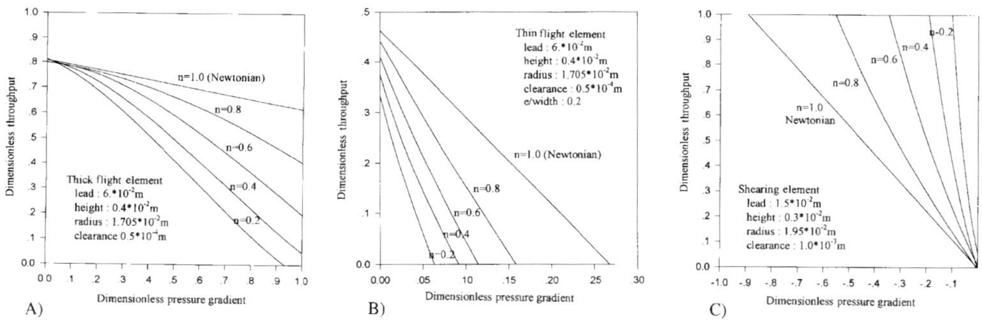

Figure 8.6: Hong and White's [24] computed screw characteristic curves for power-law fluids; (a) thick-flighted screw; (b) thin-flighted screw; (c) shearing disc

were determined. Screw characteristic curves are shown in Figure (8.6 a, b, c) for a power law fluid. They indicate that as the power law exponent decreases, the pumping ability decreases.

8.6 Melting Model

There have been relatively few investigations of melting in intermeshing counter-rotating twin screw extruders. Melting in these machines has been discussed by Janssen [2], Sakai et al. [26], Lim and White [27], and Cho and White [28] in the 1980s and 1990s. Generally, they suggest that the melting regions are much shorter here than for single screw extruders. Janssen suggested melting began in leakage gaps and not at the barrel, as found in single screw extruders. The first effort to comprehensively study melting was by Wilczynski and White [29] in 2001. They identified the calendering region between the screws as the primary location for melting initiation. A secondary location of melting was between the screw and the barrel.

The first effort to model this behavior was in a second paper by Wilczynski and White [30] in 2003. The buildup of heat by the work done by the calendering action in the inter-screw region is

$$\sigma_{xz}\,(2\,\pi\,R)\,W\,R\,\theta = G_p\,\Delta\bar{H} + h_s\,W\,R\,\theta\,(T_p - T_s) \tag{8.40}$$

If the pellets are melted, the specific enthalpy $\Delta\bar{H}$ is increased to

$$\Delta\bar{H} = C_s\,(T_m - T_s) + \lambda + C_m\,(T_f - T_m) \tag{8.41}$$

where λ is the heat of melting. The model for heat transfer from the second location of the barrel involves a heat flux of form:

$$q = h_b\,(T_b - T_m) + \frac{\eta\,(\pi\,D\,N\,\cos\theta)^2}{H_m} \tag{8.42}$$

where H_m is the thickness of melt layer.

Both melting mechanisms take place simultaneously with the calendering gap mechanism being dominant.

8.7 Composite Model of Modular Intermeshing Counter-Rotating Twin Screw Extruder

There has been little attention to composite models that combine metering, melting, and solid conveying regions. Hong and White [24] have considered this problem for the metering region of this type of machine. The throughput is specified which, together with die characteristics, allows one to determine the pressure at the end of the screws. One may then use screw characteristic curves (Q vs. Δp) that have been previously determined, to calculate the pressure gradient backward along the screw. Hong and White do not include the melting and solid conveying regions within their analysis. The flow analyses of these authors are combined with an energy balance of form:

$$\begin{aligned}\rho\,c\,\frac{\mathrm{d}\bar{\bar{T}}}{\mathrm{d}x_1} &= -[h_b\,W_b\,(\bar{\bar{T}} - T_b) + h_s\,W_s\,(\bar{\bar{T}} - T_s)] \\ &\quad + [\sigma_{12}(H)\,U_1 + \sigma_{32}(H)\,U_3]\,W_b - Q\,\frac{\mathrm{d}p}{\mathrm{d}x}\end{aligned} \tag{8.43}$$

The modular screws studied for the Leistritz modular counter-rotating twin screw extruder are shown in Figure 8.7. Typical predicted fill factors, pressure profiles, and temperature profiles along the screw are shown in Figures 8.8 a, b, c.

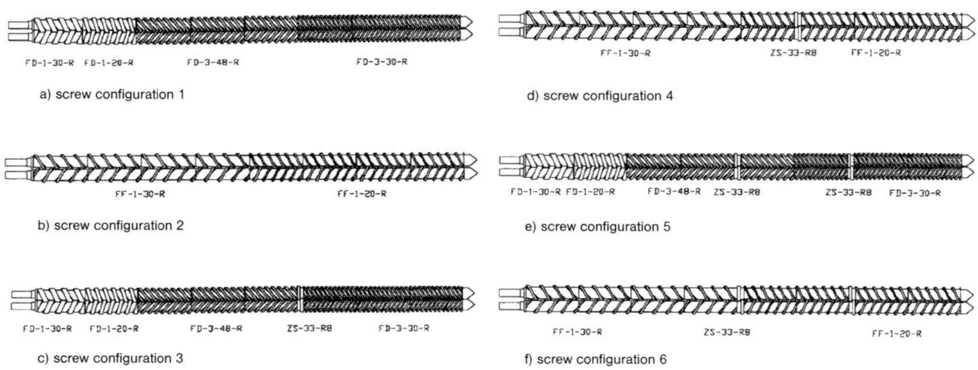

Figure 8.7: Modular screws modeled by Hong and White [24]

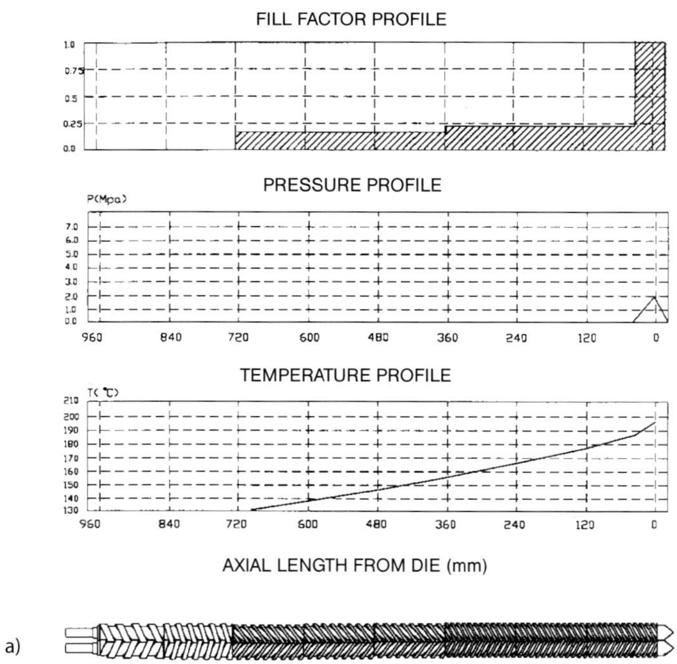

Figure 8.8: Fill factor, pressure, and temperature profiles computed by Hong and White [24] for the modular screws of Figure 8.1
(a) Predicted pressure, fill factor and temperature profiles for screw configuration 1
(b) Predicted pressure, fill factor and temperature profiles for screw configuration 2
(c) Predicted pressure, fill factor and temperature profiles for screw configuration 3

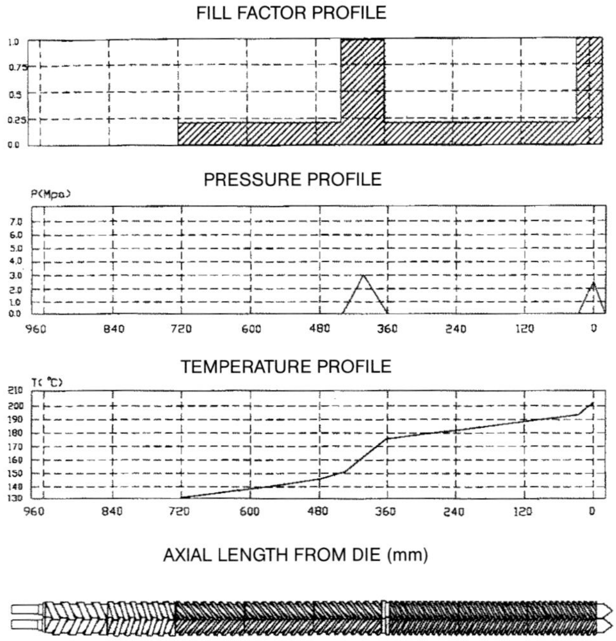

b)

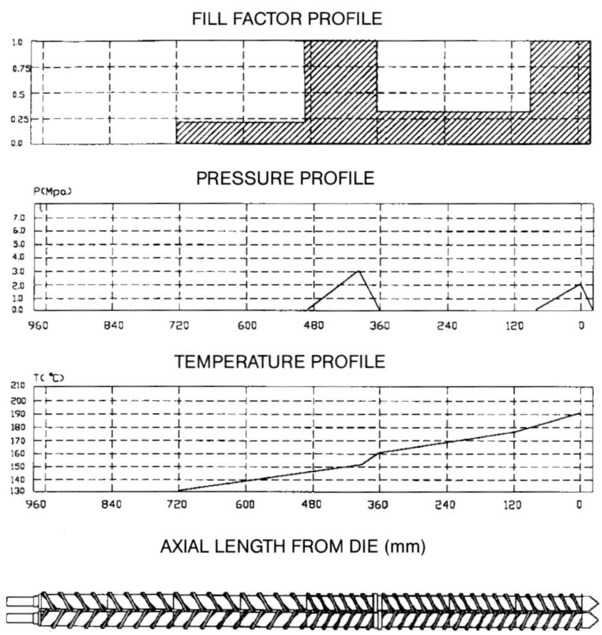

c)

Figure 8.8: (continued)

More recently, Wilczynski et al. [31] have sought to develop a composite model that includes melting and solid conveying as well as melt flow in front of the die. They presume metered starved feeding rather than flood feeding. They recognize the role of C-chambers along the length of the screw and suggest that this could control the metered feeding the pellet or powder bed as described by Doboczky [11] should have a low density. This can lead to densification of the polymer feed on melting and starved flow. This model is similar to models developed for modular co-rotating twin screw extruders [32, 33] described in Chapter 5.

The efforts of Wilczynski et al. [31] have more recently been extended by Jiang and White [34, 35], and viable computer software for calculations has been developed for both starved fed and flood fed conditions.

8.8 Screw-Screw-Barrel Interactions

It is not possible to operate intermeshing counter-rotating twin screw extruders at high speeds. This is due to the high pressure developed between the screws, which pushes them out towards the barrels. This leads to what is called the '8 o'clock-10 o'clock effect' in barrel abrasion where the barrel is abraded in this position.

White and Adewale [36] have sought to model the screw bending using Bernoulli's analysis of cantilever beams to model the problem. Consider a pair of parallel identical cantilever beams (representing twin screws) with local cross-sectional moments of inertia, $I(x)$. A pressure profile $p(x)$ exists between the parallel cantilevers. A torque balance leads to

$$M = E\,I\,\frac{\mathrm{d}^2 y}{\mathrm{d}x^2} \tag{8.44}$$

where E is the Young's modulus and y is the deflection. The deflection is obtained from solving Eq. (8.44).

$$y = \iint \frac{M}{E\,I}(x)\,\mathrm{d}x'\,\mathrm{d}x'' + C_1\,x + C_2 \tag{8.45}$$

If E and I are constant along the shafts and M is caused by a uniform pressure, so that at any position x it is $p\,C\,(L - x)$, where C is a fractional circumference and L the length of the screw.

$$y(x) = \frac{p\,C}{E\,I}\left[L\,x^2 - \frac{x^3}{2}\right] \tag{8.46}$$

The maximum deflection is at the end of the shaft

$$y(L) = \delta = \frac{p\,C\,L^3}{E\,I} \tag{8.47}$$

References

1. G. Schenkel, "Kunststoffe Extrudertechnik", Hanser, Munich (1963).
2. L. P. B. M. Janssen, "Twin Screw Extrusion", Elsevier, Amsterdam (1976).
3. S. L. Wiegand, U.S. Patent (filed April 28, 1874) 155,602 (1874).
4. W. S. Holdaway, U.S. Patent (filed June 2, 1915) 1,218,602 (1917).
5. C. O. J. Montelius, U.S. Patent (filed March 20, 1925) 1,698,802 (1929).
6. C. O. J. Montelius, U.S. Patent (filed March 1, 1929) 1,965,557 (1934).
7. S. Kiesskalt, *Zeit VDI*, **71**, 453 (1927).
8. F. Burghauser, U.S. Patent (filed November 24, 1936) 2,115,006 (1938).
9. E. Hack, German Patent (filed August 2, 1951) 815,103 (1951).
10. Z. Doboczky, *Plastverarbeiter*, **16**, 57 (1965).
11. Z. Doboczky, *Plastverarbeiter*, **16**, 395 (1965).
12. W. S. Kim, W. W. Skatschkow, and S. D. Jewmenow, *Plaste u Kautschuk*, **20**, 696 (1973).
13. L. P. B. M. Janssen, L. P. H. R. M. Mulders, and J. M. Smith, *Plast. Polym.*, **43**, 93 (1975).
14. C. E. Wyman, *Polym. Eng. Sci.*, **15**, 606 (1975).
15. K. P. Klenk, *Plastverarbeiter*, **22**, 33 (1971).
16. K. P. Klenk, *Plastverarbeiter*, **22**, 105 (1971).
17. K. P. Klenk, *Plastverarbeiter*, **22**, 189 (1971).
18. L. P. B. M. Janssen, J. J. Pelgrom, and J. M. Smith, *Kunststoffe*, **66**, 724 (1976).
19. L. P. B. M. Janssen and J. M. Smith, *Plastics and Rubber Proc.*, **44**, 90 (1976).
20. J. A. Speur, H. Mavridis, J. Vlachopoulos, and L. P. B. M. Janssen, *Adv. Polym. Tech.*, **7**, 39 (1987).
21. W. D. Mohr and R. S. Mallouk, *Ind. Eng. Chem.*, **51**, 765 (1959).
22. J. L. White and A. Adewale, *Int. Polym. Process*, **8**, 210 (1993).
23. M. H. Hong and J. L. White, *Int. Polym. Process*, **13**, 342 (1998).
24. M. H. Hong and J. L. White, *Int. Polym. Process*, **14**, 136 (1999).
25. D. Bang and J. L. White, *Int. Polym. Process*, **11**, 109 (1996); *ibid* **12**, 778 (1997).
26. T. Sakai, N. Hashimoto and N. Kobayoshi, *SPE ANTEC Tech. Papers*, **33**, 146 (1987).
27. S. Lim and J. L. White, *Int. Polym. Process*, **9**, 33 (1994).
28. J. W. Cho and J. L. White, *Int. Polym. Process*, **11**, 21 (2001).
29. K. Wilczynski and J. L. White, *Int. Polym. Process*, **16**, 257 (2001).
30. K. Wilczynski and J. L. White, *Polym. Eng. Sci.*, **43**, 1715 (2003).
31. K. Wilczynski, Q. Jiang and J. L. White, *Int. Polym. Process.*, **22**, 198 (2007).
32. S. Bawiskar and J. L. White, *Int. Polym. Process*, **12**, 331 (1997).
33. J. L. White, J. M. Keum, H. Jung, K. Ban and S. H. Bumm, *Polym Plast. Technol. Eng.*, **45**, 539 (2006).
34. Q. Jiang, Ph. D. Dissertation in Polymer Engineering, University of Akron (2008).
35. Q. Jiang and J. L. White, manuscript in preparation (2008).
36. J. L. White and A. Adewale, *Int. Polym. Process*, **10**, 15 (1995).

9 Experimental Studies of Intermeshing Counter-Rotating Twin Screw Extruders

9.1 Introduction

We now turn to experimental studies of flow in intermeshing counter-rotating twin screw extruders. There have been relatively few studies of this type in the literature. Furthermore, such experimental studies have not kept up with the development of machine technology. The experimental studies have, however, made certain aspects of the flow characteristics clear. These efforts will be reviewed in this chapter.

In Section 9.2, we will summarize flow visualization investigations which seek to determine fluid motions. We turn in Section 9.3 to studies of pumping characteristics. Investigations of residence time distributions are summarized in Section 9.4. Research on melting of solid beds is contained in Section 9.5. Dispersive mixing is considered in Section 9.6.

9.2 Flow Visualization

Flow visualization studies of intermeshing counter-rotating twin screw extruders have been reported by Jewmenow and Kim [1], Janssen and Smith [2–4], and by Sakai [5]. These investigations all confirm the positive displacement pump character of the motions.

Jewmenow and Kim [1] have reported investigations in which a window was placed on a twin screw extruder. Aluminum flake markers were placed in a polyisobutylene process fluid. Photographs were taken through the windows and were used to determine streamlines. Janssen and Smith [2–4] reported studies on an extruder with a transparent polymethyl methacrylate barrel using an aqueous polyvinylpyrolidone solution. Sakai [5] has also made similar studies.

9.3 Pumping Characteristics

There is a long history of experimental investigations of pumping characteristics of intermeshing counter-rotating twin screw pumps and extruders [4–18]. Kiesskalt [6], in a 1927 paper, studied the pumping of viscous Newtonian oils concluding that:

$$Q = V_C\, N - Q_{\text{leak}}(\Delta p) \tag{9.1}$$

i.e., there is a pressure dependent leakage between the clearances.

The earliest study of this type for a polymer melt is by Doboczky [8, 9] of Leistritz, who collected together data from a number of different twin screw machines. Doboczky compared the throughput data with Eq. (8.2b) which we repeat here:

$$Q = 2\,i\,N\,V_C \tag{9.2}$$

for the positive displacement pump output of the machine. Doboczky found that the ratio of observed to theoretical output varied from roughly 50 to 90%, with 70% being a mean value. Doboczky also argued that the output of a twin screw extruder is about three times as high as that of a single screw extruder of a similar size and screw speed.

A second study (1971) of the pumping characteristics of polymer melts was reported by Menges and Klenk [11–13] using polyvinyl chloride in a Schloemann AG Pasquetti Bitruder. It was reported that the throughput was proportional to the screw speed. The ratio of the experimental output to the theoretical predictions (Eq. (9.2) was found to be in the range of 37 to 41%. This result was only dependent on screw speed.

A third study in 1976 was published by Janssen and co-workers [17], also using a Pasquetti type machine. The screw had a diameter of 47.7 mm and an effective screw length of 360 mm.

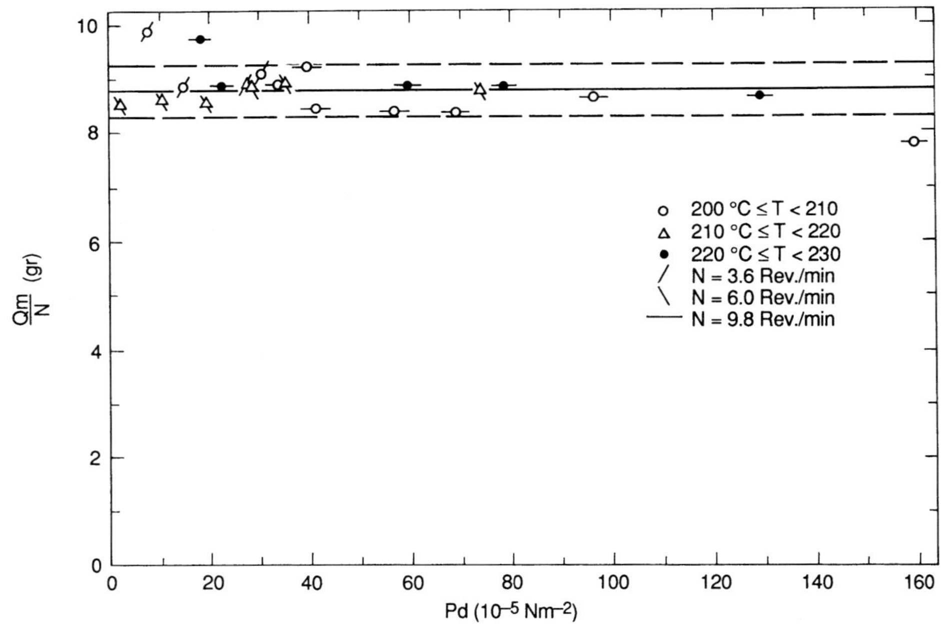

Figure 9.1: Relation of the ratio of specific extrusion rate (*Q*/*N*) to die pressure and mass throughput for the extrusion of polypropylene from Janssen et al. [17]

The material investigated was polypropylene. Die pressures from 3 to 60 atmospheres were achieved. Throughput increased linearly with screw speed *N*. The specific output *Q*/*N* seemed to be independent of pressure developed over a wide range of melt temperatures and screw speeds, as can be seen in Figure 9.1.

Janssen et al. [17, 18] hypothesized in their studies that the pressure development was associated more with the level of starvation than with the operating parameters of the machine. They were able to characterize this level of starvation by stopping the twin screw extruder while operating under steady conditions, cooling and then removing the barrel. Janssen et al.'s results are summarized in Figures 9.2 and 9.3. Data obtained at higher screw speeds and higher temperatures show lower pressure rises.

Hong et al. [19] have experimentally determined the screw pumping characteristics of three of the modules used in the Leistritz AG modular intermeshing counter-rotating twin screw extruder (see Section 7.12). These are shown in Figure 9.4 a, b, c together with the simulations for these elements of Hong and White [20] (Section 8.5). The agreement between simulation and experiment is good.

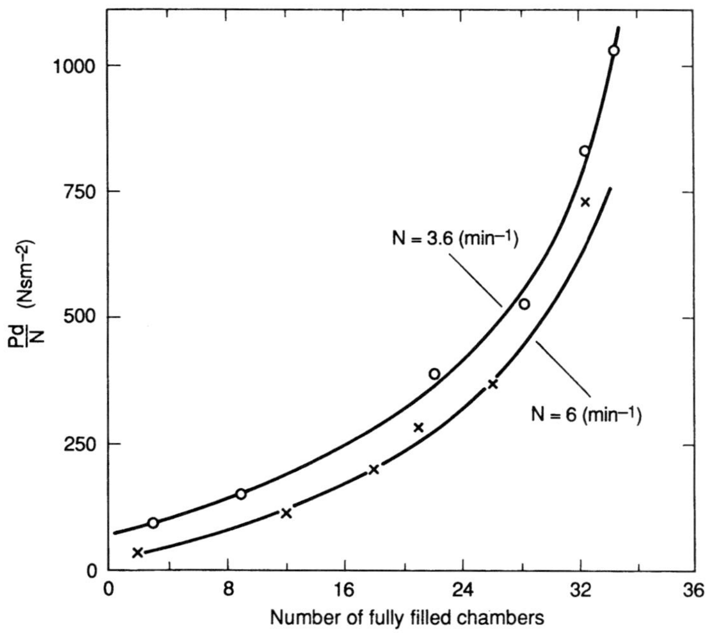

Figure 9.2: Relationship between the die pressure and the number of fully filled chambers for extrusion of polypropylene at different screw speeds from Janssen et al. [17, 18]

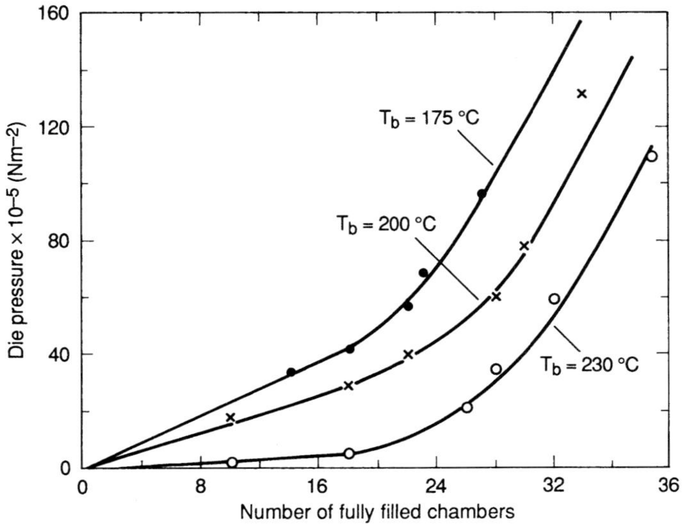

Figure 9.3: Relationship between the die pressure and the number of fully filled chambers for extrusion of polypropylene at higher temperatures from Janssen et al. [17, 18]

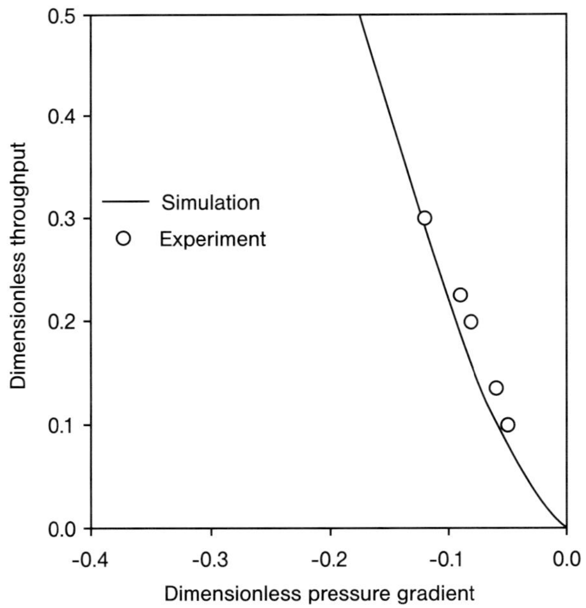

a)

Figure 9.4: Measured and simulated dimensionless screw characteristic curves for Leistritz intermeshing counter-rotating modules by Hong et al. [19, 20];
(a) thin-flighted screw

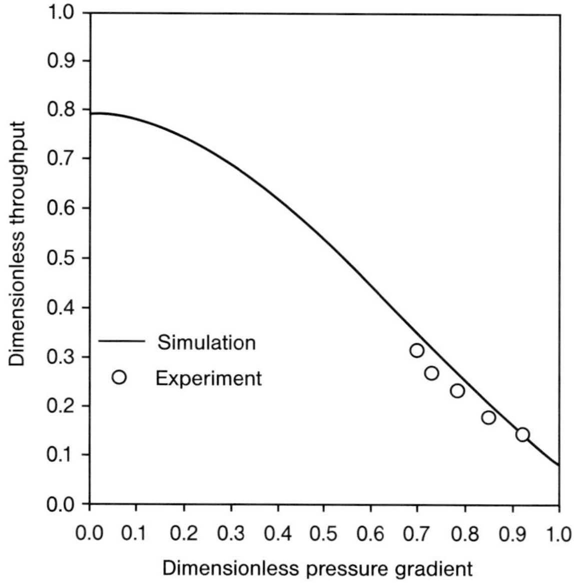

b)

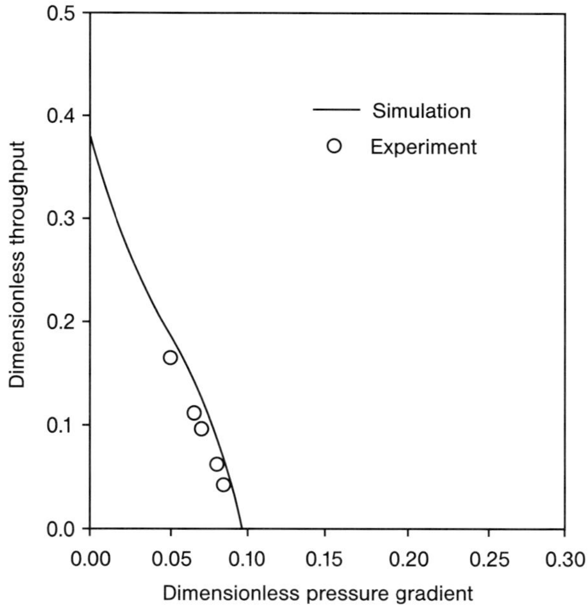

c)

Figure 9.4: Measured and simulated dimensionless screw characteristic curves for Leistritz intermeshing counter-rotating modules by Hong et al. [19, 20]; (b) thick-flighted screw; (c) shearing disc

9.4 Residence Time Distributions

Measurements of residence time distributions in intermeshing counter-rotating twin screw extruders were reported by Sakai [5,16], Janssen, Hollander, Spoor, and Smith [2,18], Rauwendaal [21], Wolf, Holen, and White [22], Potente and Schultheis [23], and Shon et al. [24].

The results of Sakai [5, 16] are summarized in Figure 9.5, which shows that the intermeshing counter-rotating twin screw extruder has a much narrower residence time distribution than the single screw extruder. Sakai associates this with both the C-chamber mechanism and the self-wiping screws. In the single screw extruder, the material is retained for much longer times associated with stagnant layers on the screw surface. Figure 9.5 also shows data for intermeshing co-rotating machines that was discussed in Chapter 6 and a non-intermeshed twin screw configuration which is not carefully defined.

In the studies of Janssen et al. [18] a pulse containing radioactive MnO_2 was injected into polypropylene in a twin screw extruder. The decay ^{56}Mn to ^{56}Fe was detected using a scintillation crystal. Attention was given to the reproducibility of results. Janssen et al. report their conclusions in terms of exit age distributions as shown in Figure 9.6.

Wolf et al. [22] also used a radioactive MnO_2 tracer and applied it to a Krauss-Maffei Model KMD 90 machine (90 mm diameter screws). The machine described by Wolf et al. was being used commercially to extrude polyvinyl chloride at Kafrit Industries (Kibbutz Kfar Aza – Israel).

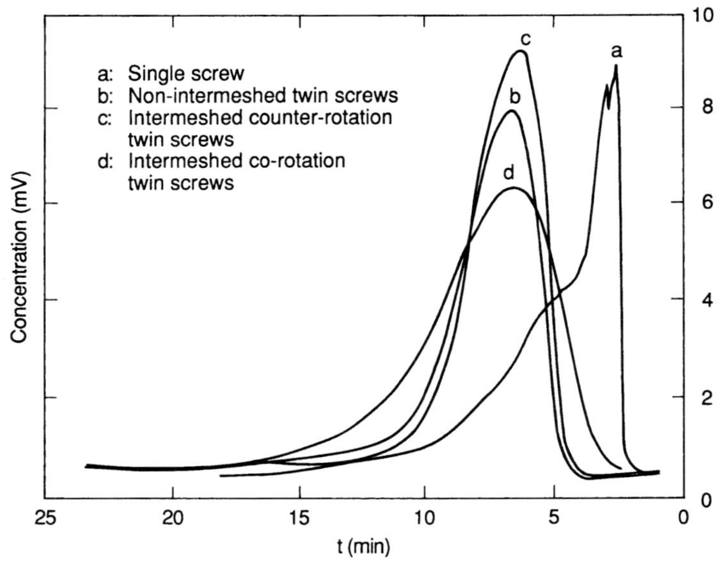

Figure 9.5: Residence time distributions of Sakai [5, 16]

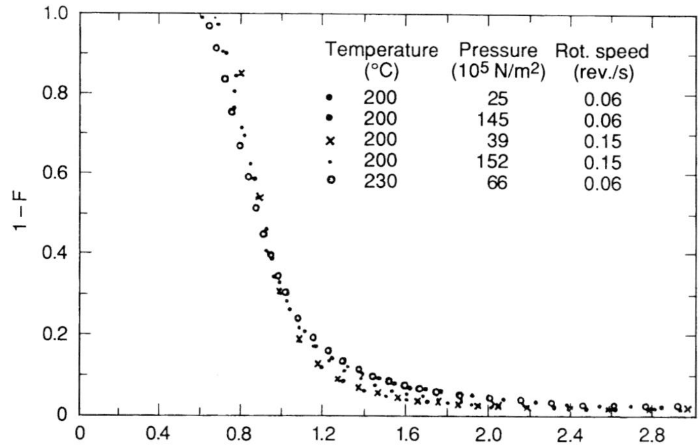

Figure 9.6: Exit age distribution of polypropylene being extruded from a Pasquetti intermeshing counter-rotating twin screw extruder after Janssen et al. [18] (reprinted with permission of the AIChE Journal)

Rauwendaal [21] reported studies using an antimony oxide tracer, which was detected by X-ray fluorescence, on a Leistritz LSM GG 30/34 machine (34 mm diameter screws). He finds a much narrower distribution of residence times than in co-rotating twin screw machines.

Shon et al. [24] in 1999 compared residence time distributions in four types of continuous mixers. The Leistritz LSM GG 30/34 intermeshing counter-rotating twin screw extruder had the narrowest residence time distribution as compared to modular co-rotating twin screw extruders and the Buss Kneader. All investigators have found the residence time distributions to be narrower than for single screw extruders or co-rotating twin screw machines.

9.5 Melting Phenomena

Experimental studies of melting in intermeshing counter-rotating twin screw extruders have been few in number. Janssen [4] seems to have been the first to study the melting process in this machine. He notes that melting occurs more rapidly than in single screw extruders and seems to begin in clearances.

Studies were subsequently made in the 1980s by Sakai [25, 26]. He also observed the more rapid melting in intermeshing counter-rotating machines as opposed to co-rotating machines.

In a later paper comparing intermeshing co-rotating and counter-rotating twin screw extrusion, Lim and White [27] also note that melting occurs much more rapidly in intermeshing counter-rotating machines than in the co-rotating machine. This was also later found by Cho and White [28].

A more recent study by Wilczynski and White [29] identified the location of the initiation of melting as being in the calendering gap between the screws. Melting induced by the heated barrel surface was a secondary location of melting initiation. They concluded that the intense mechanical energy absorbed by the pellets in the calendering region was the primary cause of the initiation of melting.

9.6 Dispersive Mixing

The relative ability of different types of intermeshing twin screw machines to disperse fillers seems first studied by Sakai, Hashimoto, and Kobayoshi [16] in the mid 1980s. They made their studies on a Japan Steel Works (JSW) TEX machine that had screws and kneading disc blocks. They studied the mixing of both talc and glass fibers. They report that the intermeshing counter-rotating does a better job of dispersing the small particles of talc but damages glass fibers to a greater extent as compared to a self-wiping co-rotating machine.

In papers published in 1983–1991, Thiele and his coworkers [30–32] with American Leistritz describe the relative compounding characteristics of an intermeshing counter-rotating twin screw machine with Leistritz elements (Figure 7.14) and a self-wiping co-rotating machine, finding them similar. In 1993, Lim and White [27] published a study of the blending characteristics of intermeshing counter-rotating twin screw extruders of Leistritz design. Polyethylene-polyamide 6 blends were studied for different modular screw configurations. They found the intermeshing counter-rotating machine gave the finest phase morphology. In a 1996 paper, Cho and White [28] compared the blending of polyethylene and polystyrene melts in a modular intermeshing counter-rotating extruder of Japan Steel Works design and a self-wiping co-rotating twin screw extruder. They reached similar conclusions.

In a 1999 paper, Shon and White [33] compared the breakage of glass fibers in a Leistritz LSM GG 30/34 intermeshing counter-rotating twin screw extruder with a modular co-rotating twin screw extruder and a Buss Kneader. The glass fibers were most seriously damaged in the intermeshing counter-rotating machine.

In 2005, Shon et al. [34] published a study comparing the dispersive mixing of calcium carbonate in polypropylene for a Leistritz LSM GG 30/34 intermeshing counter-rotating, an intermeshing co-rotating twin screw extruder, a Buss Kneader, and a Kobelco Nex-T. The most effective dispersive mixing was by the intermeshing counter-rotating twin screw machine. They represented their dispersion data through the kinetic expression of Eq. (6.7).

It seems clear that the Leistritz GG machine is the most intense and effective mixer for additives, polymer blends, and particulate fillers. It also most severely damaged glass fibers.

References

1. S. D. Jewmenow and W. S. Kim, *Plaste u Kautschuk*, **20**, 356 (1973).
2. L. P. B. M. Janssen and J. M. Smith, Proceedings of the Congress on Polymer Rheology and Plastics Processing, PRI/BSR, Loughborough, p. 160, September 1975.
3. L. P. B. M. Janssen and J. M. Smith, *Plastics and Rubber: Processing*, p. 90 (1976).
4. L. P. B. M. Janssen, Twin Screw Extrusion, Elsevier, Amsterdam (1978).
5. T. Sakai, *Góse Jushi*, **29**, 7 (1978).
6. S. Kiesskalt, *Zeit VDI*, **71**, 453 (1927).
7. H. W. Selbach, *Plastverarbeiter*, **13**, 595 (1962).
8. Z. Doboczky, *Plastverarbeiter*, **16**, 57 (1965).
9. Z. Doboczky, *Plastverarbeiter*, **16**, 395 (1965).
10. H. Markhenkel, *Kunststoffe*, **55**, 363 (1965).
11. G. Menges and K. P. Klenk, *Plastverarbeiter*, **17**, 791 (1966).
12. K. P. Klenk, *Plastverarbeiter*, **22**, 33 (1971).
13. K. P. Klenk, *Plastverarbeiter*, **22**, 105 (1971).
14. J. Stasiek, *Plaste u Kautschuk*, **21**, 207 (1974).
15. L. P. B. M. Janssen, L. P. H. R. M. Mulders, and J. M. Smith, *Plastics Polymers*, **43**, 93 (1975).
16. T. Sakai, N. Hashimoto, and N. Kobayoshi, *SPE ANTEC Tech Papers*, **33**, 146 (1987).
17. L. P. B. M. Janssen, J. L. Pelgrom, and J. M. Smith, *Kunststoffe*, **66**, 724 (1976).
18. L. P. B. M. Janssen, R. W. Hollander, M. W. Spoor, and J. M. Smith, *AIChE J.*, **25**, 345 (1979).
19. M. H. Hong, Q. Jiang and J. L. White, *Int. Polym. Process.*, **23**, 88 (2008).
20. M. H. Hong and J. L. White, *Int. Polym. Process.*, **14**, 136 (1999).
21. C. Rauwendaal, *SPE ANTEC Tech. Papers*, **38**, 618 (1981).
22. D. Wolf, W. Holen, and D. H. White, *Polym. Eng. Sci.*, **26**, 640 (1986).
23. H. Potente and S. M. Schultheis, *Int. Polym. Process.*, **4**, 247 (1989).
24. K. Shon, D. H. Chang, and J. L. White, *Int. Polym. Process.*, **14**, 44 (1999).
25. T. Sakai, Japan Steel Works (JSW) Technical Report 41.
26. T. Sakai and N. Hashimoto, *SPE ANTEC Tech. Papers*, **32**, 360 (1986).
27. S. Lim and J. L. White, *Int. Polym. Process.*, **9**, 33, (1994).
28. J. W. Cho and J. L. White, *Int. Polym. Process.*, **11**, 21 (1996).
29. K. Wilczynski and J. L. White, *Int. Polym. Process.*, **16**, 257 (2001).
30. W. C. Thiele, *SPE ANTEC Tech Papers*, **29**, 127 (1983).
31. W. C. Thiele, W. Petrozelli, and D. Lorenc, *SPE ANTEC Tech. Papers*, **36**, 120 (1990).
32. W. C. Thiele, W. Petrozelli, and C. Martin, *SPE ANTEC Tech. Papers*, **37**, 1899 (1991).
33. K. Shon and J. L. White, *Polym. Eng. Sci.*, **39**, 1157 (1999).
34. K. Shon, D. Liu, and J. L. White, *Int. Polym. Process.*, **20**, 322 (2005).

10 Technology of Non-Intermeshing Counter-Rotating Twin Screw Extruders

10.1 Introduction

The simplest of twin screw extruders from both a conceptual and a mechanical point of view are non-intermeshing (or tangential) counter-rotating machines. The non-intermeshing counter-rotating twin screw machines have developed into important commercial machines since the middle of the 20th century. It is thus logical to begin our discussion of the development of twin screw extruder technology here.

As in Chapters 4 and 7, we shall use a roughly historical perspective in this chapter to describe the development of the technology of non-intermeshing counter-rotating twin screw extruders. These machines essentially consist of right-handed and left-handed screws rotating in opposite directions inorder to feed forward the material being processed. The screws, however, are more complex than simply consisting of forward pumping constant-pitch screw-like elements. In Section 10.2, we summarize the early development of this technology. In Section 10.3, we turn, to the development of a machine that was commercialized in the late 1940s by Welding Engineers. Section 10.4 discusses applications of the Welding Engineers machine. Section 10.5 considers tapered (or conical) twin screw machines. The Werner and Pfleiderer Kammerkneter, which contained vanes in its screw flights, is considered in Section 10.6. A throttling twin screw extruder is described in Section 10.7. Machines with reversed pumping screws are described in Section 10.8.

10.2 Origins

There are several early patents for twin screw machines with parallel non-intermeshing counter-rotating pairs of screws or with screw sections [1–4]. These machines usually had rotor sections and might better be considered as continuous mixers to which we turn our attention in Chapter 13. A 1929 patent application by Loomis [4] describes machines with parallel screws that pump in opposite directions. This machine will be discussed in Section 10.8.

10.3 The Welding Engineers Machine

Pioneering efforts to develop a non-intermeshing counter-rotating twin screw extruder were carried out from the 1930s in the USA by Welding Engineers [5–8] of Norristown, PA. The Welding Engineers firm had been started in the 1920s by the Hendrickson family for metal fabrication and construction. L. J. Fuller played the key role in the development of their non-intermeshing counter-rotating machine. Machines were built initially with 2 to 6 screws and were first used for handling chicle, clay, and catalysts (for aviation gasoline). Screws with diameters up to 350 mm were produced. Plans for application to thermoplastics came about during World War II. The machines were called 'compounder-extruders'. Fuller [9] described, in a patent filed in 1943 and issued in 1948, a tangential counter-rotating twin screw extruder (Figure 10.1). The screw flights were matched according to the patent drawing. The application of the machine as a compounder was emphasized. One screw was longer than the other, which allowed for developing pressure prior to and following the twin screw region. According to the patent, twin screws only existed in the intermediate compounding section.

In a second patent filed in 1945, Fuller [10] described screw shafts containing sections of both forward and backward pumping screw elements (Figure 10.2). The relative length of the twin screw section was increased compared to the single screw section. The backward pumping elements introduced by Fuller allowed control of pressure profiles and residence times and application to devolatilization. Fuller showed a second port along the machine axis allowing removal of vapors. He noted that this port may also be used for introducing additives into the twin screw machine. One screw is shown to be longer than the other in Fuller's patent, allowing for a pressurization section prior to the die. Fuller also considered (i) semi-permeable partitions between screws, (ii) element combinations in the manner of Loomis [4] with both screws having right-handed elements, and (iii) bypass in the barrel. This is shown in Figure 10.3. This allows the fluid to move in a complex circulating manner, moving from screw to screw along the machine. Fuller's 1945 patent application contains the basis of the modern Welding Engineers machine but it contains much more than this in ingenious ideas. Fuller included sets of elements on the two screw shafts with different pitch levels at the same axial location. Fuller emphasized the application to devolatilization and discusses the use of his machine in drying synthetic rubber.

Fuller's 1943 patent application [9] generally shows matched positions of screw flights (Figure 10.1), while his 1948 patent application [10] shows staggered configurations of flights

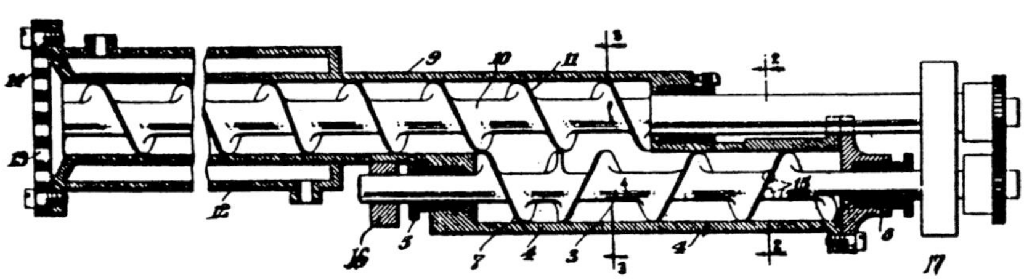

Figure 10.1: Original tangential counter-rotating twin screw extruder – 1943 patent application by Fuller of Welding Engineers [9]

(Figures 10.2 and 10.3). More recent Welding Engineers patents always indicate staggered screw flights.

A 1962 article by Schutz [11] of Welding Engineers surveyed twin screw technology at the time and described the place of the Welding Engineers machine. The screws of their machines were, at this time, made in sections and then assembled. The basic machine design has remained largely unchanged, though screw rotation speeds have become much higher.

A photograph of a modern Welding Engineers machine is shown in Figure 10.4.

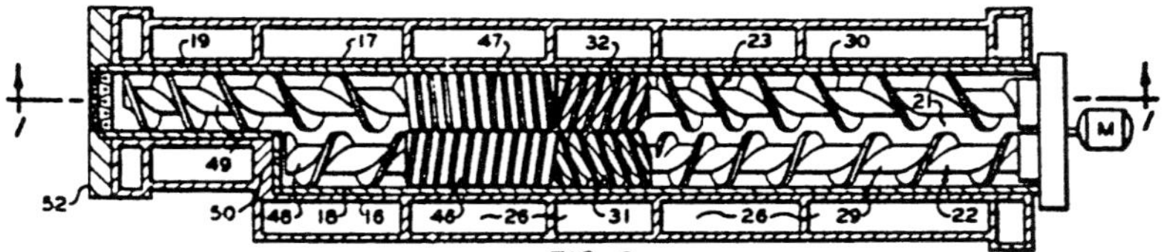

FIG. 2

Figure 10.2: Tangential counter-rotating twin screw extruder with modular screw elements – 1945 patent application by Fuller of Welding Engineers [10]

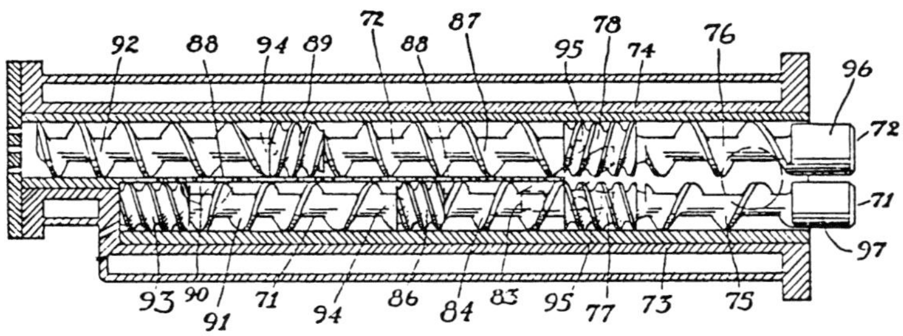

Figure 10.3: Special modular twin screw design by Fuller [10] that produces circulating flows

Figure 10.4: Modern Welding Engineers tangential counter-rotating twin screw extruder (courtesy of Welding Engineers)

10.4 Applications of the Welding Engineers Machine

There are five early published papers dealing with Welding Engineers twin screw extruders [5–7, 11, 12]. All seem to treat machines of the modular design of Fuller's second patent [10]. The two early papers [5, 6], which appeared between 1944 and 1950 discussing 'compounder-extruders', indicate the machine had three sections for feeding, compounding, and extrusion. Each section had very different screw flight designs. Applications to compounding polyvinyl chloride and its copolymers, polystyrene, and cellulose acetate were discussed. The second two papers, which appeared in 1960/1961, refer to a 'compouder-devolatilizer-extruder'. This is significant in that the 1950 and 1960 papers are both written by L. F. Street [6, 7] of Welding Engineers. These papers discuss the dewatering and drying of synthetic rubber. The 1961 paper notes that Welding Engineers equipment for this purpose had been in operation since 1955 and had been recently installed in the Esso (Exxon) butyl rubber plant in Baton Rouge, LA.

Since the key early patents of Fuller, Welding Engineers has worked extensively to improve and broaden the application of their machines. This may be seen in patents by Street [13–16] and Skidmore [17–22], who were associated with Welding Engineers. These relate to introducing multiple entry ports along the barrel for the addition of plasticizers [13] and other additives as well as modifications for application to devolatilization [17–19], coagulation [20, 22], including addition of rubber latex to polymers, and removal of insoluble liquids [17, 18] (Figure 10.5).

Skidmore's coagulation technology [20, 22] has been reviewed in the technical literature by Nichols, Senn, and Kheradi of Welding Engineers [23]. The technology involves successive steps of addition of polymer, addition of latex, addition of coagulant, steam injection, and water removal. This technology may be applied to either a product stream from a polymerization

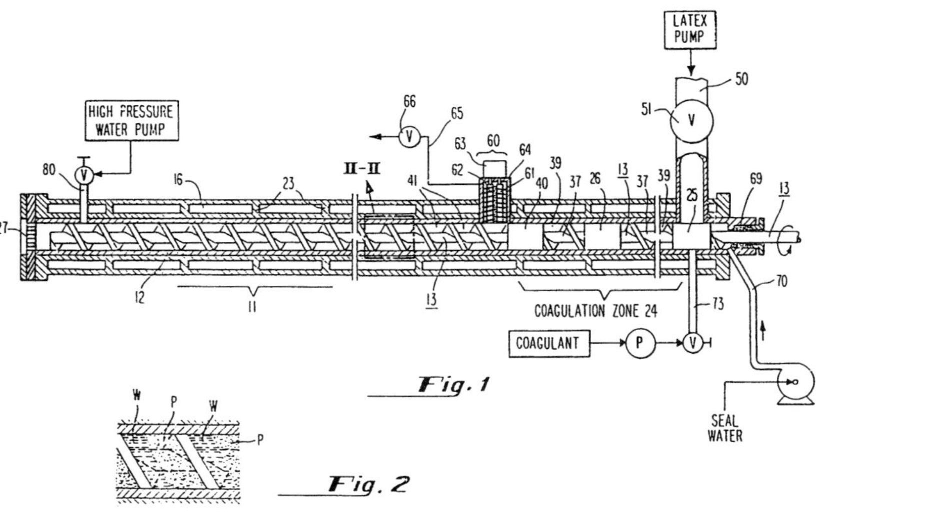

Figure 10.5: Skidmore's modified Welding Engineers tangential counter-rotating twin screw extruder showing addition of latex, coagulant, and high pressure water [22]

reactor or for the addition of rubber particles to a thermoplastic such as in the manufacture of ABS resins. The basic polymer may be added in solid form and subsequently melted or added directly in a melt form. After the addition of rubber and coagulant, the pressure is lowered so that the water separates. Further downstream, all the water is removed by again lowering the pressure and drawing a vacuum.

The Welding Engineers machine has been widely applied to reactive extrusion. In 1968, one of the first published papers on reactive extrusion by Gouinlock et al. [24] of Hooker Chemical in Niagara Falls, New York described the use of Welding Engineers machines for polycondensation using acid chloride and glycol. Terephthaloyl chloride was reacted with bisphenol A and neopentyl glycol:

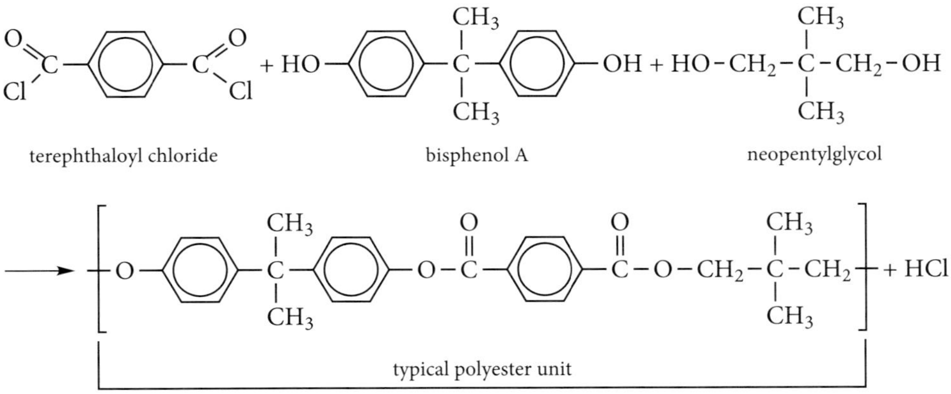

(Structure 10.1)

They carried out this polymerization in a laboratory research-type vented tangential counter-rotating twin screw extruder shown in Figure 10.6. Each vent was either closed or provided with a nitrogen purge or vacuum. The machine had a barrel diameter of 20 mm and an *L*/*D* of 37. It was operated with screw speeds of 50 to 100 rpm. They estimated residence times of 15 to 30 minutes. Generally, the extruder was fed with pre-polymer.

Further papers by Tucker and Nichols [25] of Welding Engineers described the use of a Welding Engineers machine for anionic polymerization of caprolactam to form polyamide-6:

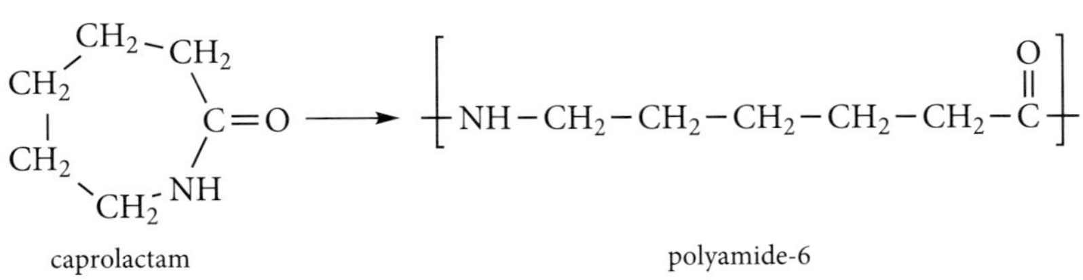

(Structure 10.2)

Production rates in excess of 9.1 kilograms per hour were obtained on a 20 mm diameter machine with speeds up to 525 rpm. The achievement of very high molecular weights was noted.

The use of tangential counter-rotating twin screw extruders for the halogenation of polymers including polyethylene, ethylene propylene copolymer/terpolymer, polyisobutylene, butyl

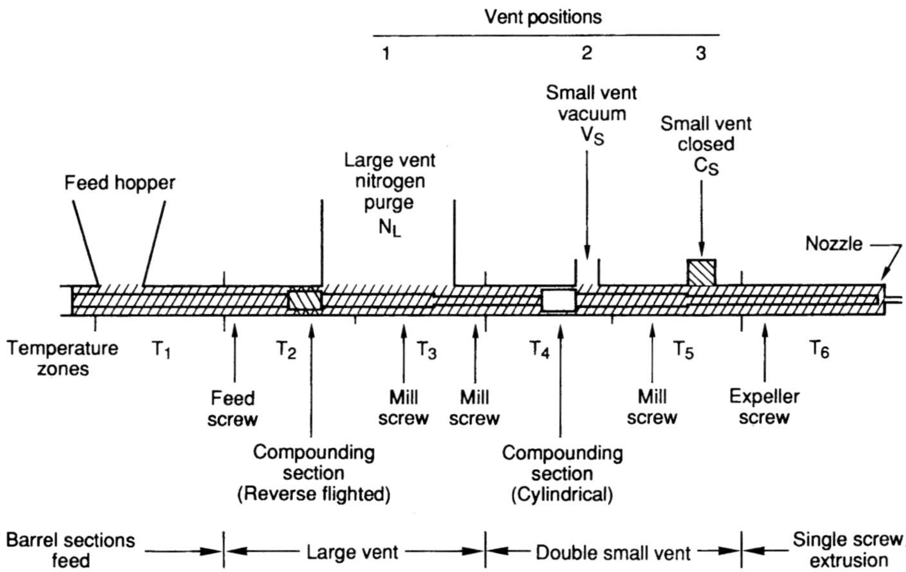

Figure 10.6: Vented research extruder for condensation polymerization of polyesters from ethylene glycol [24] by Gouinlock et al. of Hooker Chemicals (reprinted with permission of John Wiley and Sons)

rubber, ethylene vinyl acetate copolymer polyisoprene, and butadiene-styrene copolymer has been described in patents from 1984 by Kowalski and his co-workers [26–29] of Exxon. The chemical reaction has form, e.g., for polyolefins of:

$$\text{polyolefin} + \left[\frac{\text{Cl}_2}{\text{Br}_2}\right] \rightarrow \left[\frac{\text{polyolefin} - \text{Cl}_x}{\text{polyolefin} - \text{Br}_x}\right] + \left[\frac{\text{HCl}}{\text{HBr}}\right] \qquad \text{(Structure 10.3)}$$

Generally, solid polymer and perhaps a diluent were fed into the feed zone which consists of deep-flighted forward pumping screws. This region ended with a restrictive dam. The screw configuration in the reaction section can contain pins and left-handed screw elements to improve mixing. Halogen was added to the beginning of the reaction zone. The reaction zone was followed by a neutralization zone where HCl or HBr was removed, perhaps by nitrogen injection. This was followed by a scrubbing zone, where the last traces of halogen and HCl/HBr were removed. This was done by addition of co-current or counter-current gas fluxes. The scrubbing region was followed by an exit zone where stabilizers were added. Screw speeds of 200 to 400 rpm were noted. 50 mm screw diameter machines were used and the machines had a screw *L*/*D* of about 50.

One patent of Kowalski et al. [28, 29], deals specifically with the bromination of butyl rubber (isobutylene-isoprene copolymer). It is necessary to brominate isoprene so that a substantial amount of the bromine is in the primary allylic position. The brominated butyl rubber product is used as an inner liner for tubeless tires.

10.5 Tapered Twin Screw Machines

Tangential counter-rotating tapered twin screw extruders were proposed in a 1946 patent application by the American C. E. Dellenbarger [30] (Figure 10.7). A similar machine was described in a 1953 German patent application by Frohlich [31] of Hermann Berstorff GmbH of Hannover, Germany. In both patents, the patent drawing shows matched flights. The reason for this design seems to be the ability to apply high torques to the counter-rotating screws.

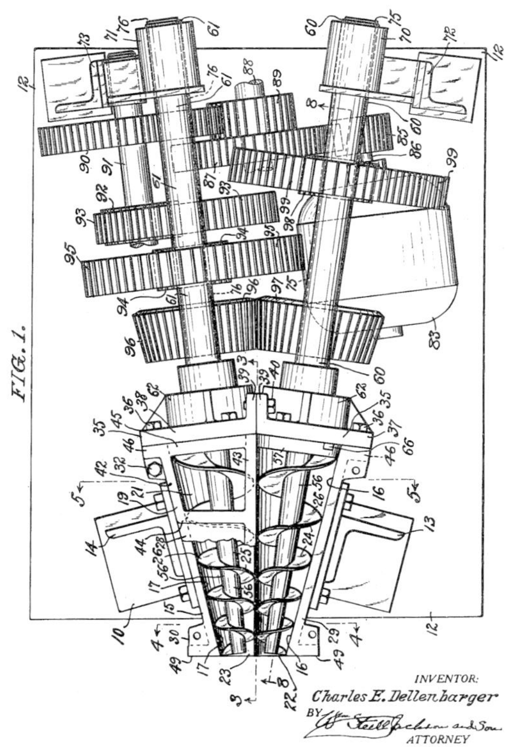

Figure 10.7: Dellenbarger tapered screw tangential counter-rotating twin screw extruder [30]

10.6 Werner and Pfleiderer Kammerkneter

In 1950, Albert Lasch [32] of Werner and Pfleiderer in Stuttgart, Germany developed a counter-rotating non-intermeshing screw kneading machine which was marketed by the firm as the Kammerkneter KK [33, 34]. The screws contained vanes in their flights, to aid in the kneading mixing action. This is shown in Figure 10.8.

In a later variant of this technology, Lasch and Stroebel [35], using specially designed screws, extended the length of selected vanes along the length so that the wipe cleans the opposite screw in their counter-rotating twin screw machine. This is shown in Figure 10.9. Werner and Pfleiderer seems to have discontinued manufacturing this machine about 1960, when they committed themselves to self-wiping co-rotating twin screw technology.

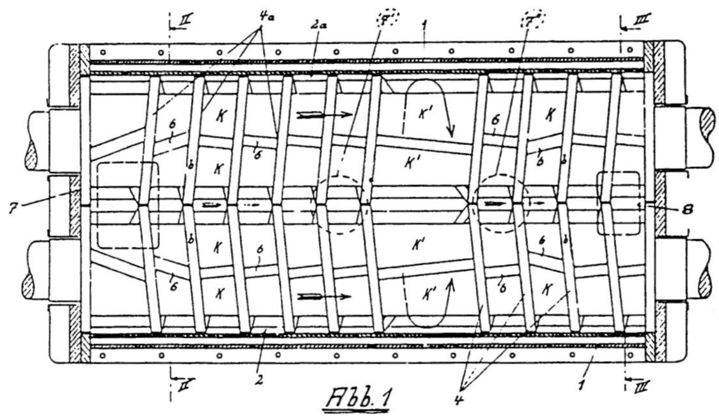

Figure 10.8: Tangential counter-rotating twin screw extruder with vanes in the flights by Lasch of Werner and Pfleiderer [32], marketed as the Kammerkneter KK

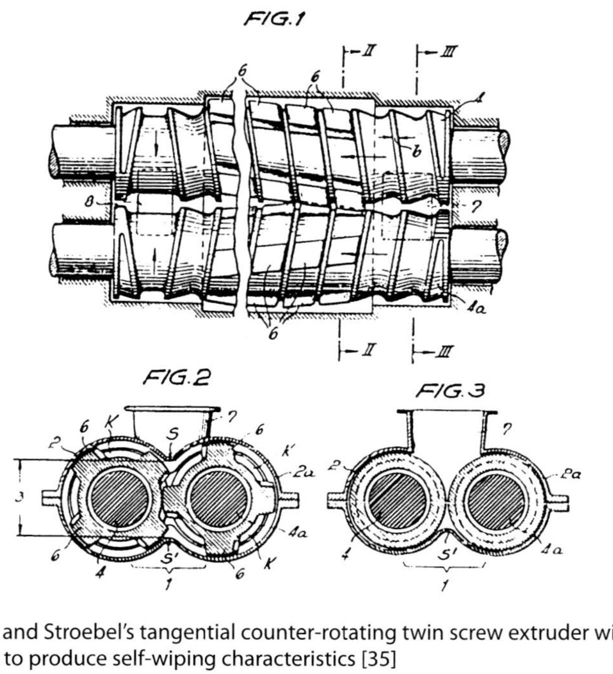

Figure 10.9: Lasch and Stroebel's tangential counter-rotating twin screw extruder with extended vanes to produce self-wiping characteristics [35]

10.7 Eisenmann Adjustable Throttling Twin Screw Extruder

In 1974, G. Eisenmann [36] of Werner and Pfleiderer received a patent for a counter-rotating twin screw extruder with separated screws. This machine was a modular twin screw machine with right-handed and left-handed screw elements. The downstream conveyance of the material was controlled by a displaceable slide, the position of which relative to the screw flights controlled the flow of the material through the extruder housing and thus the pressure buildup in the extruder. This is shown in Figure 10.10. The displaceable slide is considered a throttling device. This machine was not commercialized.

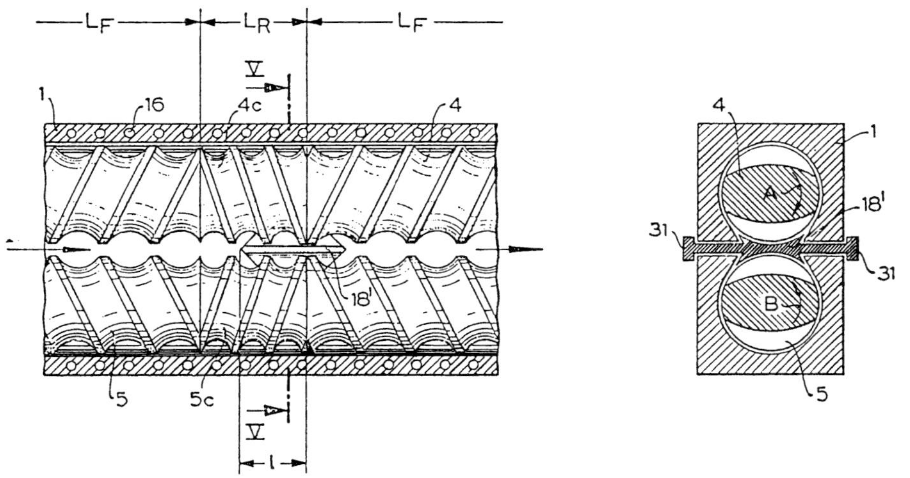

Figure 10.10: Eisenmann's adjustable throttling twin screw extruder [36]

10.8 Machines with Reversed Screws

A 1929 patent application by E. G. Loomis [4] describes a continuous mixing and kneading machine containing two parallel counter-rotating rotors and intended for polymeric materials. The machine begins with a single screw, then exhibits a region with two counter-rotating screw-like rotors, one of which is the extension of the single screw. The two rotors pump in opposite directions, as illustrated in Figure 10.11.

A 1992 patent application of Biesenberger and Todd [37] of the Polymer Processing Institute (then the Stevens Institute of Technology) proposed the machine design shown in Figure 10.12. They called the machine a "Back Mix Drag Flow Apparatus". They emphasized its advantages as a continuous mixing device. The patent also discussed a co-rotating version (screws with opposite hand) and four screw versions.

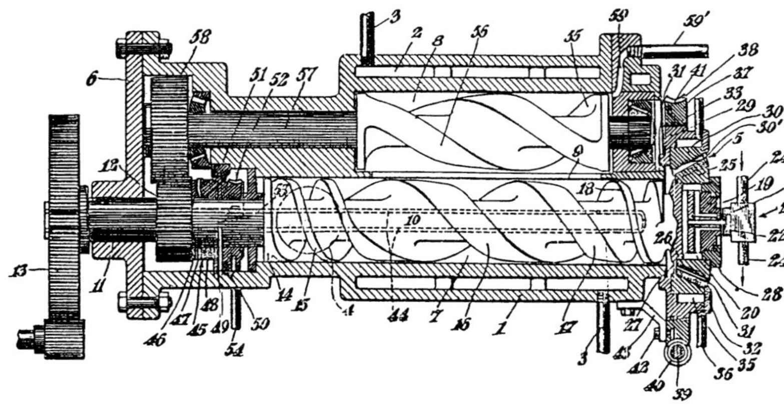

Figure 10.11: Loomis reversed tangential twin screw machine [4]

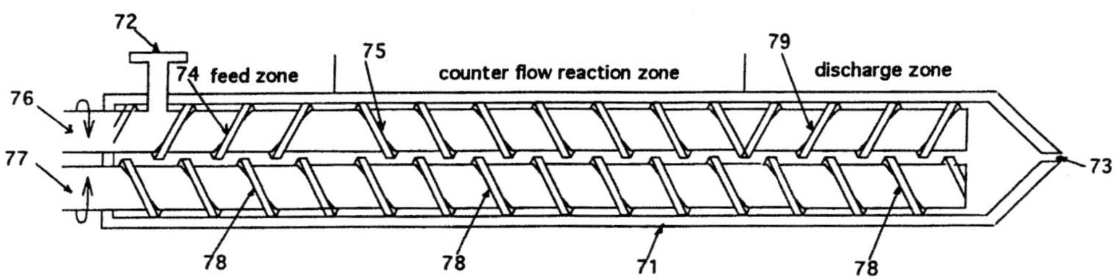

Figure 10.12: Biesenberger-Todd reversed tangential twin screw machine [37]

References

1. A. W. Hale, U.S. Patent 72,393 (1867).
2. P. Pfleiderer, German Patent (filed April, 14 1879) 10,164 (1881).
3. H. Ahnhudt, German Patent (filed May 29, 1923) 397,961 (1924).
4. E. G. Loomis, U.S. Patent (filed Oct. 11, 1929) 1,990,555 (1935).
5. Anonymous, *Modern Plastics*, (June) 87 (1949).
6. L. F. Street, *Rubber World*, **123**, 58 (1950).
7. L. F. Street, *Rubber and Plastics Age*, 1519 (1960).
8. J. L. White, *Int. Polym. Process.*, **8**, 98 (1993).
9. L. J. Fuller, U.S. Patent (filed Oct. 10, 1943) 2,441,222 (1948).
10. L. J. Fuller, U.S. Patent (filed May 15, 1945) 2,615,199 (1952).
11. F. C. Schutz, *SPE Journal*, **18**, (Sept.) 1147 (1962).
12. Anonymous, *Rubber and Plastics Age*, 1227 (1961).

13. L. F. Street, U.S. Patent (filed Aug. 16, 1952) 2,733,051 (1956).
14. L. F. Street, U.S. Patent (filed Feb. 23, 1954) 3,068,514 (1962).
15. L. F. Street, U.S. Patent (filed Aug. 15, 1957) 3,085,288 (1963).
16. L. F. Street, U.S. Patent (filed March 31, 1960) 3,078,511 (1963).
17. R. H. Skidmore, U.S. Patent (filed Dec. 28, 1959) 3,082,816 (1963).
18. R. H. Skidmore, U.S. Patent (filed April 8, 1970) 3,742,093 (1973).
19. R. H. Skidmore, U.S. Patent (filed Sept. 27, 1973) 3,917,507 (1975).
20. R. H. Skidmore, U.S. Patent (filed Dec. 13, 1974) 3,993,292 (1976).
21. R. H. Skidmore, U.S. Patent (filed Aug. 29, 1978) 4,110,843 (1978).
22. R. H. Skidmore, U.S. Patent (filed Feb. 13, 1978) 4,148,991 (1979).
23. R. J. Nichols, R. K. Senn, and F. Kheradi, *Adv. Polym. Technol.*, **3**, 62 (1983).
24. F. T. Gouinlock, H. W. Marciniak, M. H. Schatz, E. J. Quinn, and R. H. Hindersinn, *J. Appl. Polym. Sci.*, **12**, 2403 (1968).
25. C. S. Tucker and R. J. Nichols, *Plastics Eng.*, **43** (May) 27 (1987).
26. R. C. Kowalski, W. M. Davis, N. F. Newman, Z. A. Foroulis, and F. P. Baldwin, European Patent Application (filed March 30, 1984), 0,124,278 (1984).
27. R. C. Kowalski, W. M. Davis, N. F. Newman, Z. A. Foroulis, and F. P. Baldwin, U.S. Patent (filed Sept. 17, 1984) 4,548,995 (1985).
28. R. C. Kowalski, W. M. Davis, N. F. Newman, Z. A. Foroulis, and F. P. Baldwin, U.S. Patent (filed Oct. 1, 1984) 4,563,506 (1986).
29. R. C. Kowalski, *Chem. Eng. Prog.*, **85**, (May) 67 (1989).
30. C. E. Dellenbarger, U.S. Patent (filed Jan. 5, 1946) 2,466,934 (1948).
31. H. Frohlich, German Patent (filed April 10, 1953) 1,030,555 (1958).
32. A. Lasch, German Patent (filed April 29, 1950) 807,186 (1951).
33. H. Herrmann, Schneckenmaschinen in der Verfahrenstechnik, Springer, Berlin (1972).
34. H. Herrmann in Kunststoffe-Ein Werkstoff Macht Karriere, edited by W. Glenz, Hanser, Munich (1985).
35. A. Lasch and A. Stroebel, U.S. Patent (filed Aug. 15, 1955) 2,778,482 (1957).
36. G. Eisenmann, U.S. Patent (filed Oct. 2, 1973) 3,856,278 (1974).
37. J. A. Biesenberger and D. B. Todd, U.S. Patent (filed Nov. 19, 1992) 5,372,418 (1994).

11 Flow Mechanisms and Modeling of Non-Intermeshing Counter-Rotating Twin Screw Extruders

11.1 Introduction

In this chapter, we will turn to understand flow in non-intermeshing counter-rotating twin screw extruders. Generally, the literature considers two identical screws of opposite hand, i.e., one basically right-handed and the other left-handed. The efforts to date have been limited, as the reader may see from the relative scarcity of references at the end of this chapter. In recent years, there has been enhanced activity. The key early paper, which initiated research in this area, was published by Kaplan and Tadmor [1] in 1974. It presumed that the two screws may be analyzed independently and each pumps forward through drag flow caused by the relative motion between the screw and barrel. In the inter-screw region, there is no drag- but only pressure-driven flow. A later series of papers by Nichols and his co-workers [2–5] of Welding Engineers, the primary manufacturer of this type of machine, emphasized the importance of backward leakage flow. In succeeding years, various research groups have undertaken numerical simulation approaches to flow in tangential counter-rotating twin screw extruders. There has been little effort to simulate flow in other non-intermeshing counter-rotating twin screw machines. This may be caused by the fact that the Welding Engineers tangential machine has the simplest geometry and it is necessary to establish the theory of flow in this geometry first.

We begin this chapter by critically describing the Kaplan-Tadmor model (Section 11.2), then move on to consider the contributions of Nichols and his team to analyze flow in this geometry (Section 11.3). More recent numerical simulation models are then considered with the lubrication theory based approach and finite element simulations being described in successive sections (Section 11.4 and 11.5). Non-Newtonian numerical models are then described (Section 11.6). The role of backward pumping elements in modular twin screw machines is considered in Section 11.7. Then, we will analyze flow in modular machines and starvation effects (Section 11.8). We finally discuss scale-up of the Welding Engineers machine (Section 11.9).

11.2 Kaplan-Tadmor Model of Flow in Tangential Counter-Rotating Screw Elements

A model for interpreting the behavior of a non-intermeshing counter-rotating twin screw extruder was published in 1974 by A. Kaplan and Z. Tadmor [1]. These authors understood and presumed that the screw channel was open to the die and they limited their considerations to matched screw flights. They used a 3-plate model as shown in Figure 11.1. They presumed, as in the classical model of rectilinear screw extrusion (see Chapter 3), that the screw is stationary and the barrel moves. This is valid except in the region where the screws are in contact with each other, where a separate model is required.

In Zone I, where the screws are close to the barrel, the velocity field is determined by Eqs. (3.4). This may be solved by using Eq. (3.6) for $v_1(x_2)$. The throughput rate Q_I is given by Eq. (3.7). It is to be understood that Q_I is the throughput of the screw in this region. We may write from Eq. (3.7):

$$Q_I = \frac{1}{2} W\, H\, U_1 - \frac{W\, H^3}{12\,\eta} \frac{\Delta p_I}{(x_1)_I} \tag{11.1a}$$

$$\Delta p_I = \frac{12\,\eta}{W\, H^3}\left(\frac{1}{2} W\, H\, U_1 - Q_I\right)(x_1)_I \tag{11.1b}$$

In the region, where the screws are in contact, there is no drag flow but only pressure flow. For tangential screws $(\partial v_1/\partial x_2)\,(H)$ would be expected to be zero. Using this together with 'no slip' on the screws, Kaplan and Tadmor [1] show that:

$$Q_{II} = -\frac{W\, H^3}{3\,\eta} \frac{\Delta p_{II}}{(x_1)_{II}} \tag{11.2}$$

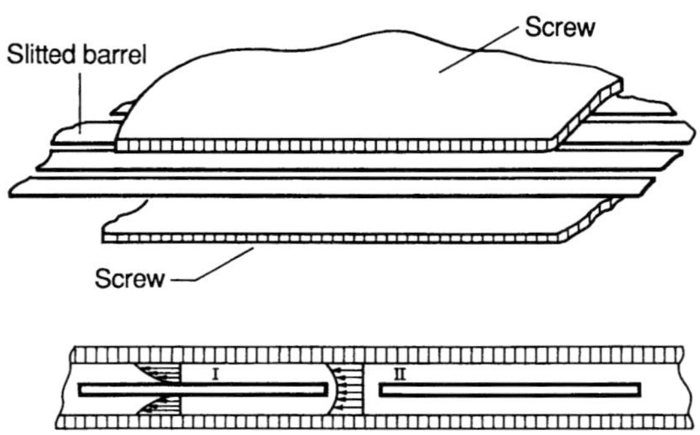

Figure 11.1: Three-plate model of Kaplan and Tadmor [1]
(reprinted with permission of the Society of Plastics Engineering)

Combining Eqs. (11.1) and (11.2) through:

$$\Delta p = \Delta p_{\mathrm{I}} + \Delta p_{\mathrm{II}} \tag{11.3}$$

they found that:

$$Q = \frac{1}{2} W\, H\, U_1 \left(\frac{4\, f}{3\, f + 1} \right) - \frac{W\, H^3}{12\, \eta} \left(\frac{\Delta p}{\Delta x_1} \right)_{\text{turn}} \left(\frac{4}{3\, f + 1} \right) \tag{11.4a}$$

where

$$f = \frac{(x_1)_{\mathrm{I}}}{(x_1)_{\mathrm{I}} + (x_1)_{\mathrm{II}}} \tag{11.4b}$$

Kaplan and Tadmor determined the influence of the f-factor on the $Q - \Delta p$ characteristic curves which they find is to decrease drag flow, increase pressure flow, and reduce the net throughput per screw. This reduction in pumping effectiveness is due to the reduction of drag flow.

The above formulation is oversimplified because it neglects leakage flow between the screws. This was noted by Kaplan and Tadmor [1], who considered leakage to occur between the screw flight tips in Zone II. They conceived this region to consist of a roughly rectangular region of depth 2 δ_{f} and length W^*. This channel extends backward over a length b, representing screw flight thickness. Presuming a rectangular leakage region and using arguments of the same form as leads to Eq. (11.2), they obtain:

$$Q_{\mathrm{L}} = \frac{2}{3} \frac{W^*}{\eta} \delta_{\mathrm{f}}^3 \left(\frac{\Delta p_{\mathrm{T}}}{b} \right) \tag{11.5}$$

where Δp_{T} is the pressure drop per turn, W^* is $(1 - f)\, \pi\, D$. The total flow rate Q_{T} should be:

$$Q_{\mathrm{T}} = 2\, Q - Q_{\mathrm{L}} \tag{11.6}$$

This yields:

$$Q_{\mathrm{T}} = W\, H\, U_1 \left(\frac{4\, f}{3\, f + 1} \right) - \frac{W\, H^3}{6\, \eta} \left(\frac{\Delta p}{\Delta x_1} \right)_{\mathrm{T}} \left(\frac{4}{3\, f + 1} \right) - \frac{2}{3} \frac{W^*\, \delta_{\mathrm{f}}^3}{\eta} \left(\frac{\Delta p_{\mathrm{T}}}{b} \right) \tag{11.7a}$$

or

$$Q_{\mathrm{T}} = W\, H\, U_1 \left(\frac{4\, f}{3\, f + 1} \right) - \frac{W\, H^3}{6\, \eta} \left(\frac{\Delta p}{\Delta x_1} \right)_{\mathrm{T}} \left(\frac{4}{3\, f + 1} \right) \cdot \left[1 + 4 \left(\frac{3\, f + 1}{4} \right) \left(\frac{\delta_{\mathrm{f}}}{H} \right)^3 \frac{W^*}{W} \frac{\Delta x_1}{b} \right] \tag{11.7b}$$

11.3 Theories of Nichols and his Co-Workers

At the 1982 Society of Plastics Engineers Annual Technical Conference (ANTEC), R. J. Nichols and J. Yao [3] of Welding Engineers expressed concern that the Kaplan-Tadmor analysis described in the previous section was under-estimating the backward leakage flow. They specifically noted that their model for the area allowing leakage was not correct. Specifically, the apex was not rectangular and was larger. Nichols and Yao proposed the expression for the apex area:

$$\begin{aligned} A_\mathrm{T} &= A_1 + A_2 \\ &= 4\left\{\frac{x\,s}{2} - \left[\frac{r^2}{2}(\overline{\alpha} - \sin\overline{\alpha})\right]\right\} + 2\,W^*\,\delta_\mathrm{f} \end{aligned} \tag{11.8}$$

The first term, A_1, represents the apex area, the second term, A_2, the correction of Kaplan and Tadmor. The origin of the A_1 term may be seen from Figure 11.2. The $x\,s/2$ term represents the area of a half apex triangle. The quantity 4 $(x\,s/2)$ represents the triangular area of the upper apex and lower apex. The term $(r^2/2)(\overline{\alpha} - \sin\overline{\alpha})$ represents a correction due to the area of the segment at the screw periphery.

Nichols and Yao proposed to use the Kaplan-Tadmor model of Eq. (11.4) and (11.6), but with a leakage flow term based on the area defined by Eq. (11.8). The Nichols-Yao model predicts much weaker pumping and greater backward leakage than Kaplan and Tadmor's analysis.

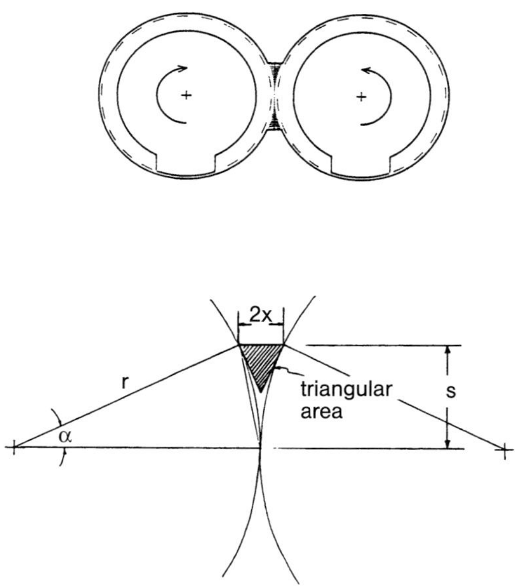

Figure 11.2: Nichols et al. apex areas allowing leakage [3] (reprinted with permission of the Society of Plastics Engineers)

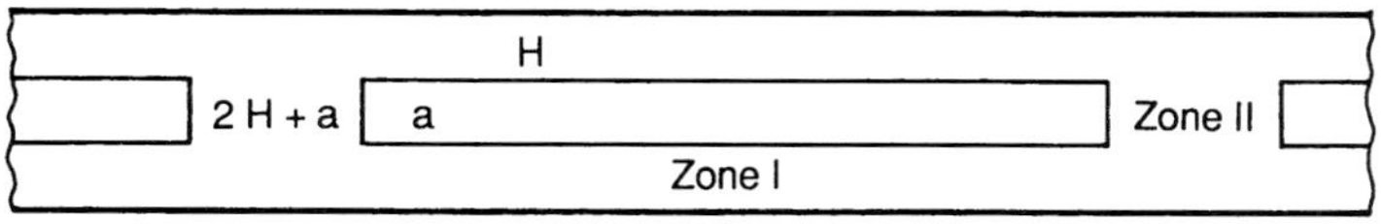

Figure 11.3: Modified 3-plate model of Nichols [5]
(reprinted with permission of the Society of Plastics Engineers)

Nichols returned to the issue at the 1984 Society of Plastics Engineers ANTEC and presented an analysis developed with J. T. Lindt [5, 6]. The 3-plate model of the previous section is retained as is the treatment of flow in Zone I. The flow treatment in Zone II is modified by recognizing that the apex area makes the region for the pressure flow to be larger (see Figure 11.3). Equation (11.2) was replaced with:

$$Q_{\mathrm{II}} = -\frac{W\left(H+\frac{1}{2}a\right)^3}{3\,\eta}\,\frac{\Delta p_{\mathrm{II}}}{(x_1)_{\mathrm{II}}} \tag{11.9}$$

The leakage flow through the apex was taken as:

$$Q_{\mathrm{L}} = K\,\frac{x^3}{\eta}\,\frac{\Delta p_{\mathrm{T}}}{b} \tag{11.10}$$

which is equivalent to the Newtonian flow pressure drop-extrusion rate relationship for a triangular-shaped conduit.

Nichols with Lindt now combined Eqs. (11. 1) and (11.9) using Eq. (11.3):

$$Q = \frac{1}{2}\,W\,H\,U\,F_{\mathrm{DCRT}} - \frac{W\,H^3}{12\,\eta}\,\frac{\Delta p_{\mathrm{T}}}{(x_1)_{\mathrm{T}}}\,F_{\mathrm{PCRT}} \tag{11.11}$$

Here:

$$F_{\mathrm{DCRT}} = \frac{4\,f}{4\,f + \left[\frac{1-f}{\left(1+\frac{a}{H}\right)^3}\right]} \tag{11.12a}$$

$$F_{\mathrm{PCRT}} = \frac{4}{4\,f + \left[\frac{1-f}{\left(1+\frac{a}{H}\right)^3}\right]} \tag{11.12b}$$

When the parameter '*a*' goes to zero, Eq. (11.11) becomes Kaplan and Tadmor's Eq. (11.4). If one includes the leakage term, Eq. (11.10), we obtain Eq. (11.13):

$$Q_{\mathrm{T}} = W\,H\,U\,F_{\mathrm{DCRT}} - \frac{W\,H^3}{6\,\eta}\left[\frac{\Delta p}{(x_1)}\right]_{\mathrm{T}} F_{\mathrm{PCRT}}\left[1 + \frac{6\,K\,x^3}{F_{\mathrm{PCRT}}\,W\,H^3}\,\frac{(\Delta x_1)_{\mathrm{T}}}{b}\right] \tag{11.13}$$

The Nichols et al. models have been critically discussed in the monograph by Rauwendaal [7].

11.4 Lubrication Theory Numerical Model of Flow in Tangential Counter-Rotating Screw Elements

M. H. Kim, W. Szydlowski, J. L. White, and K. Min [8] and subsequently Kim and White [9, 10] sought to develop a model of flow in tangential counter-rotating twin screw extruders based on Cartesian coordinate hydrodynamic lubrication theory. More recently Bang and White [11, 12] have modeled the flow with a cylindrical coordinate lubrication formulation.

M. H. Kim et al. begin using the hydrodynamic lubrication theory as developed in Chapter 2 (Section 2.8). As before, the screw is unwrapped and treated in Cartesian coordinates. The coordinate system is embedded in one of the screws with coordinates: "1" along the screw axis, "2" perpendicular to the screw channel, and "3" with the circumferential direction. One proceeds by integration of the equations of motion through the "2" direction between the screw and barrel. A balance is made with integrated fluxes (see Eq. (2.65)):

$$q_1 = \int_0^H v_1\,\mathrm{d}x_2 = \frac{1}{2}U_1\,H - \frac{H^3}{12\,\eta}\frac{\partial p}{\partial x_1} \tag{11.14a}$$

$$q_3 = \int_0^H v_3\,\mathrm{d}x_2 = \frac{1}{2}U_3\,H - \frac{H^3}{12\,\eta}\frac{\partial p}{\partial x_3} \tag{11.14b}$$

The function $H(x_1,x_3)$ defines both the distance between the screw and the barrel chamber and the intermeshing region. The velocities U_1 and U_3 should be seen to have values.

$$U_1 = 0$$
$$U_3 = \begin{cases} \pi\,D\,N & \text{(between screw and barrel)} \\ 0 & \text{(intermeshing region)} \end{cases} \tag{11.15}$$

This may be viewed in Figure 11.4.

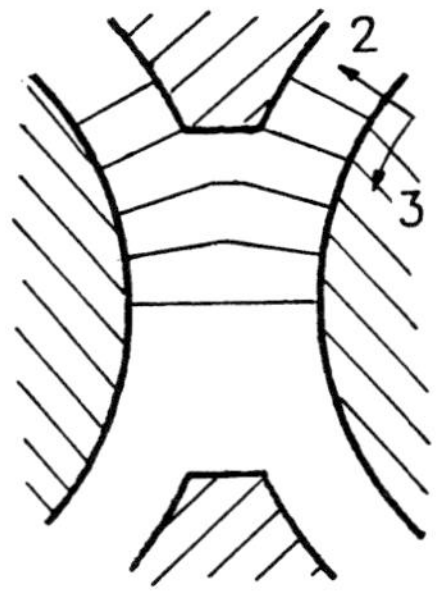

Figure 11.4: Cross-sectional view of twin screw geometry and FAN elements [8, 9]

This problem was solved by M. H. Kim et al. [8, 9] using the FAN method developed by Tadmor and his co-workers [13, 14] and modified by Szydlowski et al. [15] for inclusion of a moving boundary. In the screw-barrel region, we have the balance:

$$\begin{aligned} &q_1^{\mathrm{I}}(x_1 - \Delta x_1, x_3)\,\Delta x_3 + q_3^{\mathrm{I}}(x_1, x_3 - \Delta x_3)\,\Delta x_1 \\ &= q_1^{\mathrm{I}}(x_1 + \Delta x_1, x_3)\,\Delta x_3 + q_3^{\mathrm{I}}(x_1, x_3 + \Delta x_3)\,\Delta x_1 \end{aligned} \tag{11.16a}$$

where the q_1^{I} are fluxes in the "1" direction or the 'hopper' and 'die' ends of the element and the q_3^{I} are fluxes in the 3-direction in the 'circumferential direction'. The same general form is valid for the screw contact region (Figure 11.4). However, at the entrance and exit to the region, we have different forms. Specifically, at the entrance:

$$\begin{aligned} &q_1^{\mathrm{II}}(x_1 - \Delta x_1, x_3)\,\Delta x_3 + q_3^{\mathrm{IR}}(x_1, x_3 - \Delta x_3)\,\Delta x_1 + q_3^{\mathrm{IL}}(x_1, x_3 - \Delta x_3) \\ &= q_1^{\mathrm{II}}(x_1 + \Delta x_1, x_3)\,\Delta x_3 + q_3^{\mathrm{II}}(x_1, x_3 + \Delta x_3)\,\Delta x_1 \end{aligned} \tag{11.16b}$$

At the exit from the inter-screw region, we have:

$$\begin{aligned} &q_1^{\mathrm{II}}(x_1 - \Delta x_1, x_3)\,\Delta x_3 + q_3^{\mathrm{II}}(x_1, x_3 - \Delta x_3)\,\Delta x_1 \\ &= q_1^{\mathrm{II}}(x_1 + \Delta x_1, x_3)\,\Delta x_3 + q_3^{\mathrm{IR}}(x_1, x_3 + \Delta x_3)\,\Delta x_1 + q_3^{\mathrm{IL}}(x_1, x_3 + \Delta x_3)\,\Delta x_1 \end{aligned} \tag{11.16c}$$

The actual integration is applied from the bottom screw-barrel combination through the inter-screw region to the top screw-barrel combination of the same screw. Symmetry arguments are then applied. This analysis may be compared to the Cartesian coordinate analysis for co-rotating twin screw machines given in Sections 5.3 and 5.5.

Equation (11.16) should be considered as a series of finite difference equations in pressure analogous to Reynolds' equations of lubrication theory (see Section 2.8 and Eqs. (2.65) and (2.66)). This follows from substitution of Eq. (11.14) into Eq. (11.16).

The method described above is immediately more powerful than the analytical methods described earlier. It is able to not only make computations for matched flight screw configurations as done by Kaplan and Tadmor [1] and by Nichols [2, 5], but also for the staggered flight configurations, which seem to be more widely used (see Chapter 10 (Sections 10.3 and 10.4)).

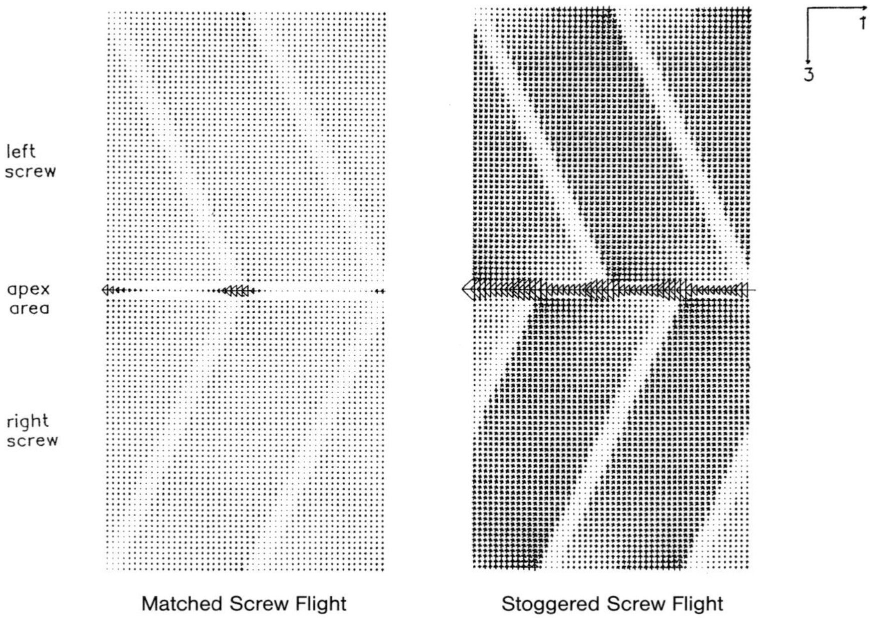

Figure 11.5: Local flux vector map for matched and staggered screw configurations [8, 9]

Calculations were made by Kim et al. based on the above model using dimensions supplied by R. J. Nichols of Welding Engineers, which are representative of their machines. A flux vector map indicating mean velocity fields in the 1–3 plane for matched and staggered screw flight configurations is shown in Figure 11.5. This shows that the backflow is much higher for the staggered flight configurations. It also indicates that the level of distributive mixing should be much greater for the staggered screw flight. Kim et al. put this in quantitative form by computing the relative extent of backflow Q_b to net forward flow '$Q - Q_b$':

$$G = \frac{Q_b}{Q - Q_b} \tag{11.17}$$

G is much larger for the staggered flight configuration. The situation for the intermediate configurations is shown in Figure 11.6.

Calculations for extruder throughput versus pressure development for matched and staggered screw configurations are shown in Figure 11.7. These indicate that the matched flight screw configuration is a much better pump than the staggered flight configuration.

Comparisons can be made for selected geometries between the Kaplan-Tadmor [1], Nichols and Lindt [5], and Kim et al. [8, 9] approaches for the case of matched screw flights. This is shown in Figure 11.8. The Kaplan-Tadmor model predicts inferior pumping characteristics for two single screw machines. The Nichols et al. and the Kim et al. models predict worse pumping behavior. The latter authors find the highest level of backflow and worst pumping characteristics.

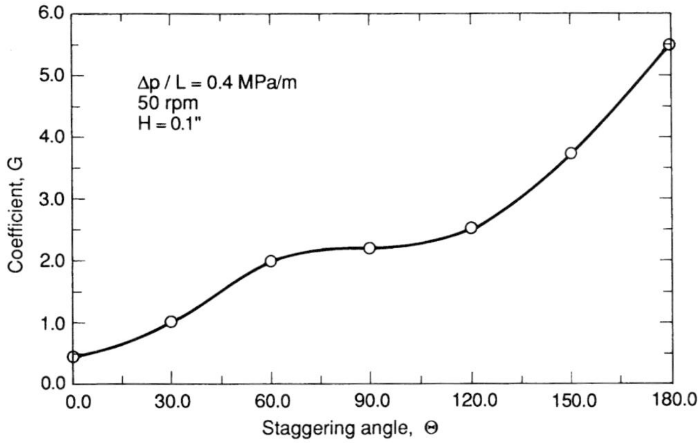

Figure 11.6: Effect of screw flight staggering on fractional throughput G [8, 9]

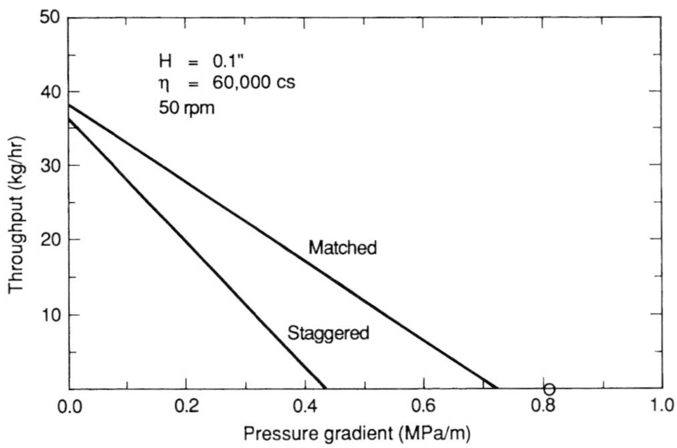

Figure 11.7: Screw characteristic curves for matched and staggered screw flight configuration [8, 9]

Subsequently, Bang and White [11, 12], using a cylindrical coordinate FAN analysis, developed by David et al. [16], re-solved the tangential counter-rotating twin screw extruder problem analyzed by Kim et al. [8, 9] presented above. First consider the velocity field:

$$\underset{\sim}{v} = v_z(r)\,\underset{\sim}{e}_z + v_\theta(r)\,\underset{\sim}{e}_\theta + 0\,\underset{\sim}{e}_r \qquad (11.18)$$

The equations of motion in cylindrical coordinates for a Newtonian fluid in the lubrication approximation are

$$0 = -\frac{\partial p}{\partial z} + \eta \frac{1}{r}\frac{\partial}{\partial r}\left(r\frac{\partial v_z}{\partial r}\right)$$
$$0 = -\frac{1}{r}\frac{\partial p}{\partial \theta} + \eta \frac{1}{r}\frac{\partial}{\partial r}\left[r^3 \frac{\partial}{\partial r}\left(\frac{v_\theta}{r}\right)\right] \qquad (11.19\text{ a, b})$$

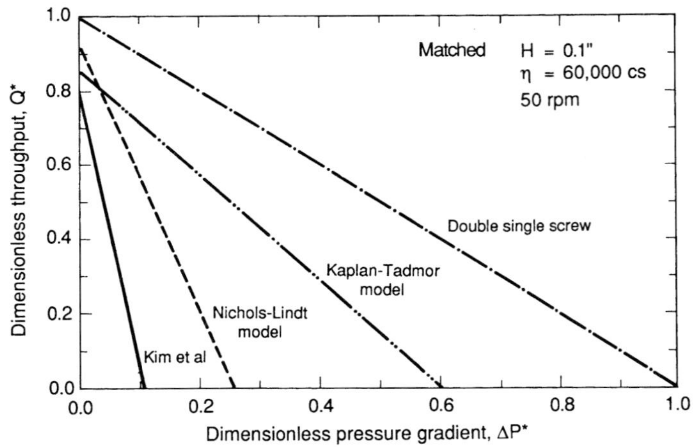

Figure 11.8: Screw characteristic curves predicted by Kaplan-Tadmor [1], Nichols-Yao [3], and Kim et al. [8] models

with the boundary conditions for $v_z(R_i)$ and $v_z(R_0)$ are 0. We also have $v_\theta(R_i) = 0$ and $v_\theta(R_0)$ is $v_\theta = \pi D N$. Equations (11.19a) and (11.19b) lead to fluxes

$$q_z = \frac{1}{R^*} \int_{R_i}^{R_0} r\, v_z \, dr \tag{11.20a}$$

$$q_\theta = \int_{R_i}^{R_0} v_\theta \, dr \tag{11.20b}$$

where R^* is a characteristic length. For a Newtonian fluid these fluxes are

$$q_z = -\frac{R_i^4}{16\,\eta\, R^*} \frac{\partial p}{\partial z} \left[-1 + \frac{1}{\kappa^4} + \frac{\left(\frac{1}{\kappa^2} - 1\right)^2}{\ln \frac{1}{\kappa}} \right] \tag{11.21a}$$

$$q_\theta = \frac{U_\theta\, R_i}{2} \frac{\kappa}{1-\kappa^2} \left(\frac{1}{\kappa^2} - 1 + 2 \ln \kappa \right) - \frac{R_i^2}{8\,\eta} \frac{\partial p}{\partial \theta} \left(1 - \frac{1}{\kappa^2} \right) \left[-1 + \left(\frac{2 \frac{1}{\kappa} \ln \frac{1}{\kappa}}{1 - \frac{1}{\kappa^2}} \right)^2 \right] \tag{11.21b}$$

The flux balance between the screw and barrel is

$$q_z(\theta, z - \Delta z)\, R^*\, \Delta\theta + q_\theta(\theta - \Delta\theta, z)\, \Delta z = q_z(\theta, z + \Delta z)\, R^*\, \Delta\theta \tag{11.22}$$

For the inter-screw region we must distinguish between fluxes from the left and right screws when the fluids enter this region. This leads to:

$$\begin{aligned} & q_\theta^{IR}(\theta - \Delta\theta, z)\,\Delta z + q_\theta^{IL}(\theta - \Delta\theta, z)\,\Delta z + q_z^{int}(\theta, z - \Delta z)\,R^*\,\Delta\theta \\ &= q_\theta^{int}(\theta + \Delta\theta, z)\,\Delta z + q_z^{int}(\theta, z + \Delta z)\,R^*\,\Delta\theta \end{aligned} \tag{11.23}$$

where q_θ^{IR} and q_θ^{IL} are input fluxes from the right and left screws and q_z^{int} is the inter-screw axial flux. In the midst of the inter-screw region we have

$$\begin{aligned} & q_\theta^{int}(\theta - \Delta\theta, z)\,\Delta z + q_z^{int}(\theta, z - \Delta z)\,R^*\,\Delta\theta \\ &= q_\theta^{int}(\theta + \Delta\theta, z)\,\Delta z + q_z^{int}(\theta, z + \Delta z)\,R^*\,\Delta\theta \end{aligned} \tag{11.24}$$

The flux balance at the exit of the inter-screw region may be represented by a balance similar to Eq. (11.23) (compare the cylindrical coordinate analysis for intermeshing screws given in Section 8.5).

Bang and White [11] computed machine pumping characteristics for a Newtonian fluid in the cylindrical coordinate formulation given above. The results of the computations are very similar to those of Kim et al. [8] for the Newtonian fluid formulation.

11.5 Finite Element Models of Flow in Tangential Counter-Rotating Screw Elements

The use of finite element analysis [17, 18] to model flow in a tangential counter-rotating twin screw extruder has been described by Bigio and his coworkers [19, 21] and by Nguyen and Lindt [22–24]. The work has only been described incompletely. The solutions are based on creeping flow (inertia-less) solutions of the equations of motion for a Newtonian fluid. The full set of viscous components is used, including extensional as well as shearing stresses. (This differs from the previously described work of Kim et al. [8, 9], which considers shearing stress alone.) Both Bigio et al. [19–21] and Nguyen and Lindt [22–24] also consider the full curvature inherent in the system, including the inter-screw region.

Bigio and Zerafati [19, 20] present a 2-dimensional simulation of flow perpendicular to the screw axes. The intention of these authors was to study the mixing in the nip region. They find that increasing the size of the apex region improves the mixing. Increasing the aspect ratio (outer screw radius divided by inner radius) improves the mixing. The screw speed did not affect the mixing performance. These efforts were extended by Conner and Bigio [21].

Nguyen and Lindt [22–24] consider 3-dimensional flow of a Newtonian fluid. They both consider mixing and calculate screw characteristic curves. They study the influence of screw staggering and apex design on screw pumping and mixing. Both, introducing screw staggering and increasing the size of the apex region, improves mixing and reduces pumping characteristics.

11.6 Non-Newtonian Flow Modeling

Non-Newtonian models of flow in tangential counter-rotating twin screw extruders have been described by Kim and White [10] and subsequently by Bang and White [12]. These authors represent the non-Newtonian viscosity in terms of the power law and use hydrodynamic lubrication theory.

The earlier study by Kim and White [10] takes

$$\eta = K\left[\left(\frac{\partial v_1}{\partial x_2}\right)^2 + \left(\frac{\partial v_3}{\partial x_2}\right)^2\right]^{\frac{n-1}{2}} \tag{11.25}$$

and solves the equations

$$\begin{aligned} 0 &= -\frac{\partial p}{\partial x_1} + K\frac{\partial}{\partial x_2}\left[\left(\frac{\partial v_1}{\partial x_2}\right)^2 + \left(\frac{\partial v_3}{\partial x_2}\right)^2\right]^{\frac{n-1}{2}}\frac{\partial v_1}{\partial x_2} \\ 0 &= -\frac{\partial p}{\partial x_3} + K\frac{\partial}{\partial x_2}\left[\left(\frac{\partial v_3}{\partial x_2}\right)^2 + \left(\frac{\partial v_3}{\partial x_2}\right)^2\right]^{\frac{n-1}{2}}\frac{\partial v_3}{\partial x_2} \end{aligned} \tag{11.26 a, b}$$

using the FAN method. The coordinate axis

"3" is along the screw channel, "1" is transverse to the channel and "2" is perpendicular to the screw root.

Fluxes are defined in the manner of Eq. (11.14 a, b). In the screw barrel section

$$\begin{aligned} q_1 &= -\frac{H^3}{12\,\overline{\eta}}\frac{\partial p}{\partial x_1} \\ q_3 &= \frac{1}{2}U_3\,H - \frac{H^3}{12\,\overline{\eta}}\frac{\partial p}{\partial x_3} \end{aligned} \tag{11.27 a, b}$$

where the shear viscosity $\overline{\eta}$ of Eq. (11.27 a, b) was expressed as

$$\overline{\eta} = K\left[\left(\overline{\frac{\partial v_1}{\partial x_2}}\right)^2 + \left(\overline{\frac{\partial v_3}{\partial x_2}}\right)^2\right]^{\frac{n-1}{2}} \tag{11.27 c}$$

To determine $(\overline{\partial v_1}/\partial x_2)$ and $(\overline{\partial v_3}/\partial x_2)$, they consider three different cases, i.e., (i) drag flow only, (ii) pressure flow only, (iii) both drag and pressure flow but with zero output. We consider the basic nature of the flow in question and choose appropriately the Newtonian fluid forms. In the region between the screws, the balances of Eq. (11.16 a, b, c) were applied.

Bang and White [11, 12] model this problem in cylindrical coordinates with the axis being coincident with one of the screws. The non-Newtonian shear viscosity was taken as

$$\eta = K\left\{\left[r\frac{\partial}{\partial r}\left(\frac{v_\theta}{r}\right)\right]^2 + \left(\frac{\partial v_z}{\partial r}\right)^2\right\}^{\frac{n-1}{2}} \tag{11.28}$$

being related to shear stresses through

$$\sigma_{\theta r} = \eta\, r\frac{\partial}{\partial r}\left(\frac{v_\theta}{r}\right) \tag{11.29 a}$$

$$\sigma_{zr} = \eta\frac{\partial v_z}{\partial r} \tag{11.29 b}$$

The equations of motion are

$$\begin{aligned} 0 &= -\frac{1}{r}\frac{\partial p}{\partial \theta} + \frac{1}{r^2}K\frac{\partial}{\partial r}\left\{r\left[\left(r\frac{\partial}{\partial r}\left(\frac{v_\theta}{r}\right)\right)^2 + \left(\frac{\partial v_z}{\partial r}\right)^2\right]^{\frac{n-1}{2}}\frac{\partial}{\partial r}\left(\frac{v_\theta}{r}\right)\right\} \\ 0 &= -\frac{\partial p}{\partial z} + \frac{1}{r}K\frac{\partial}{\partial r}\left\{r\left[\left(r\frac{\partial}{\partial r}\left(\frac{v_\theta}{r}\right)\right)^2 + \left(\frac{\partial v_z}{\partial r}\right)^2\right]^{\frac{n-1}{2}}\frac{\partial v_z}{\partial r}\right\} \end{aligned} \tag{11.30 a, b}$$

Again, a FAN procedure was used with

$$\begin{aligned} q_\theta &= \frac{1}{2}\int_{R_i}^{R_0}\left[\frac{1}{m}(R_0^2 - r^2)\frac{1}{r}\left(\frac{1}{2}\frac{\partial p}{\partial \theta} + \frac{C_\theta}{r^2}\right)\right]\mathrm{d}r \\ q_z &= \frac{1}{2}\int_{R_i}^{R_0}\left[\frac{1}{m}(R_0^2 - r^2)\left(\frac{1}{2}\frac{\partial p}{\partial z}r + \frac{C_z}{r}\right)\right]\mathrm{d}r \end{aligned} \tag{11.31 a, b}$$

with

$$m = K\left\{\left[r\frac{\partial}{\partial r}\left(\frac{v_\theta}{r}\right)\right]^2 + \left(\frac{\partial v_z}{\partial r}\right)^2\right\}^{\frac{n-1}{2}} \tag{11.32a}$$

$$m = K\left[\left(\frac{1}{2}\frac{\partial p}{\partial \theta} + \frac{C_\theta}{r^2}\right)^2 + \left(\frac{1}{2}\frac{\partial p}{\partial z}r + \frac{C_z}{r}\right)^2\right]^{\frac{n-1}{2}} \tag{11.32b}$$

The boundary conditions are

$$\pi D N = R_0 \int_{R_i}^{R_0} \frac{1}{m} \frac{1}{r} \left(\frac{1}{2} \frac{\partial p}{\partial \theta} + \frac{C_\theta}{r^2} \right) \mathrm{d}r \tag{11.33 a}$$

$$0 = \int_{R_i}^{R_0} \frac{1}{m} \left(\frac{1}{2} \frac{\partial p}{\partial z} r + \frac{C_z}{r^2} \right) \mathrm{d}r \tag{11.33 b}$$

The results of calculations are shown in Figures 11.9 a and b and 11.10 a and b for two screw elements of the modular Leistritz twin screw extruder for matched flight and staggered screw flight conditions. It can be seen that increasing non-Newtonian character (decreasing power law exponent, n) reduces the pumping characteristics. Similarly, screw staggering reduces pumping ability.

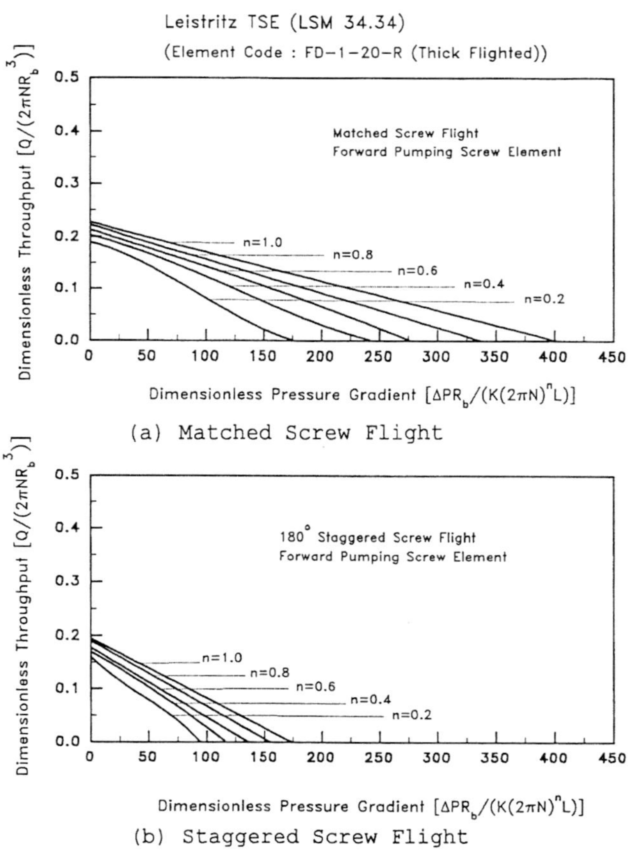

Figure 11.9 a, b: Screw characteristic curves calculated by Bang and White [12] for power law fluids in thick-flighted Leistritz screw elements

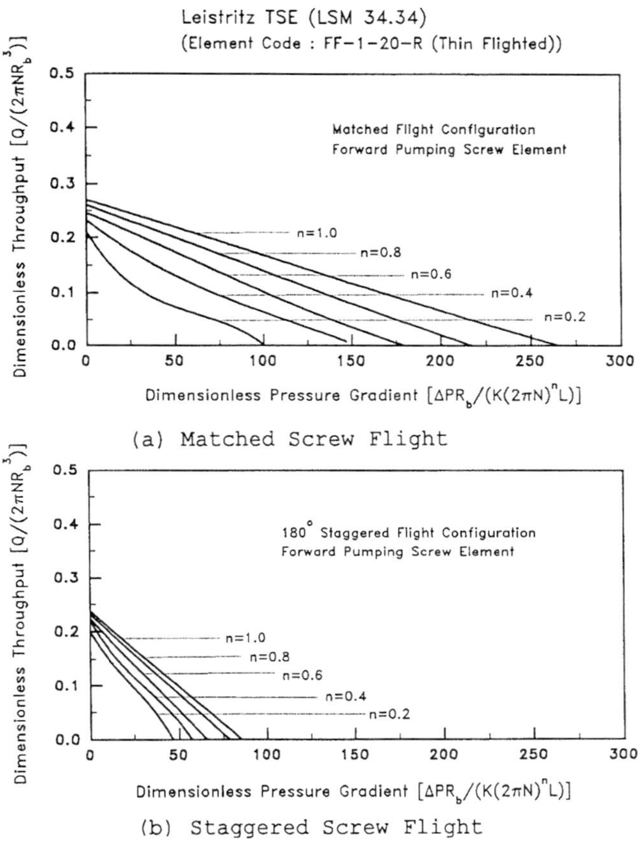

Figure 11.10 a, b: Screw characteristic curves calculated by Bang and White [12] for power law fluids in thin-flighted screw elements

11.7 Backward Pumping Tangential Counter-Rotating Screw Elements

The design of commercial tangential counter-rotating twin screw extruders is following the patent of Fuller [25]. They are generally modular in character, involving forward pumping and reversed or backward pumping screw elements. In single screw or co-rotating twin screw machines, we distinguish right-handed and left-handed screws. However, such a distinction cannot be made here because the configurations on opposite counter-rotating screws are different. We will speak here instead of forward pumping and backward pumping screw elements.

The simulation method developed in Section 3.3 may be directly applied to a backward pumping screw element (Figure 3.13). If one considers an unwound flattened screw channel, Eqs. (3.6) and (3.8) may still be considered to apply for the velocity components $v_1(x_2)$ and $v_3(x_2)$. If the coordinate axes are placed on the screw, one finds that the boundary condition on the barrel changes sign. In place of velocity U_1, which drags the fluid forward, there is a velocity $(-U_1)$, which drags the fluid backward to the hopper.

The velocity components $v_1(x_2)$ and $v_3(x_2)$ in the region (I) between the screw and barrel are found to be

$$v_1(x_2) = -U_1\left(\frac{x_2}{H}\right) - \frac{H^2}{2\,\eta}\frac{\mathrm{d}p}{\mathrm{d}x_1}\left[\left(\frac{x_2}{H}\right)^2 - \left(\frac{x_2}{H}\right)\right] \tag{11.34}$$

$$v_3(x_2) = -U_3\left[2\left(\frac{x_2}{H}\right) - 3\left(\frac{x_2}{H}\right)^2\right] \tag{11.35}$$

The throughput of a backward pumping tangential counter-rotating twin screw extruder element in the Kaplan-Tadmor approximation of Section 11.2 is given by

$$Q = -W\,H\,U_1\left(\frac{4\,f}{3\,f+1}\right) - \frac{W\,H^3}{6\,\eta}\left(\frac{\Delta p}{\Delta x_1}\right)_{\text{turn}}\left(\frac{4}{3\,f+1}\right) \tag{11.36}$$

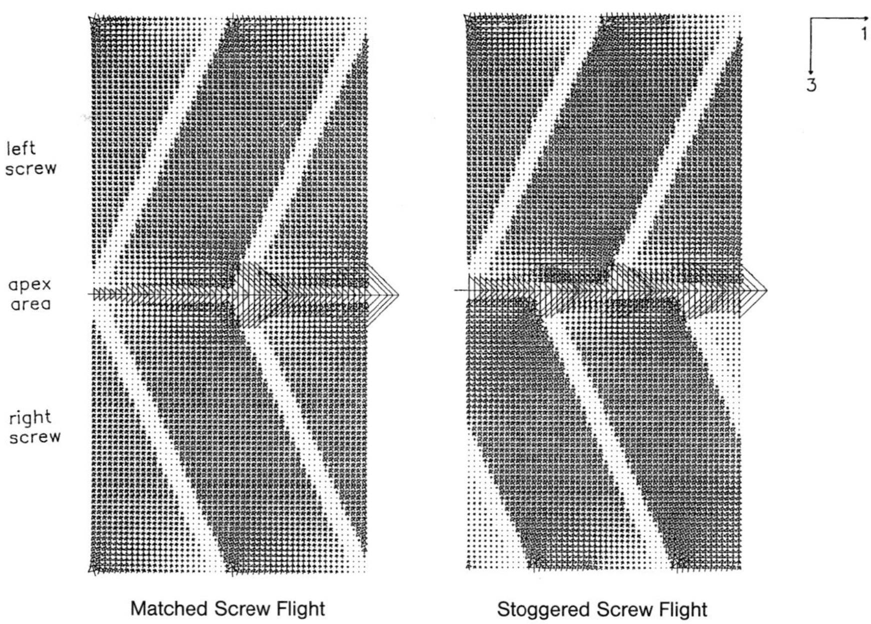

Figure 11.11: Local flux vector map for backward pumping screws from Kim and White [10]

where we have omitted leakage flows. Both drag flow and a positive pressure gradient tend to push the fluid back to the hopper. A positive value of Q requires a negative pressure gradient. Thus backward pumping screw elements result in pressure reductions. This will also change the direction of the leakage flow between the screws. The predicted forward flux calculated by M. H. Kim and White [10] between the screws is shown in Figure 11.11.

11.8 Non-Isothermal Modeling

For the velocity profile of Eq. (11.18), which we used to represent the flow in a tangential counter-rotating twin screw extruder, the energy equation of Eq. (2.56) may be written:

$$\rho\, c\left(\frac{v_\theta}{r}\frac{\partial T}{\partial \theta}+v_z\,\frac{\partial T}{\partial z}\right)=k\,\frac{1}{r}\frac{\partial}{\partial r}\left(r\,\frac{\partial T}{\partial r}\right)+\sigma_{r\theta}\,r\,\frac{\partial}{\partial r}\left(\frac{v_\theta}{r}\right)+\sigma_{zr}\,\frac{\partial v_z}{\partial r} \tag{11.37}$$

If Eq. (11.37) is integrated following Bang and White [12] in the radial direction, we obtain

$$\begin{aligned}\rho\, c\left(\frac{q_\theta}{r}\frac{\partial \bar{T}}{\partial \theta}+q_z\,\frac{\partial \bar{T}}{\partial z}\right)&=\frac{1}{R_{ch}}\left[R_0\, h_b\,(T_b-\bar{T})+R_i\, h_s\,(T_s-\bar{T})\right]\\&\quad+\frac{1}{R_{ch}}\,R_0\,\sigma_{r\theta}\,(R_0)\,U_\theta-\left(\frac{\partial p}{\partial \theta}\right)\frac{q_\theta}{R_{ch}}-\left(\frac{\partial p}{\partial z}\right)q_z\end{aligned} \tag{11.38}$$

Here q_z and q_θ are defined in Eq. (11.20) and

$$\frac{\partial \bar{T}}{\partial \theta}=\frac{\int_{R_0}^{R_i} v_\theta\,\frac{\partial T}{\partial \theta}\,\mathrm{d}r}{\int_{R_0}^{R_i} v_\theta\,\mathrm{d}r} \tag{11.39a}$$

$$\frac{\partial \bar{T}}{\partial z}=\frac{\int_{R_0}^{R_i} r\,v_z\,\frac{\partial T}{\partial z}\,\mathrm{d}r}{\int_{R_0}^{R_i} r\,v_z\,\mathrm{d}r} \tag{11.39b}$$

We may integrate Eq. (11.38) with respect to θ, which leads to a differential equation for the cup mixing temperature $\bar{\bar{T}}(z)$.

$$\rho\, c\, Q \frac{\mathrm{d}\bar{\bar{T}}}{\mathrm{d}z} = 2\,\pi\,[R_0\; h_b\; (T_b - \bar{\bar{T}}) + R_i\; h_s\; (T_s - \bar{\bar{T}})]_{\text{left screw}}$$
$$+\, 2\,\pi\,[R_0\; h_b\; (T_b - \bar{\bar{T}}) + R_i\; h_s\; (T_s - \bar{\bar{T}})]_{\text{right screw}} \qquad (11.40)$$
$$+ \int_0^{2\pi(1+f)} R_0\; \sigma_{\theta r}\; (R_0)\, U_\theta\; \mathrm{d}\theta - Q\left(\frac{\mathrm{d}p}{\mathrm{d}z}\right)$$

The energy balance of Eq. (11.40) must be solved moving forward from the hopper to the die.

11.9 Composite Model of Modular Tangential Counter-Rotating Twin Screw Extruder

Tangential counter-rotating twin screw extruders have been modular since the time of Fuller [25]. The basis of analysis and design of modular tangential twin screw extruders is contained in a 1987 paper by White and Szydlowski [26] and developed in detail a decade later by Bang and White [12].

The approach is based on the machine having metered starved feeding. In such a twin screw extruder, the output of the die is known. From the die characteristics, the pressure level at the exit to the screw region may be calculated. One may then use the screw characteristic curves for the various elements along the modular screw to calculate the pressure profile backward along the machine axis. Typical screw characteristic curves are shown in Figures 11.7, 11.8, 11.9 and 11.10. Such calculations generally predict starved flow as well as fully filled screw elements. This is shown in Figure 11.12 a, b, c, where calculations are carried out back to the probable position of melting.

The analysis described in the above paragraph serves for isothermal machines. However, typically tangential counter-rotating twin screw extruders are quite non-isothermal. We must solve the backward fluid mechanical calculations together with forward calculations of an energy balance.

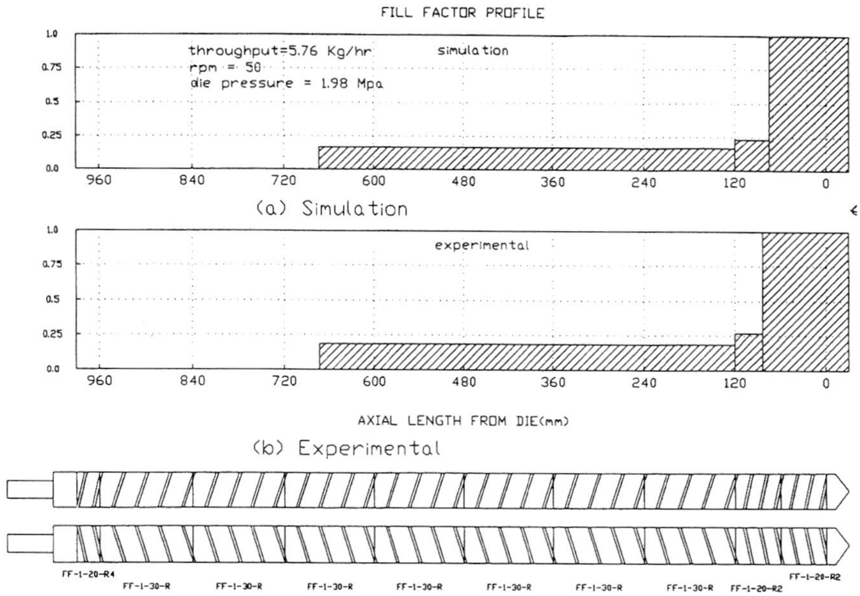

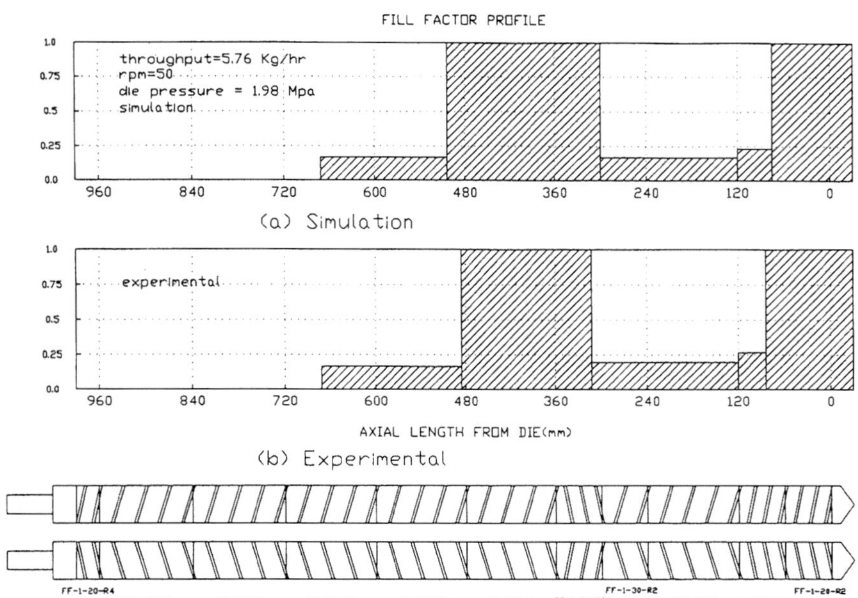

Figure 11.12 a, b, c: Calculated fill factor profiles for different modular screws [12]
(a) Calculated and measured fill factor profiles for Screw Configuration 1
(b) Calculated and measured fill factor profiles for Screw Configuration 2
(c) Calculated and measured fill factor profiles for Screw Configuration 3

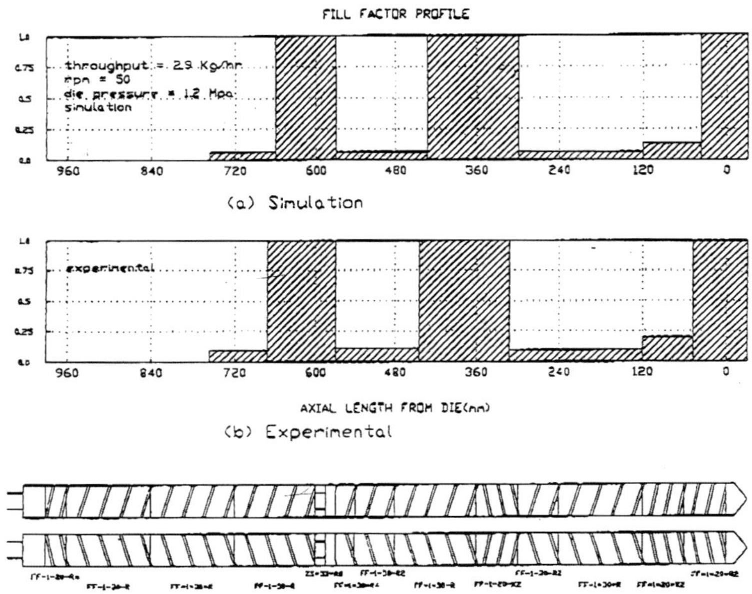

Figure 11.12 a, b, c: (continued)

11.10 Scale-Up of Tangential Counter-Rotating Twin Screw Extruders

The scale-up of tangential counter-rotating twin screw extruders has received little attention in the literature. The first significant consideration of this problem was by Nichols and Lindt [27]. This scale-up issue was subsequently discussed by Bash [28].

Nichols and Lindt [27] argue that the principles involved in scale-up of single screw extruders should be used for tangential counter-rotating machines. They argue that the early method of Carley and Mckelvey [29], which maintains shear rate and residence time, is suitable and they suggest it scales mixing. This approach leads to scaling

$$\frac{Q}{Q_0} = \left(\frac{D}{D_0}\right)^{\alpha} \tag{11.41}$$

where Q is the extrusion rate, D is screw diameter, and α is 3.0 while maintaining L/D and screw geometry. Nichols and Lindt [27] note that industrial practice leads to the exponent α being 2.5. This view is generally accepted by Bash [28].

As noted by both Nichols and Lindt [27] and Bash [28], the Carley-Mckelvey approach neglects heat transfer. If viscous dissipation and heat transfer are included as is done in the modeling of Section 3.8, the value of α will certainly be reduced.

References

1. A. Kaplan and Z. Tadmor, *Polym. Eng. Sci.*, **14**, 58 (1974).
2. R. J. Nichols, *Kunststoffe*, **72**, 506 (1982).
3. R. J. Nichols and J. Yao, *SPE ANTEC Tech. Papers*, **28**, 416 (1982).
4. R. J. Nichols, J. C. Golba, and P. K. Shete, Paper presented at AIChE Diamond Jubilee Meeting, Washington, Oct.-Nov. (1983).
5. R. J. Nichols, *SPE ANTEC Tech. Papers*, **30**, 6 (1984).
6. J. T. Lindt, Personal Communication (1988).
7. C. Rauwendaal, Polymer Extrusion, 4th ed. Hanser, Munich (2001).
8. M. H. Kim, W. Szydlowski, J. L. White, and K. Min, *Adv. Polym. Technol.*, **9**, 87 (1989).
9. M. H. Kim and J. L. White, *SPE ANTEC Tech. Papers*, **35**, 49 (1989); *Int. Polym. Process.*, **5**, 201 (1990).
10. M. H. Kim and J. L. White, *J. Non. Newt. Fluid Mech.*, **37**, 37 (1990).
11. D. S. Bang and J. L. White, *Int. Polym. Process.*, **11**, 109 (1996).
12. D. S. Bang and J. L. White, *Int. Polym. Process.*, **12**, 278 (1997).
13. Z. Tadmor, E. Broyer and C. Gutfinger, *Polym. Eng. Sci.*, **14**, 660 (1974).
14. E. Broyer, C. Gutfinger and Z. Tadmor, *Trans. Soc. Rheology*, **19**, 423 (1975).
15. W. Szydlowski and J. L. White, *Adv. Polym. Technol.*, **7**, 177 (1987);
W. Szydlowski, R. Brzoskowski and J. L. White, *Int. Polym. Process.*, **1**, 207 (1987).
16. B. David, T. Sapir, A. Nir and Z. Tadmor, *Int. Polym. Process.*, **7**, 204 (1992).
17. O. C. Zienkiewicz, The Finite Element Method, McGraw Hill, London (1979).
18. A. J. Baker, Finite Element Computational Fluid Mechanics, McGraw-Hill, NY (1983).
19. D. I. Bigio and S. Zerafati, *SPE ANTEC Tech. Papers*, **34**, 85 (1988).
20. D. I. Bigio and S. Zerafati, *Polym. Eng. Sci.*, **31**, 1900 (1991).
21. J. H. Conner and D. I. Bigio, *SPE ANTEC Tech. Papers*, **38**, 31 (1992).
22. K. Nguyen and J. T. Lindt, *SPE ANTEC Tech. Papers*, **34**, 93 (1988).
23. K. Nguyen and J. T. Lindt, *SPE ANTEC Tech. Papers*, **35**, 154 (1989).
24. K. Nguyen and J. T. Lindt, *Polym. Eng. Sci.*, **29**, 709 (1989).
25. L. J. Fuller, U.S. Patent (filed May 15, 1945) 2,615,199 (1952).
26. J. L. White and W. Szydlowski, *Adv. Polym., Technol.*, **7**, 419 (1987).
27. R. J. Nichols and J. T. Lindt, *SPE ANTEC Tech. Papers*, **34**, 80 (1988).
28. T. F. Bash, in Plastics Compounding: Equipment and Processing, edited by D. B. Todd, Hanser, Munich (1998).
29. J. F. Carley and J. M. McKelvey, *Ind. Eng. Chem.*, **45**, 985 (1953).

12 Experimental Studies of Non-Intermeshing Counter-Rotating Twin Screw Extruders

12.1 Introduction

In this chapter we will describe experimental studies of the mechanisms of flow in non-intermeshing counter-rotating twin screw extruders. These studies have been limited in number, as we shall see, and much still needs to be done. However, certain essential characteristics are clear.

In Section 12.2, we describe flow visualization investigations of fluid motions. Section 12.3 summarizes studies of pumping characteristics of tangential counter-rotating twin screw extruders based on forward pumping screw designs. We turn in Section 12.4 to investigations of pumping characteristics carried out on modular screw systems, which contain left-handed screws and right-handed screws, as well as non-screw elements. Section 12.5 summarizes studies of residence time distributions for this type of machine. Melting studies are described in Section 12.6.

12.2 Flow Visualization of Fluid Motions

Flow visualization investigations for tangential counter-rotating twin screw extruders were pioneered in the late 1970s in a series of papers by T. Sakai [1, 2] of Japan Steel Works. The first paper [1] contrasts flow patterns in tangential counter-rotating twin screw extruders with intermeshing co-rotating and intermeshing counter-rotating machines. Sakai used suspended pigments in high-viscosity (300 Stokes) silicone oil in his studies. The twin screws were mounted in a polymethyl methacrylate barrel. He found that in these machines the flow is primarily along a single screw but occasionally, the marker strays over to the second screw (Figure 12.1). Sakai's second paper [2] investigated streamlines in different mixing sections of a complex character. There was again only occasional movement of fluid streamlines from screw to screw.

A third flow visualization study using low viscosity fluids was reported by Bigio and Penn [3] (Figure 12.2). Their procedure was similar to Sakai's in using a viscous liquid in a transparent plastic barrel. They proceeded by following a neutralization reaction (potassium hydroxide and phenolphthalein in corn syrup) of a colored liquid. This reaction occurs much more rapidly, i.e., over shorter lengths, in staggered twin screw configurations. Fully-staggered, 50/50 configurations are more effective than 90/10 configurations. Subsequently, Bigio et al. [4] have described an image processing technique applied to this experiment.

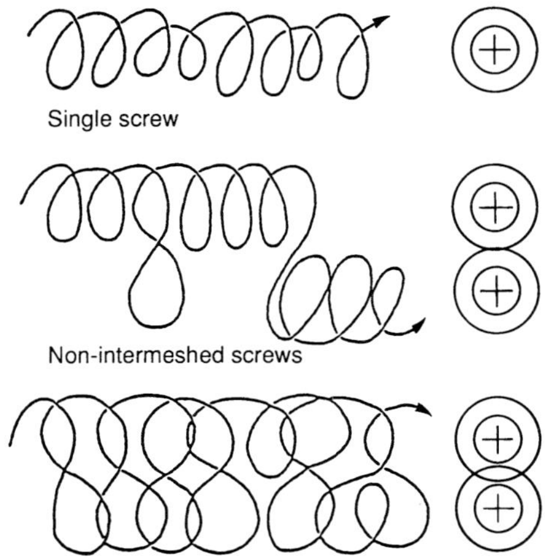

Figure 12.1: Sakai's [1] investigation of streamline motion in a tangential twin screw extruder

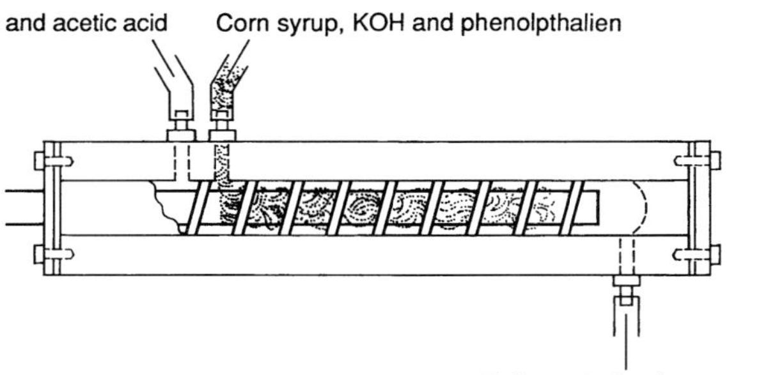

Figure 12.2: Flow visualization apparatus of Bigio and Penn [3] (reprinted with permission of the Society of Plastics Engineers)

Flow visualization studies showing the motions of a polyethylene melt in a tangential counter-rotating twin screw extruder in a chamber with glass windows have been reported by Min, Kim, and White [5]. The screw channels were filled and the motions observed are generally similar to those described by Sakai. Little inter-screw fluid movement was observed.

Nichols [6] reported that most applications using tangential counter-rotating twin screw extruders involve starved flow conditions. In starved flow machines with staggered screw flights, he finds that there is a pencil of fluid which stretches from flight to flight along the length of the screw in this region.

Bigio and Wigginton [7] have reported flow visualization studies of starved flow. They find that little inter-screw transfer occurred with matched screw flights. However, with staggered flights, substantial inter-screw transfer is observed. Bigio and Wigginton find the maximum inter-screw transfer occurring at 50% stagger with 70% levels of inter-screw transfer.

12.3 Pumping Characteristics of Simple Screw Assemblies in Tangential Counter-Rotating Twin Screw Machines

The first basic experimental study of flow in this geometry was presented by Kaplan and Tadmor [7] using a Bausano 25 mm diameter tangential counter-rotating machine. The machine had two identical screws (one right-handed and one left-handed) with the same pitch and arranged with matched screw flights. They studied a low-molecular weight polyisobutylene oil and determined its pumping characteristics. Screw characteristic curves were determined for the twin screw machine and for a single screw extruder formed from the Bausano twin screw machine by removal of one screw and plugging the barrel with a solid cylindrical rod (Figure 12.3). They reported that their model (described in Section 11.2) worked well.

A second experimental study using a Newtonian oil was reported by Nichols [9]. He contrasted the behavior of two Newtonian oils in a series of Welding Engineers tangential counter-rotating twin screw machines. He presented screw characteristic curves, Q vs Δp, which indicate that the pumping characteristics of these machines are significantly less than two equivalent single screw extruders. A most interesting result is that the pumping characteristics of matched screw flight configurations (also used by Kaplan and Tadmor [8]) are much superior to staggered flight configurations (Figure 12.4). This result suggests a much greater leakage flow with staggered screw configurations.

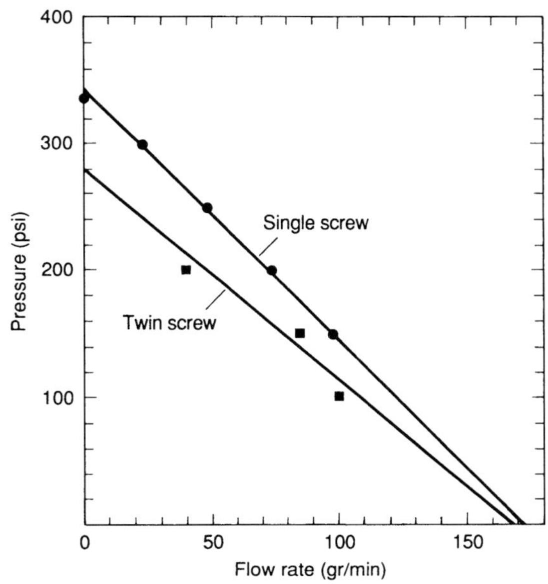

Figure 12.3: Screw characteristic curves comparing a tangential counter-rotating twin screw extruder and an equivalent single-screw extruder formed by removing one screw and plugging the barrel (after Kaplan and Tadmor [8]) (reprinted with permission of the Society of Plastics Engineers). The data is normalized per screw

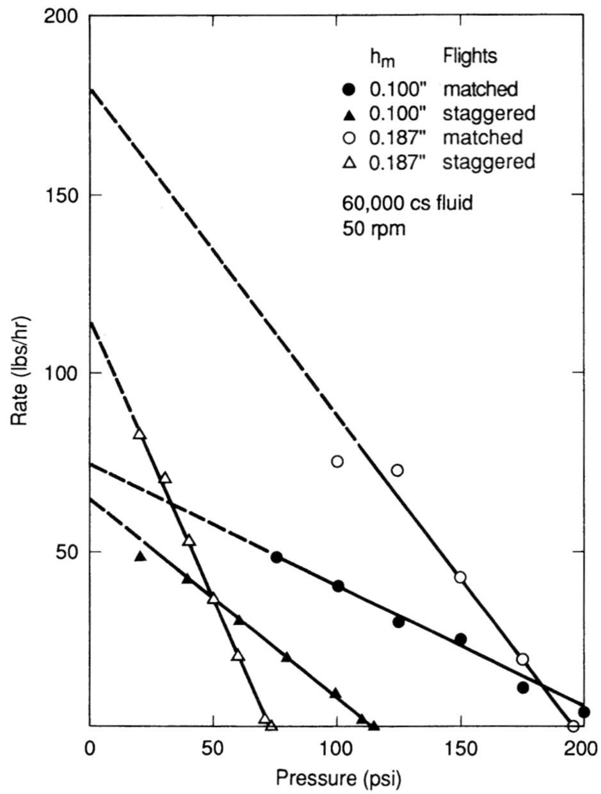

Figure 12.4: Screw characteristic curves comparing single screw and tangential counter-rotating twin screw extruders for polyethylene with matched and staggered screw flights (after Nichols [9]) (reprinted with permission of the Society of Plastics Engineers)

Subsequently Min, Kim, and White [5] built a laboratory tangential counter-rotating twin screw extruder that is modular in design and has transverse windows allowing for flow visualization. Min et al. determined the screw characteristic curves for their machine using forward pumping screw configurations for a polyethylene melt. Their machine was used for plastication, melting, and pumping. Comparisons were made of matched and staggered flights as shown in Figure 12.5. The results indicate that the matched flight configurations have much better pumping characteristics than the staggered flight machines. It is to be noted that a survey of Welding Engineers patents [10–13] through the years indicates that staggered flight configurations are generally used in their machines.

Kim, Szydlowski, White, and Min [14] have compared the experimental data of Nichols [9] described above with their own simulation of flow. This was described in Section 11.4. The comparison is displayed in Figure 12.6. Quantitatively, the prediction of superior pumping characteristics of the matched flight relative to the staggered flight configuration is observed. However strikingly, quantitative agreement is found between theory and experiment. Earlier

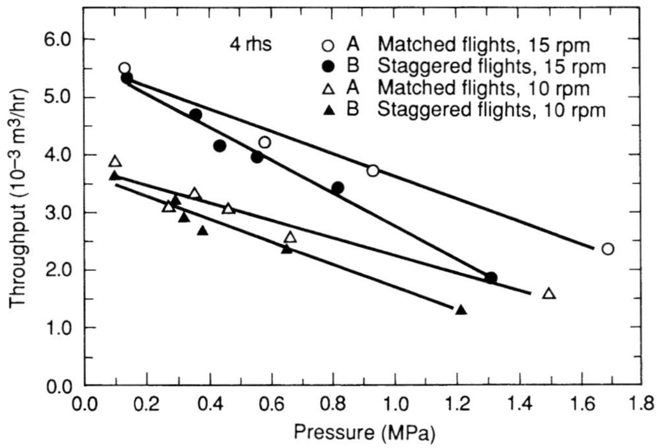

Figure 12.5: Screw characteristic curves comparing tangential counter-rotating twin screw extruders for polyethylene with matched and staggered screw flights (after Min, Kim, and White [5])

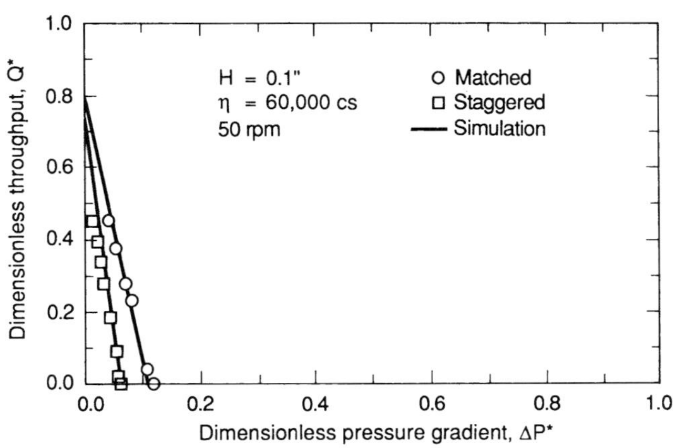

Figure 12.6: Comparison of data of Nichols et al. with the theory of Kim, Szydlowski, White, and Min [14]

models by Kaplan and Tadmor [8] (Section 11.2) and Nichols and Yao [15] (Section 11.3) tend to overpredict the pumping characteristics and underestimate leakage.

Nguyen and Lindt [16, 17] made a similar comparison for the matched flight configuration data of Nichols with their theory. Again, good agreement was found.

In a 1998 paper, Bang et al. [18] determined experimentally screw characteristic curves for modular elements in a tangential counter-rotating twin screw extruder manufactured by Leistritz AG. This was done by characterizing die and screw exit pressure losses and comparing them with fill lengths in front of the die (see Figure 12.7). Typical experimental screw characteristic curves are shown in Figure 12.8 together with the results of simulation (Section 11.6)

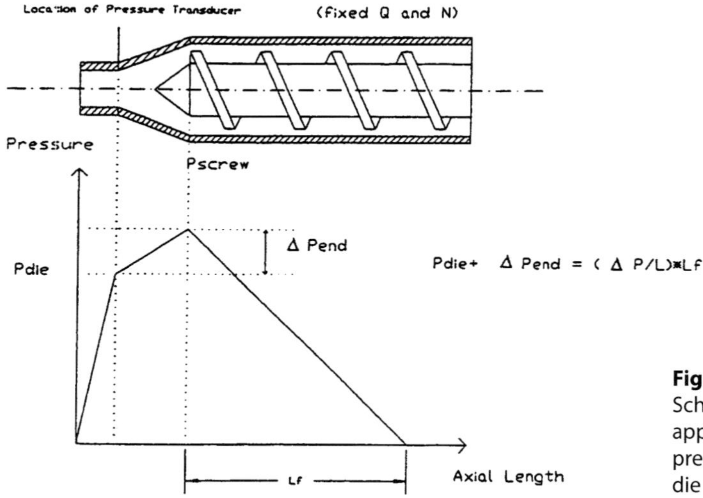

Figure 12.7:
Schematic of experimental apparatus with details of the pressure profiles around the die [18]

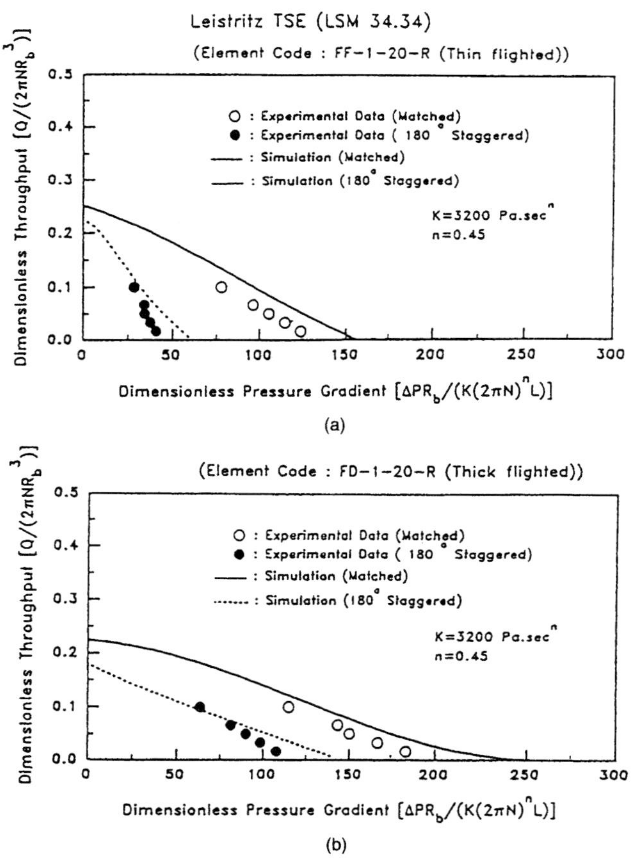

Figure 12.8:
Dimensionless screw characteristic curves for forward pumping screw elements (comparison of simulation with experiment);
a) FF-1-20-R element, and b) FD-1-20-R element (after Bang et al. [18])

12.4 Pumping Characteristics of Modular Screws in Tangential Counter-Rotating Twin Screw Machines

Modular screw configurations in non-intermeshing counter-rotating twin screw extruders date to the 1945 patent application of Fuller [10] of Welding Engineers. They are described in various later Welding Engineers patents [11–13]. Typical Welding Engineers designs using right-handed and left-handed screw elements are shown in Figures 10.4 and 10.5.

The only study showing the influence of modular screws with both right-handed and left-handed screw elements is that of Min et al. [5]. These authors contrasted the pumping performance of the two sets of modular screw systems shown in Figure 12.9. These screws contain four sections. The 1st, 2nd, 4th elements are forward pumping screw pairs, while the 3rd element is backward pumping. The modular screws were compared with all forward pumping screw element designs in both matched and staggered flight configurations. The screw characteristic curves observed using a polyethylene melt are found in Figure 12.10. These results show that screw configurations with backward pumping screw elements have lower pumping capacity and develop smaller pressures than the configurations with all forward pumping screw elements. This is found for both matched and staggered screw flights.

Min et al. [5] have also determined pressure profiles along the length of their twin screw extruder. Typical results are shown in Figure 12.11. These results indicate that pressure increases along the length of the screw for forward pumping screw elements. However, the introduction of backward pumping screws causes the pressure to decrease in the flow direction.

The results summarized above generally agree with the view given, for instance, by White and Szydlowski [19] and others (compare Sections 5.4 and 11.7) that backward pumping screw elements exhibit pressure reductions and thus weaker overall pumping performance. They are, of course, positioned in Fuller design [10] Welding Engineers machines to produce pressure decreases for applications such as devolatilization.

Bang et al. [18] have also compared predicted fill factor distributions (Section 11.9) with experimental observations on modular screws. This is shown in Figure 12.12.

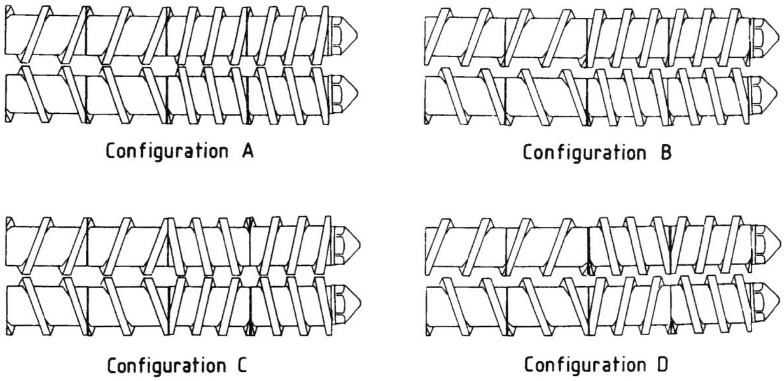

Figure 12.9: Modular screw designs of Min, Kim, and White [5]

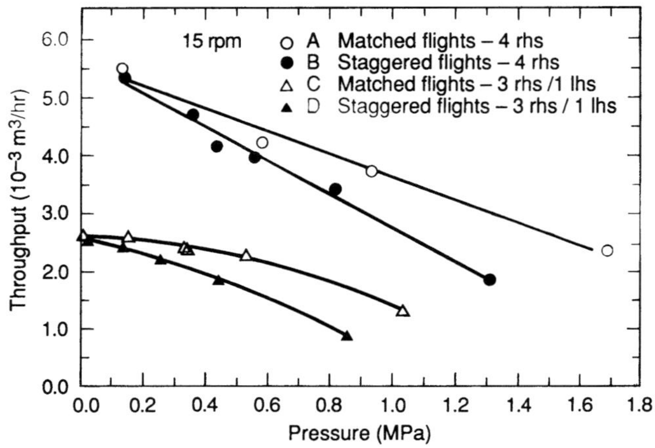

Figure 12.10: Screw characteristic curves comparing the modular screws of Figure 12.9 [5]. Screw configurations C and D have backward pumping screw elements

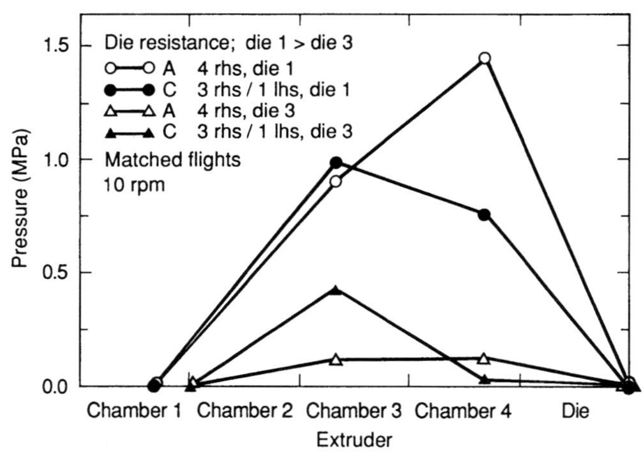

Figure 12.11: Pressure profile measurements along the length of a tangential intermeshing twin screw extruder with modular elements [5]

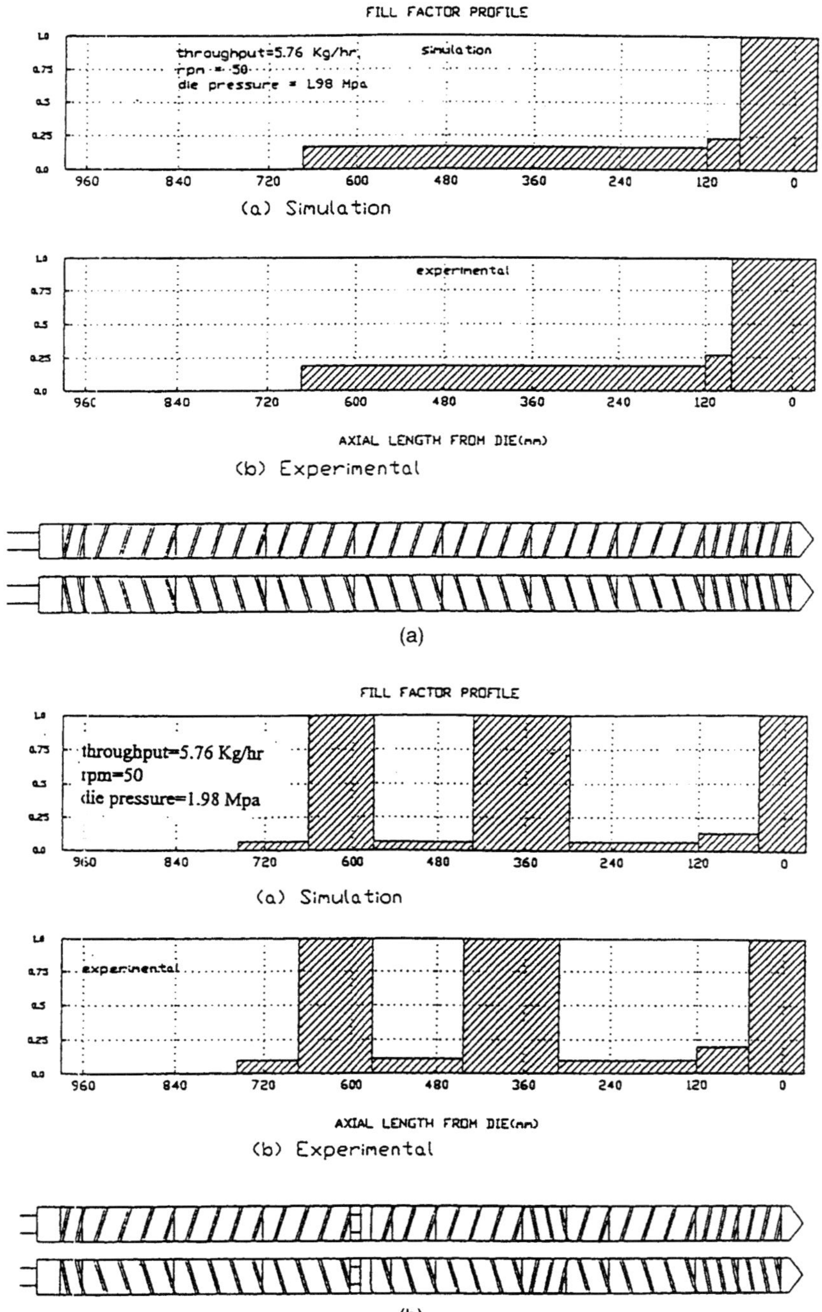

Figure 12.12: Comparison of predicted fill factor profiles with measured fill factor profiles [18]; a) screw configuration 1, and b) screw configuration 4

12.5 Residence Time Distributions in Tangential Counter-Rotating Twin Screw Machines

Three experimental studies of residence time distributions in tangential counter-rotating twin screw extruders have been reported in the literature. The earliest of these investigations was by Sakai [1, 2] of Japan Steel Works. More recently, residence time distributions have been determined by Walk [20], Nichols, Golba, and Shete [21], and Chen et al. [22].

Sakai [1] used a tracer technique to measure residence time distributions using an emission diode and a glass fiber conductor to measure a voltage which depended on the concentration of the particles. Walk [20] used a cadmium selenide trace in polymethyl methacrylate, which was detected using X-ray fluorescence. In the work of Nichols, Golba, and Shete [21], a conductive carbon black tracer and capacitance/dielectric constant was used as the sensing method. In a 1995 paper, Chen et al. [22] made tracer experiments using polystyrene ethylbenzene solution. They used 1-aminoanthraquinone as a tracer and detected it with an ultraviolet spectrophotometer.

Walk [20] found that residence times decreased with increasing screw speed or throughput. Nichols et al. [21] find mean residence times are generally longer for staggered flight as compared to matched flight configurations. They also found that residence time decreases with increasing screw speed and extrusion rate but increases with decreasing channel depth. The behavior was similar with matched flight and staggered flight pairs of screws.

Generally, the residence time distributions for the tangential twin screw extruders are broader than for single screw extruders and are narrower than the behavior of a series of continuous stirred tank reactors.

12.6 Mixing Studies in Tangential Counter-Rotating Twin Screw Machines

Howland and Erwin [23] studied the striation behavior of black and white silicone rubber, which were added to a twin screw extruder and subsequently cured. Samples were removed and cut out at different angles. They found that mixing was much more rapid than in equivalent single screw extruders. They attributed this to backflow between the screws.

During the 1990s, Bigio and Baim [24, 25] published studies of mixing in tangential counter-rotating twin screw extruders. In one study they used a pigment solution in corn syrup (a non-Newtonian fluid), into which they injected a red dye. It was found that with increasing fill factor, stagger between the screws led to increasing inter-screw dye transfer. Increasing screw speed also increased the amount of inter-screw dye transfer. Generally, they found the level of inter-screw transfer to be less than for Newtonian fluids, especially at lower screw speeds.

In a second study, Baim and Bigio [25] studied the mixing of two pigmented silicone oils (one black and one white) with different viscosity levels.

12.7 Melting Behavior in Tangential Counter-Rotating Twin Screw Machines

The first experimental study of melting in a tangential counter-rotating twin screw extruder was presented by Kaplan and Tadmor [8] using low density polyethylene (LDPE). Following Maddock's experiment [26], they fed mixed pigmented and pure particles of LDPE to the extruder, processed them until achieving the steady state, and then cooled the machine down. The screws were pushed out and the solidified plastic was unwound and sectioned. This revealed a solid bed adjacent to the pushing flight. This was in contrast to studies of single screw extruders, where the solid bed is generally adjacent to the neutral flight [26]. Kaplan and Tadmor thought this result was not due to the machine being a twin screw extruder, but because of the large flight clearances.

Observations on melting in non-intermeshing counter-rotating twin screw extruders were also presented by Sakai [27]. The studies were carried out in a twin screw machine in which the screws are separated, but which had mixing sections at the end of the screw. Sakai evaluated the decrease in width of the solid bed with axial position along the screw.

A third investigation was presented by Nichols and Kheradi [28]. They found the details of melting to be determined by the configuration of the screw flights for matched screw flights. When a matched screw flight configuration was used, the melting behavior was similar to a single screw extruder with a melt pool on the leading flight. When the screw flights were staggered, no melt pool was formed and pellets were found scattered in a melt matrix. This may be associated with pressure equilibration between the screws in the transverse direction. The pressure equilibration of one screw channel in screw A with two channels in screw B is disrupting.

Min, Kim, and White [5] presented flow visualization studies of the melting process. They generally observed a solid bed floating in the midst of a melt pool for all conditions.

Certainly no consistent picture has developed of melting in tangential counter-rotating twin screw machines and more work is needed.

References

1. T. Sakai, *Góse Jushi*, **29**, 7 (1978).
2. T. Sakai, *Kobunshi Ronbunshu*, **38**, 279 (1984).
3. D. I. Bigio and D. Penn, *SPE ANTEC Tech. Papers*, **34**, 59 (1988).
4. D. I. Bigio, J. H. Conner, and A. Vashihat, *SPE ANTEC Tech. Papers*, **35**, 133 (1989).
5. K. Min, M. H. Kim, and J. L. White, *Int. Polym. Process.*, **3**, 165 (1988).
6. R. J. Nichols, *Personal Communication*, June (1988).
7. D. I. Bigio and M. Wigginton, *SPE ANTEC Tech. Paper*, **36**, 1905 (1990).
8. A. Kaplan and Z. Tadmor, *Polym. Eng. Sci.*, **14**, 58 (1974).
9. R. J. Nichols, *SPE ANTEC Tech. Papers*, **29**, 130 (1983).
10. L. J. Fuller, U.S. Patent (filed May 15, 1945) 2,615,199 (1952).
11. L. F. Street, U.S. Patent (filed August 16, 1952) 2,733,051 (1956).
12. R. H. Skidmore, U.S. Patent (filed December 28, 1959) 3,082,816 (1963).
13. R. H. Skidmore, U.S. Patent (filed April 8, 1970) 3,742,093 (1973).
14. M. H. Kim, W. Szydlowski, K. Min, and J. L. White, *Adv. Polym. Technol.*, **9**, 87 (1989).
15. R. J. Nichols and J. Yao, *SPE ANTEC Tech. Papers*, **28**, 416 (1982).
16. K. Nguyen and J. T. Lindt, *SPE ANTEC Tech. Papers*, **34**, 93 (1988).
17. K. Nguyen and J. T. Lindt, *SPE ANTEC Tech. Papers*, **35**, 154 (1989); *Polym. Eng. Sci.*, **29**, 709 (1989).
18. D. S. Bang, M. H. Hong and J. L. White, *Polym. Eng. Sci.*, **38**, 485 (1998).
19. J. L. White and W. Szydlowski, *Adv. Polym. Technol.*, **7**, 419 (1987).
20. C. J. Walk, *SPE ANTEC Tech. Papers*, **28**, 423 (1982).
21. R. J. Nichols, J. C. Golba, and P. K. Shete, Paper presented at AIChE Diamond Jubilee Meeting, Washington, D. C., November (1983).
22. L. Chen, G. H. Hu and J. T. Lindt, *Polym. Eng. Sci.*, **35**, 598 (1995).
23. C. Howland and L. Erwin, *SPE ANTEC Tech. Papers*, **29**, 113 (1983).
24. D. I. Bigio and W. Baim, *SPE ANTEC Tech. Papers*, **38**, 1809 (1992).
25. W. Baim and D. I. Bigio, *SPE ANTEC Tech. Papers*, **40**, 266 (1994).
26. B. H. Maddock, *SPE Journal*, **15**, 303 (1959).
27. T. Sakai, *Nippon Kikai Gakkai Ronbunshu*, **B47**, 420 (1981).
28. R. J. Nichols and F. Kheradi, *Mod. Plastics*, **61** (February), 70 (1984).

13 Technology of Continuous Mixers

13.1 Introduction

Clearly classify twin screw processing machines presents a problem to. One method considers relative directions of screw rotation, a second the level of screw intermesh. Here, we have decided to also separate out machines that are the outgrowth of an attempt to upgrade known established batch mixers to continuous mixers. These machines contain mixing rotor sections along their screw axes. This is to some extent artificial, but this presentation is quite useful in understanding the development of continuous mixing technology.

13.2 Batch and Internal Mixers

Processing of elastomers (and thus polymer melts) on counter-rotating twin rotor machines dates to E. M. Chaffee's [1] 1836 American patent for the mill and calender. This was intended for compounding natural rubber and coating it onto fabrics.

The next stage in the development took place in Stuttgart, Germany. Here, in 1877, Paul Freyburger [2] developed a two-rotor kneading machine with elliptical disc rotors. The significance of this invention was realized by Paul Pfleiderer, who developed this machine further. He brought Hermann Werner into partnership in 1879, and the firm of Maschinenfabrik Werner and Pfleiderer [4] was created to manufacture twin rotor internal mixers [3]. These were used at first for foodstuffs, and later for early industrial polymeric materials such as rubber. The Freyburger rotors did not work out. Pfleiderer [5] subsequently redesigned and substantially improved Freyburger's internal mixer rotors, shown in Figure 13.1. They represent the beginning of the 'sigma blade' mixer.

It was agreed by Hermann Werner and Paul Pfleiderer that Werner would operate the company in Stuttgart and Pfleiderer would move to London, England. This move would allow him to market the Werner and Pfleiderer products throughout the British empire [5]. Pfleiderer remained in England until his death in 1903.

In the 1890s, Hermann Werner took the lead of establishing Werner and Pfleiderer in the United States with manufacturing facilities for mixing equipment in Saginaw, Michigan [6]. This facility was partly owned by Werner, Pfleiderer, and Perkins [6].

The early 20th century saw developments in the design of batch internal mixers, which were to have a profound effect on the design of counter-rotating twin screw extruders. The first batch mixers used for rubber were Werner and Pfleiderer machines with specially designed masticating rotors [7] devised by Kempter (Figure 13.2). This is illustrated by Herrmann [3] who reprinted a 1910 Werner and Pfleiderer advertisement for a "Gummi Kneter". These machines were open at the top and were rotated to discharge the mixed products.

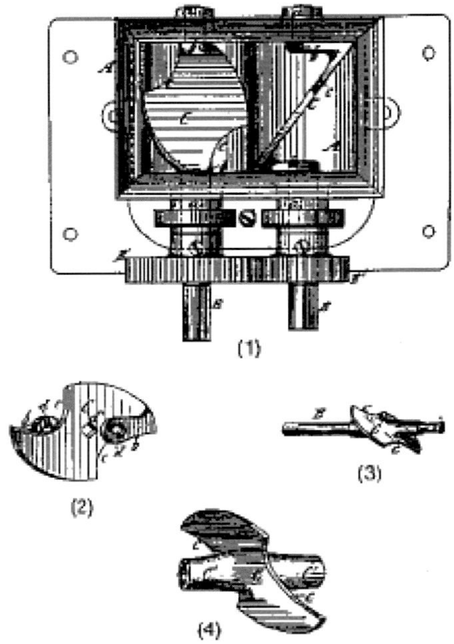

Figure 13.1: Pfleiderer's 1879 batch mixer rotors [5]

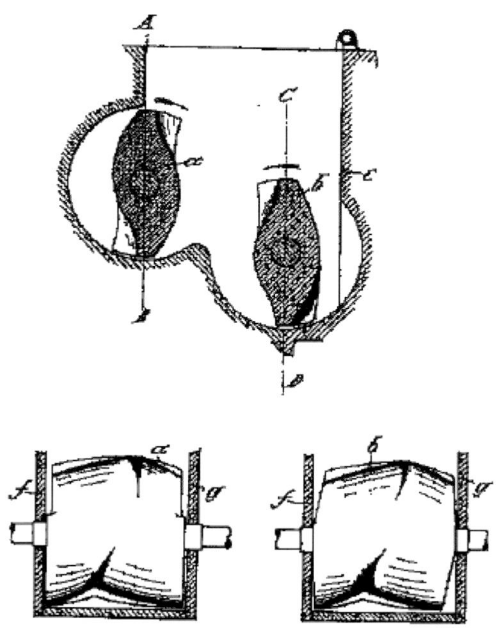

Figure 13.2: Kempter's 1913 batch mixer rotors [7]

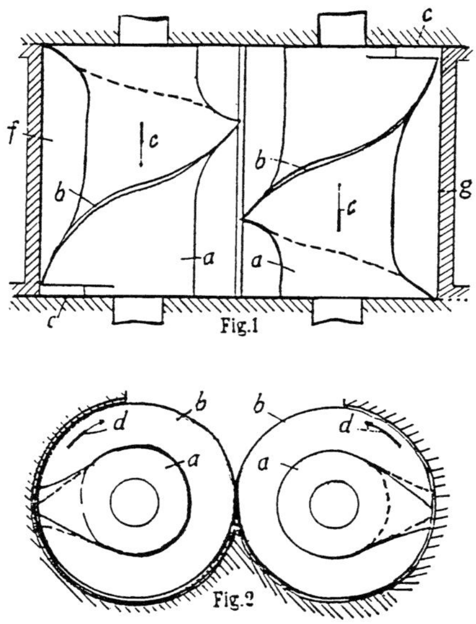

Figure 13.3: Pointon's 1915 batch mixer rotors [8]

A 1915 patent of John E. Pointon [8], a protégé of Paul Pfleiderer, who had become Chief Engineer of Werner, Pfleiderer, and Perkins described an internal mixer for rubber based on two similarly posed (right-handed) counter-rotating screws shown in Figure 13.3.

In 1913, Kurt Pfleiderer of Werner, Pfleiderer, and Perkins met and convinced Fernley H. Banbury of the importance of internal mixers for rubber compounding and induced him to join Werner and Pfleiderer in Saginaw, Michigan [9]. Between 1913 and 1916, F. H. Banbury as an employee of Werner and Pfleiderer in Saginaw, developed a modified design of a true internal mixer with a ram and with rotors with two sets of displaced winged flights on each rotor. In Banbury's machine, as in Pfleiderer's and Kempter's, the two rotors operated at different angular velocities. In a dispute with the management in Saginaw, Banbury resigned from Werner and Pfleiderer in possession of his invention, for which he personally obtained a patent [10]. A drawing of Banbury's invention from his patent is shown in Figure 13.4.

Banbury later joined forces with H. F. Wanning and F. D. Wanning and his internal mixer design [9] was developed by their firm, the Birmingham Iron Foundry of Derby, Connecticut [9]. Banbury's internal mixer was to dominate the compounding of rubber in succeeding years. Birmingham Iron Foundry introduced the trademark of the Banbury® Mixer.

In 1927, the Birmingham Iron Foundry was merged into the Farrel Foundry and Machine Company of Ansonia, Connecticut to form the Farrel-Birmingham Corp., which subsequently became as Farrel Corp. [9, 11]. The Banbury mixer would, in the 1960s, be the starting place for Farrel in their development of the Farrel Continuous Mixer (Section 13.8).

During World War I, Werner, Pfleiderer, and Perkins of Peterborough, England, changed its name to Perkins Engineers Ltd. under the pressure of anti-German feeling which ravaged in England and the Anglo-Saxon parts of their empire. The Pfleiderer family changed their name

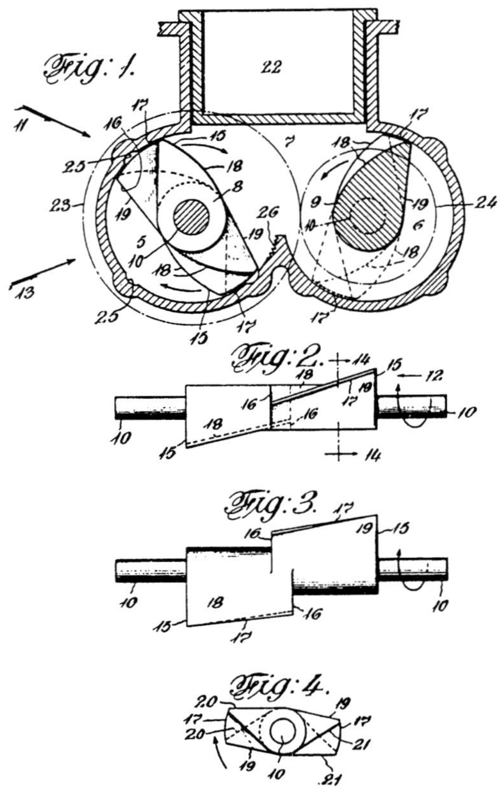

Figure 13.4: Banbury's 1916 internal mixer with a ram [10]

to Pelmore. In 1917, at the behest of the English government and the Pfleiderer-Pelmores, the Werners of Stuttgart renounced their stockholdings. Perkins Engineers was subsequently taken over by Joseph Baker and Sons Ltd. of England, becoming first Joseph Baker Sons and Perkins Ltd. and later Baker Perkins Ltd. [6]. The Baker family had the major control, and the first Chairman of the Board was Allan R. Baker, who held this position until 1942. Initially, both the Joseph Baker and Sons and Werner, Pfleiderer, and Perkins facilities were maintained. In 1933, these were consolidated in the old Werner, Pfleiderer, and Perkins Petersborough site.

During World War I, the Werner and Pfleiderer Saginaw facility was seized by the Alien Custodian Authority of the U.S. government. It was later sold to Baker Perkins Ltd., becoming Baker Perkins Inc. in the US [6].

In the 1920s, Baker Perkins Ltd. continued to market Werner and Pfleiderer machinery in England. In 1927, Baker Perkins obtained a major (but minority) financial interest in Werner and Pfleiderer [6] and remained a major stockholder until the 1980s. Werner and Pfleiderer and Baker Perkins would be competitors in manufacturing intermeshing co-rotating twin screw extruders a half century later (see Chapter 10). In 1987, Baker Perkins was merged into APV Chemical Machinery Ltd.

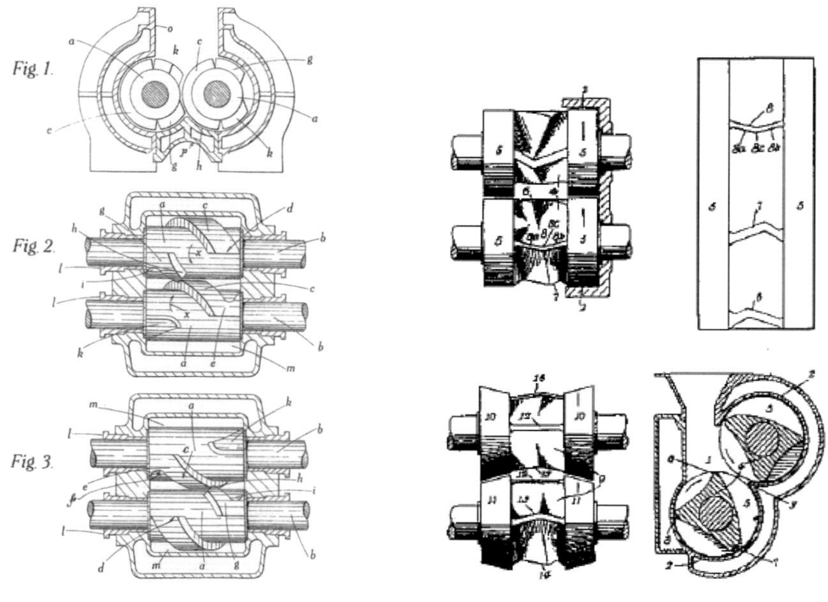

(a) Cooke (b) Lasch and Stromer

Figure 13.5: Intermeshing internal mixers (1934); a) Cooke [13]; b) Lasch and Stromer [14]

The Birmingham Iron Foundry/Farrel-Birmingham and Werner and Pfleiderer companies remained as competitive manufacturers of mixing machinery in the 1920s and 1930s. It was not until the late 1930s that Werner and Pfleiderer GmbH introduced Banbury's innovations in mixing chamber design. The David Bridge Company of Manchester, England was licensed to manufacture Banbury mixers. In the 1930s, Friederich Krupp of Essen, Germany became a licensee of Farrel-Birmingham and manufactured Banbury mixers in their Grusonwerk in Magdeburg (see Section 13.4) [12].

A major innovation made in the 1930s was the invention of internal mixers with intermeshing rotors by Rupert Cooke of Francis Shaw and Company of Manchester, England [13] and by Albert Lasch and Ernst Stromer [14] of Werner and Pfleiderer. These machines allowed additional dispersive mixing between the rotors (Figure 13.5). The Francis Shaw machine was commercialized in the 1930s and became successful because of its superior dispersive mixing capability.

In the post-war period, Farrel-Birmingham (later Farrel Corp.) was still competing with Werner and Pfleiderer. They sought to establish licensees in Japan and on the European continent replacing Krupp. Kobe Steel became a licensee of Farrel to produce internal mixers in 1965. The same year, the Italian machinery manufacturer Pomini became a licensee, and a new firm Pomini-Farrel, based in Castellanza, Italy was created. These relationships lasted until 1985. In the United States, the firm Stewart Bolling of Cleveland, Ohio began to manufacture internal mixers.

13.3 Early Continuous Mixer Designs

The earliest continuous mixer design seems to occur in a patent of Paul Pfleiderer. In 1881, Paul Pfleiderer [15] designed and patented a continuous tangential counter-rotating twin screw mixing machine. He proposed twin counter-rotating shafts which included both mixing rotors and screw elements (Figure 13.6). The material first moved vertically downward through twin pairs of mixing rotors, and was then picked up and pumped by twin screws to another set of mixing elements. A second set of twin screws pumped the material to the exit of the machine. Pfleiderer also envisioned a three-screw machine. The screw and mixing elements on the different screws were tangential. Much of the technology of the next 85 years is implicit in Pfleiderer's patent. This machine, however, was not commercialized.

Following Pfleiderer's [15] continuous mixing machine, the next development was a 1924 patent by H. Ahnhudt [16] of Charlottenburg (Berlin), Germany, for continuous mixing rubber and plastic materials. This involved shafts of two non-intermeshing counter-rotating screw elements which led to a pair of effectively intermeshing conical mixing rotors. One of these rotor shafts became a single screw that built pressure and pumped material out of the machine. The concept is shown in Figure 13.7.

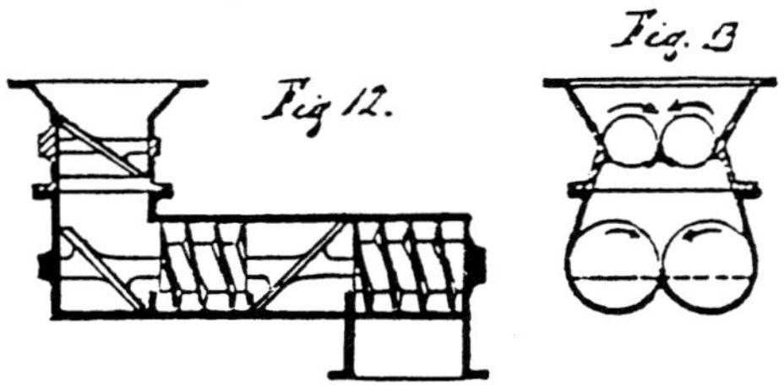

Figure 13.6: Pfleiderer's 1881 continuous mixer [15]

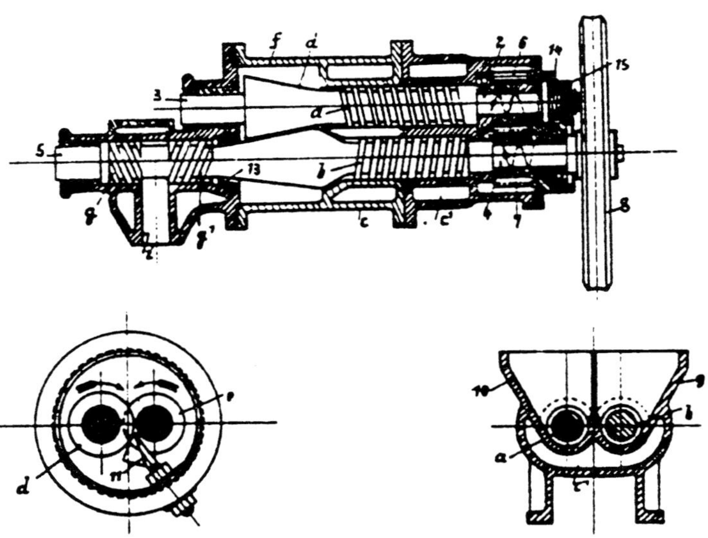

Figure 13.7: Ahnhudt's 1924 continuous mixer [16]

13.4 Krupp Knetwolf

In the 1930s, Farrel-Birmingham licensed German steel and machinery manufacturer Friederich Krupp [12] to manufacture and market Banbury design internal mixers in Europe. In his biography of F. H. Banbury, Killheffer [9] describes a visit of Banbury in the 1930s to the Krupp Grusonwerk in Magdeburg, Germany, where he met Wilhelm Ellermann. This was where Krupp manufactured Banbury mixers. Together they then visited Continental Gummiwerke AG in Hannover, and the Kuebler factory in Berlin. They inspected the first Krupp Banbury mixer with anti-friction bearings for the rotors.

About 1940, Wilhelm Ellermann of Krupp developed a new intermeshing counter-rotating twin rotor continuous mixer, perhaps at the suggestion of the I.G. Farbenindustrie. This was built by Krupp in their Grusonwerk and made commercially available in 1941 [17, 18]. A patent was filed in 1941 and issued in 1945 [18], but the inventor and assignee are not indicated on the patent[1]. Krupp obtained a license from the I.G. Farbenindustrie, because they considered to use the patent technology developed by Kiesskalt et al. described in Section 7.3. The Krupp machine was called the *Knetwolf* and was described as a 'Geschlossene Knetmaschine für organische plastische Massen'. It was the first commercially produced continuous mixer. The screws of the Krupp Knetwolf had diameters of 400 mm and the machine was used in the production of polyisobutylene. This application is explicitly discussed in the patent.

The Knetwolf contained intermeshing screws near the extruder feed port, which pumped material to mixing elements. These elements are longitudinal blades or flights. The rotors again become intermeshing screws as they approached the discharge port (Figure 13.8).

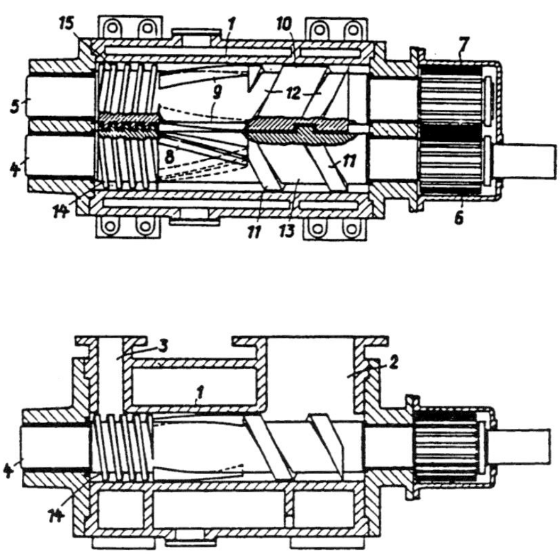

Figure 13.8: Ellermann's 1941 Krupp Knetwolf [18]

[1] This was common practice in Germany toward the end of World War II.

13.5 Eck Mixtruder

The 'Knetwolf' of Section 13.4, which was manufactured by Krupp, was discontinued at the end of World War II. Ellermann was able to move to Western Germany and worked with Josef Eck und Söhne Maschinenfabrik of Düsseldorf [3, 17, 19]. Ellermann worked with Eck and modified the design of this continuous mixing machine. In a patent application filed in 1951, Ellermann [19] described a design in which the two existing screws became tangential near the die rather than intermeshing. The mixing section has the form of a backward pumping screw. This is shown in Figure 13.9. This machine was called the 'Eck Mixtruder'.

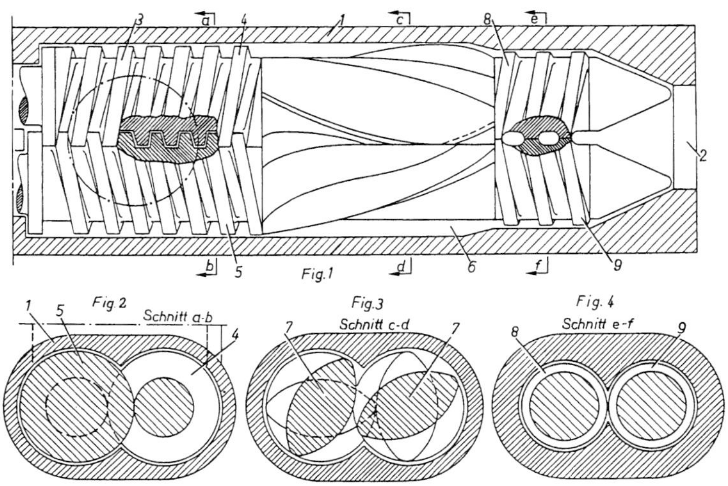

Figure 13.9: Ellermann's 1951 Eck Mixtruder [19]

13.6 Ellermann Krauss-Maffei DSM

In 1953, Ellermann [17, 20, 21] filed a patent for a self-wiping co-rotating intermeshing twin screw extruder with a similar intermediate kneading section. This extruder had two separate shafts with screw cuts in the barrel (Figure 13.10).

The patent was assigned to Joseph Eck und Söhne of Düsseldorf, who had built the Eck mixtruder. However, the technology was licensed and commercially developed by Krauss-Maffei of Munich. The machine was called the DSM (DSM meant Doppel Schnecken Mischer). The Krauss-Maffei machine was similar but contained threads cut in the screw rather than the housing (Figure 13.11). The Krauss-Maffei DSM machine is described in detail in a 1964 paper by Proksch [22]. The purpose of this machine, as with its predecessors, was continuous mixing and kneading.

Subsequently, this machine (DSM) was licensed to Japan Steel Works [23] in Hiroshima, who continue to manufacture it.

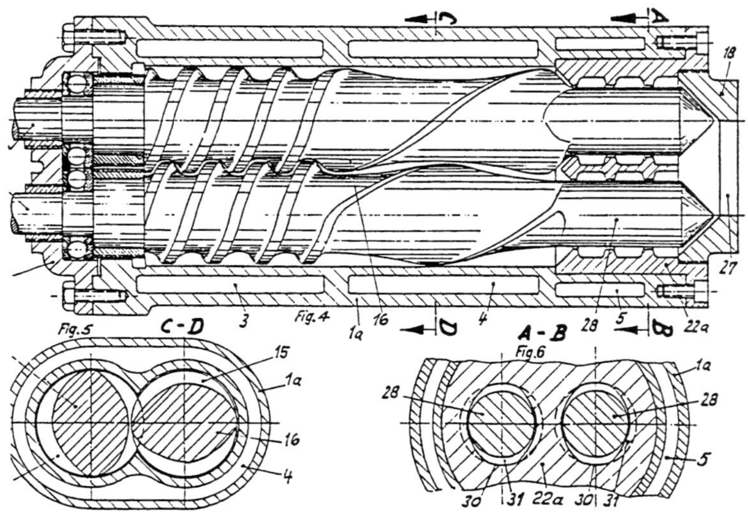

Figure 13.10: Ellermann's 1953 intermeshing co-rotating internal mixer [20, 21]

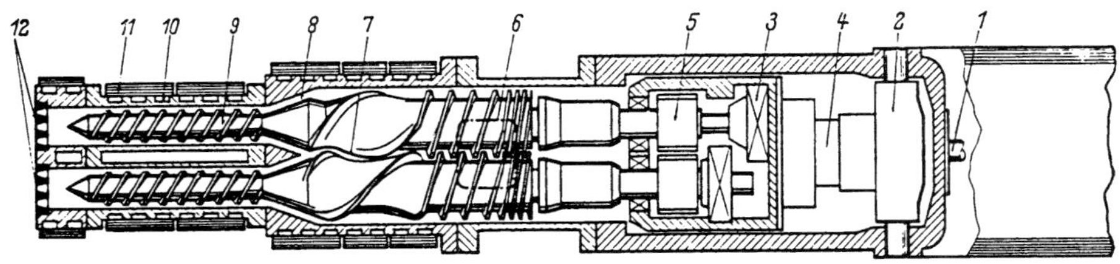

Figure 13.11: Krauss Maffei DSM [22]

13.7 Farrel Continuous Mixer

The development of the Banbury mixer and its association with the Birmingham Iron Foundry and later Farrel-Birmingham and Farrel, was described in Section 13.2. The 1960s saw the development of the Farrel Continuous Mixer by E. H. Ahlefeld, A. J. Baldwin, P. Hold, W. A. Rapetzki, and H. R. Scharer [24, 25] of the Farrel Corporation of Ansonia, Connecticut. The Farrel Continuous Mixer is in the tradition of the patents of Pfleiderer [15] and Ahnhudt [16]. However, the design of the mixing elements is different. They follow the tradition of the rotors designed by Banbury [10] for his internal mixer (long manufactured by Farrel (Section 13.2)). The Farrel Continuous Mixer (FCM) (Figure 13.12) involves the combination of screw elements with elements similar to Banbury's internal mixer rotors, which had long been used in the rubber industry. The mixing elements first involve flights of opposite hand, which induce further forward pumping. They are followed by sections of reversed pitch, which

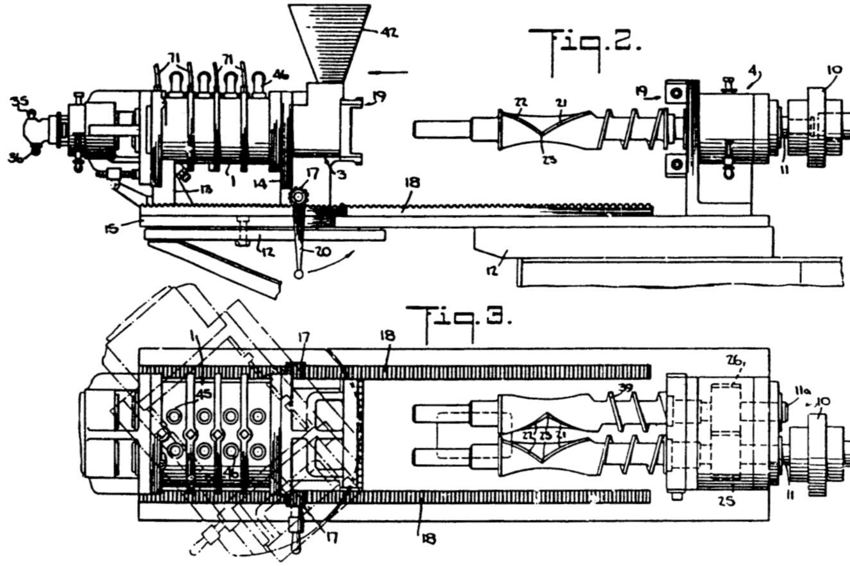

Figure 13.12: Farrel Continuous Mixer 1962 patent application [24]

induce backward pumping. It would be expected, as in the Banbury mixer, that there would be significant flow over the mixing element flight tips, which would induce dispersive mixing. This is in turn followed by the machine outlet. The screws and mixing sections of the FCM are machined on individual shafts. In early machines, the rotors had different rotation speeds as in internal mixers. More recently, both rotors have had the same velocity.

The original intent of the Farrel Continuous Mixer was for the continuous compounding of rubber and carbon black, with powdered rubber being fed to the machine inlet. However, there have been only limited applications for this type of machine. It has been more widely used in post-reactor processing and for compounding of thermoplastics.

The Farrel Continuous Mixer was later described in papers by Hold [26] and Scharer [27]. They discuss the operation of the machine, indicating that with metered flow and low discharge pressures, starved flow conditions result. The screw shafts rotate at different speeds and only modest pressure is developed. The polymer product exits through a sizeable 'discharge' orifice gate, usually into the hopper of a screw extruder.

There were various efforts to improve the design of the Farrel Continuous Mixer. These are found in patents of the 1960s and 1970s by Gagliardi [28] and Mosher, Treat, and Drab [29] of the same organization.

Farrel licensed Kobe Steel of Kobe, Japan, for several years to manufacture Farrel Continuous Mixers in Japan. Similarly, Farrel licensed Pomini (Pomini-Farrel) in Castellanza, Italy for some years for the same purpose. The licenses expired in 1985 but both Kobe Steel and Pomini continued to manufacture machines and to develop new designs. The efforts of Kobe Steel are described in Section 13.10.

13.8 Matsuoka and Bolling Mixtrumat

Patents filed by J. T. Matsuoka [30–35] of Intercole Automation (Stewart Bolling) of Cleveland, Ohio between 1968 and 1974 led to the development of a new continuous mixer. The first patent [30] deals with mixers containing two non-intermeshing rotors consisting of a screw type feeding portion and a mixing section (Figure 13.13). The mixing section of one of the rotors has two flights, each of which twists in opposite directions, with the first pumping in the same direction as the screw and the second pumping in the opposite direction. The second rotor has four flights, arranged in pairs, which also twist in opposite directions.

The later patents of Matsuoka et al. [33–35] deal with interfacing a single screw extruder with the mixing section as shown in Figure 13.14. The discharge from the twin screw mixer consists of a single screw extruder arranged in a transverse manner to a second screw section, which follows the mixing section. This machine was commercialized by Bolling as the 'Mixtrumat'.

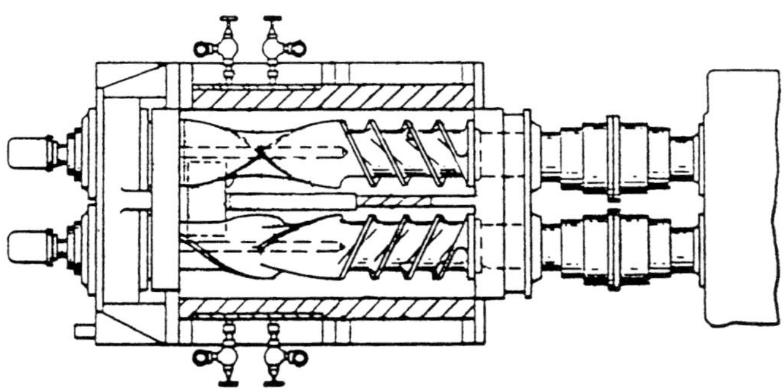

Figure 13.13: Matsuoka continuous mixer [30]

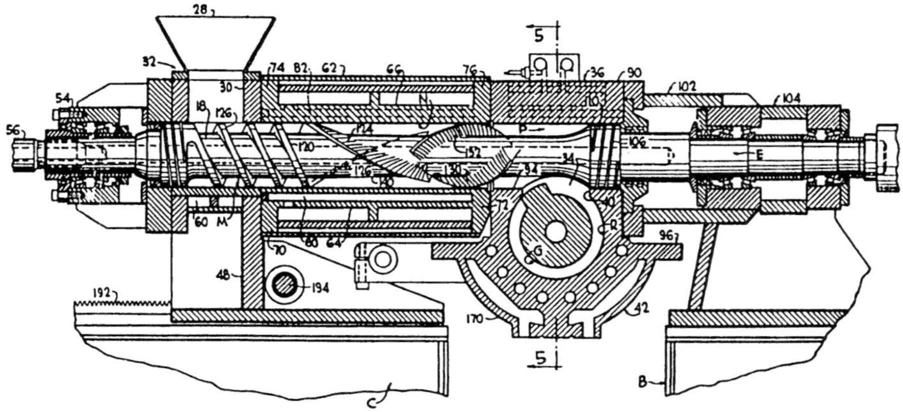

Figure 13.14: Bolling Mixtrumat [33]

13.9 Japan Steel Works Activities

Japan Steel Works, Ltd [23] based in Hiroshima, Japan has been involved in the development of twin screw extruders and continuous mixers since the 1960s. We noted in Section 13.6 that they licensed and manufactured the Krauss-Maffei DSM. A 1971 U.S. patent application of Okada, Taniguchi, and Kamimori [36] described a machine similar to the Farrel continuous mixer, but with a special barrel design and kneading region at the exit (Figure 13.15).

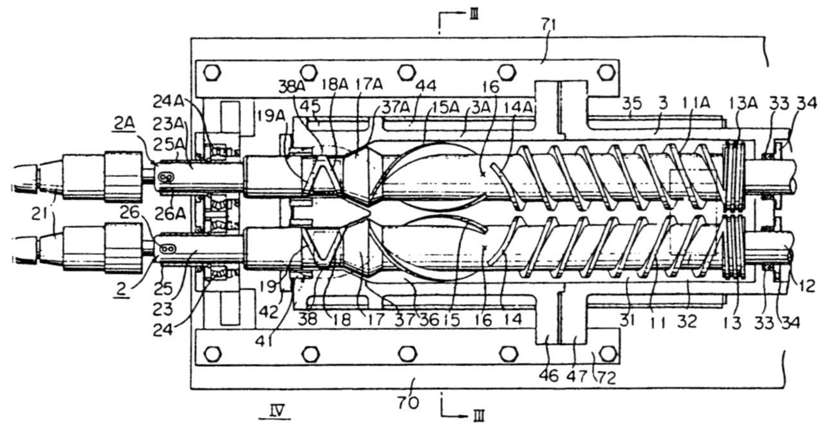

Figure 13.15: JSW continuous mixer by Okada et al. [36]

This type of machine was described in a 1988 paper by Sakai [37] of Japan Steel Works. It was used for the grafting of maleic anhydride/styrene monomer onto ethylene vinyl acetate copolymer.

13.10 Kobe Steel L-Series Continuous Mixer

Kobe Steel Ltd. [38] of Kobe, Japan, originally became involved in the manufacture of counter-rotating twin screw machines as a licensee of Farrel of Ansonia, Connecticut (see Section 13.7), for the Farrel continuous mixer (as well as the Farrel Banbury mixer). They have since become involved in the development of new machine designs. One of the new machines was the Kobe Steel L-Series continuous mixer developed by K. Inoue, K. Ogawa, T. Fukui, T. Asai, and S. Hashizume [39], shown in Figure 13.16. It consists of a machined shaft with three sections. First, there is a screw section, followed by a mixing element with forward and backward pumping flights. At the end of the shaft is a second set of mixing rotors with forward and backward pumping flights. Generally, in the screw section of modern Kobe Steel machines, the screws are intermeshing. Some commercial machines place additional screw elements between the mixing rotors. There is also a throttle between the mixing rotors.

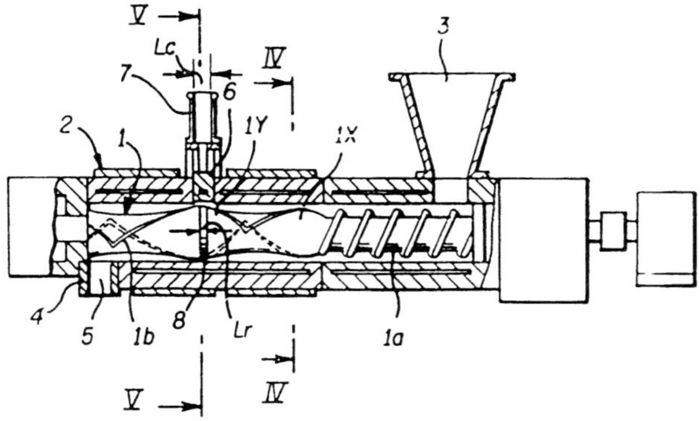

Figure 13.16: L-Series continuous mixer by Inoue et al. of Kobe Steel [39]

The compounded polymer exits the machine after the second mixing rotor. The throttle can be used to influence the residence time in the mixing chamber. A fairly similar machine was developed and marketed by Pomini-Farrel.

In 1989, Kobe Steel purchased Stewart Bolling (Sections 13.2 and 13.9) and formed the new company Kobelco-Stewart Bolling, located in Hudson (Akron), Ohio.

13.11 Subsequent Farrel Developments

Farrel has also sought continuous improvements of their continuous mixer. This has included the development of new rotor designs in which multiple screw and rotor sections occur. Typical Farrel designs are shown in Figure 13.17 [40].

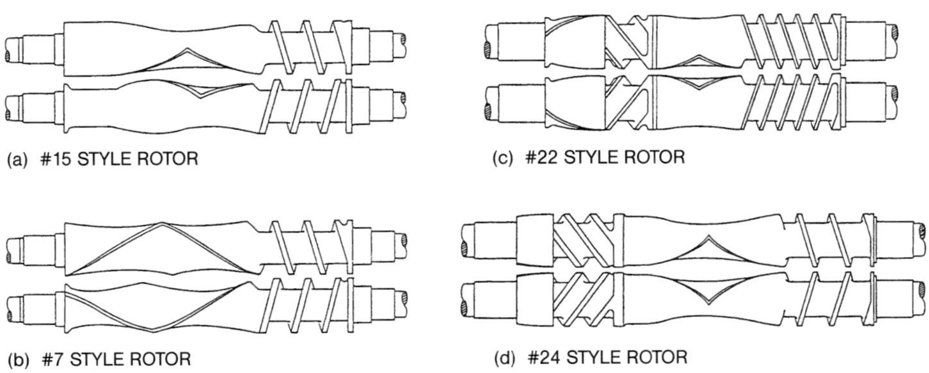

Figure 13.17: Farrel continuous mixer rotors [40];
a) No. 15 style rotor; b) No. 7 style rotor; c) No. 22 style rotor; d) No. 24 style rotor

Another innovation was the use of gear pumps to control the polymer exiting the machine [41, 42] (see Figure 13.18) discussed by Scharer [27]. The gear discharge allows the pressurization of the material in the mixing rotor portion of the machine. This allows for improved mixing. The patent of Valenzky and Markhardt [42] describes a control system in which the gear pump discharge rate is determined by the torque on the mixing rotors.

Farrel Corporation manufactures a combination of a Farrel Continuous Mixer (FCM) and a single screw extruder as the Compact Processor (CP) [43]. Several models are offered with capacities between 20 and 2000 kg/h.

Farrel markets a variety of continuous mixers with the FCM designation. As an example, the VMSD (Variable Mixer, Side Discharge) line combines a continuous mixer with a gear pump and optionally a pelletizer.

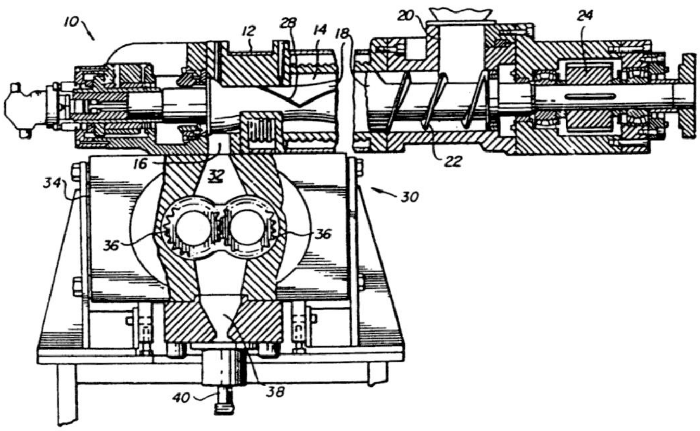

Figure 13.18: Gear pump discharge controlled FCM [41]

Figure 13.19: Modern Farrel Continuous Mixer (courtesy of Farrel Corporation)

Scharer [27] reported that machines with screw diameters from 50 mm to 375 mm are produced by Farrel with screw RPM on the larger sizes generally in the range of 300. Screw *L*/*D* based on screw diameters are generally less than 10.

Farrel reported various applications of their Continuous Mixer. These include its use in reactive extrusion including viscracking (visbreaking) of polypropylene with organic peroxides to lower molecular weight and narrow its molecular weight distribution [43]. Another application is for devolatilization. Valsamis and Canedo [43] describe the removal of volatiles from partially degraded viscracked polypropylene. Figure 13.19 shows a modern Farrel Continuous Mixer.

References

1. E. M. Chaffee, U.S. Patent 16 (1836).
2. P. Freyburger, German Patent 1,454 (1877).
3. H. Herrmann, in Kunststoffe-Ein Werkstoff Macht Karriere, edited by W. Glenz, Hanser, Munich (1985).
4. J. L. White, *Int. Polym. Process.*, **8**, 2 (1993).
5. P. Pfleiderer, German Patent (filed April 4, 1879) 10,164 (1880).
6. A. Muir, The History of Baker Perkins, Heffer, Cambridge (1968).
7. F. Kempter, German Patent (filed November 16, 1913) 279,649 (1914).
8. J. E. Pointon, U.S. Patent 1,138,410 (1915).
9. D. H. Killeffer, Banbury, The Master Mixer, Palmerton, NY (1962).
10. F. H. Banbury, U.S. Patent (filed January 15, 1916) 1,200,070 (1916).
11. J. L. White, *Int. Polym. Process.*, 7, 2 (1992).
12. J. L. White, *Int. Polym. Process.*, **9**, 198, 290 (1994).
13. R. T. Cooke, British Patent (filed June 14, 1934) 431,012 (1935).
14. A. Lasch and E. Stromer, German Patent (filed October 16, 1934) 641,685 (1937).
15. P. Pfleiderer, German Patent (filed June 10, 1881) 18,797 (1882).
16. H. Ahnhudt, German Patent (filed May 29, 1923) 397,961 (1924).
17. H. Herrmann, Schneckenmaschinen in der Verfahrenstechnik, Springer, Berlin (1972).
18. Anonymous (W. Ellermann), German Patent (filed January 31, 1941) 750,509 (1945).
19. W. Ellermann, German Patent (filed July 4, 1951) 879,164 (1953).
20. W. Ellermann, German Patent (filed July 31, 1953) 935,634 (1955).
21. W. Ellermann, U.S. Patent (filed October 12, 1953) 2,693,348 (1954).
22. W. Proksch, *Kunststoff und Gummi*, **3**, 426 (1964).
23. J. L. White, *Int. Polym. Process.*, 7, 194 (2002).
24. E. H. Ahlefeld, A. J. Baldwin, P. Hold, W. A. Rapetzki, and H. R. Scharer, U.S. Patent (filed May 15, 1962) 3,154,808 (1964).
25. E. H. Ahlefeld, A. J. Baldwin, P. Hold, W. A. Rapetzki, and H. R. Scharer, U.S. Patent (filed July 24, 1964) 3,239,878 (1966).

26. P. Hold, *Adv. Polym. Technol.*, **4**, 281 (1984).
27. H. R. Scharer, *Adv. Polym. Technol.*, **5**, 65 (1985).
28. G. R. Gagliardi, U.S. Patent (filed July 24, 1964) 3,237,241 (1966).
29. D. E. Mosher, C. H. Treat, and E. H. Drab, U.S. Patent (filed May 28, 1970) 3,704,866 (1972).
30. J. T. Matsuoka, U.S. Patent (filed September 5, 1968) 3,565,403 (1971).
31. J. T. Matsuoka and A. Cantarutti, U.S. Patent (filed July 19, 1969) 3,700,374 (1972).
32. J. T. Matsuoka and A. Cantarutti, U.S. Patent (filed January 4, 1971) 3,723,039 (1972).
33. J. T. Matsuoka, U.S. Patent (filed February 17, 1972) 3,764,118 (1973).
34. J. T. Matsuoka, U.S. Patent (filed May 7, 1973) 3,829,067 (1974).
35. J. T. Matsuoka and A. Cantarutti, U.S. Patent (filed September 12, 1974) 3,923,291 (1975).
36. T. Okada, K. Taniguchi, and K. Kamimori, U.S. Patent (filed October 12, 1971) 3,802,670 (1974).
37. T. Sakai, *SPE ANTEC Tech. Papers*, **34**, 1853 (1988).
38. J. L. White, *Int. Polym. Process.*, **8**, 190 (1993).
39. K. Inoue, K. Ogawa, T. Fukui, T. Asai, and S. Hashizume, U.S. Patent (filed March 27, (1980).
40. L. N. Valsamis and E. L. Canedo, *Int. Polym. Process.*, **4**, 247 (1989).
41. H. R. Scharer, D. A. D'Amato, P. Hold, and M. Hobner, U.S. Patent (filed November 5, 1976) 4,310,251 (1982).
42. D. Valenzky and G. T. Markhardt, U.S. Patent (filed November 22, 1985) 4,707,139 (1987).
43. E. L. Canedo and L. N. Valsamis, in Plastics Compounding: Equipment and Processing, edited by D. B. Todd, Hanser, Munich (1998).

14 Flow Mechanisms and Modeling in Continuous Mixers

14.1 Introduction

In this chapter, we will describe the flow modeling of continuous mixers. This is a subject that has received very little attention in the literature. All studies relate to the Farrel Continuous Mixer® and none to the machines of Sections 13.3 to 13.6. We first consider in Section 14.2 the earliest modeling by Valsamis and Canedo [1–5] of Farrel Corporation. In Section 14.3, we consider a subsequent model of M. H. Kim and White [7], which is based on Cartesian coordinate hydrodynamic lubrication theory and FAN calculations. The more realistic cylindrical coordinate model of Bang and White [8] is dicussed in Section 14.4.

14.2 Canedo-Valsamis Model

The first efforts of modeling flow in a continuous mixer were made by Valsamis and Canedo [1–3] between 1989 and 1991. These studies do not consider the fluid mechanics of pumping but rather emphasize residence times and mixing.

The modeling is essentially 2-dimensional and neglects the forward flow of the polymer compound along the machine axis in the continuous mixer. This effort was expanded upon in later papers by Canedo and Valsamis [4, 5] and by Tadmor and Gogos [6] in the second edition of their book. These authors basically consider a shearing flow motion into a wedge, where one of the surfaces in motion represents the mixing of the machine. The relative frequency of the movement of elements of the fluid through the gap between the rotor tip and the barrel is considered.

14.3 Cartesian Coordinate Lubrication Theory Model of Flow in a Farrel Continuous Mixer

In 1992, M. H. Kim and White [7] published a study using the Cartesian coordinate lubrication theory of Eqs. (5.11) and (5.12) to analyze flow in a fully filled machine resembling a Farrel Continuous Mixer®. The analyses are similar to those of the same authors for tangential counter-rotating twin screw extruders, which we described in Chapter 11 (Section 11.4).

They presume the machine is locally fully filled with polymer compound that is metered into the machine.

Hydrodynamic lubrication theory was used to compute screw characteristic curves for the various sections of the machine. The machine was considered to consist (in the direction from the hopper) of forward pumping screws, a (more poorly) forward pumping rotor section and a backward pumping rotor section. Pressure profiles were computed along the axis of the continuous mixer.

14.4 Cylindrical Coordinate Lubrication Theory Model of Flow in a Farrel Continuous Mixer

In a 1997 paper, Bang and White [8] modeled flow in this same type of machine, using cylindrical coordinate lubrication theory [9] described by Eqs. (11.30) to (11.33). This is the most realistic and sophisticated model to date. The rotors analyzed in this paper are given in Figure 14.1. It involves a forward pumping screw and forward pumping, backward pumping and neutral rotor sections. The cross-sectional mesh considered in this model is shown in Figure 14.2. The problem of starvation is considered.

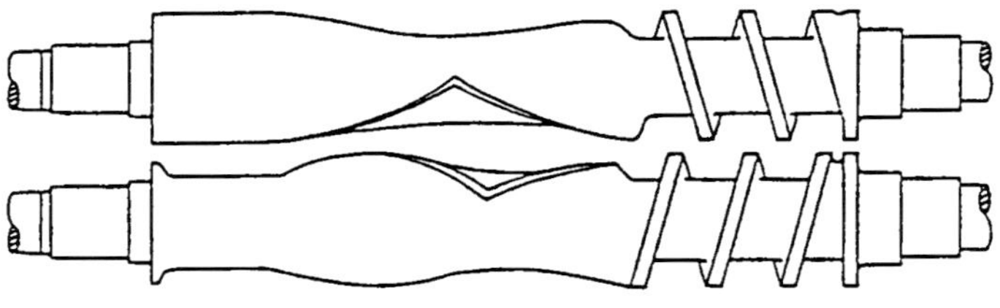

Figure 14.1: Rotors of machine described by Bang and White [8]

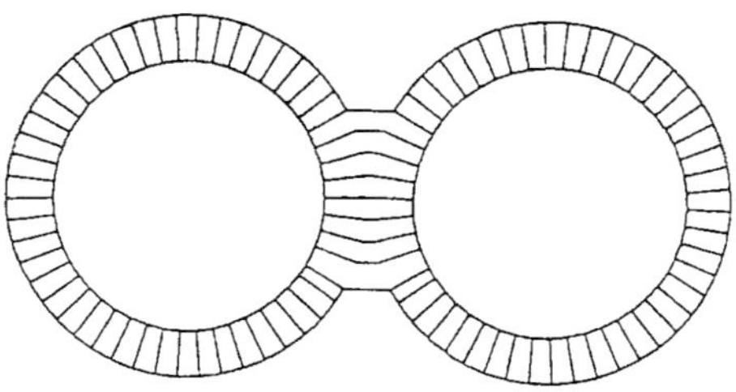

Figure 14.2: Cross-sectional mesh used by Bang and White [8] to model a continuous mixer

The basic flow and equations of motion in the fully filled melt region are:

$$\underline{v} = v_z(r)\,\underline{e}_z + 0\,\underline{e}_r + v_\theta(r)\,\underline{e}_\theta \tag{14.1}$$

$$\begin{aligned} 0 &= -\frac{\partial p}{\partial z} + \frac{1}{r}\frac{\partial}{\partial r}(r\,\sigma_{zr}) \\ 0 &= -\frac{1}{r}\frac{\partial p}{\partial \theta} + \frac{1}{r^2}\frac{\partial}{\partial r}(r^2\,\sigma_{\theta r}) \end{aligned} \tag{14.2}$$

The shear stresses σ_{zr} and $\sigma_{\theta r}$ are:

$$\sigma_{zr} = \eta\,\frac{\partial v_z}{\partial r} \qquad \sigma_{\theta r} = \eta\, r\,\frac{\partial}{\partial r}\left(\frac{v_\theta}{r}\right) \tag{14.3}$$

The shear viscosity η was represented by the power law:

$$\eta = K\left\{\left(\frac{\partial v_z}{\partial r}\right)^2 + \left[r\,\frac{\partial}{\partial r}\left(\frac{v_\theta}{r}\right)\right]^2\right\}^{\frac{n-1}{2}} \tag{14.4}$$

Solving these equations leads to the expressions for the shear stresses:

$$\begin{aligned} \sigma_{zr} &= \frac{r}{2}\frac{\partial p}{\partial z} + \frac{C_z}{r} = \eta\,\frac{\partial v_z}{\partial r} \\ \sigma_{\theta r} &= \frac{1}{2}\frac{\partial p}{\partial \theta} + \frac{C_\theta}{r^2} = \eta\, r\,\frac{\partial (v_\theta / r)}{\partial r} \end{aligned} \tag{14.5 a, b}$$

and for the velocity field:

$$\begin{aligned} v_z(r) &= \frac{1}{K^{1/n}}\int_R^r \left[\left(\frac{r}{2}\frac{\partial p}{\partial z} + \frac{C_z}{r}\right)^2 + \left(\frac{1}{2}\frac{\partial p}{\partial \theta} + \frac{C_\theta}{r^2}\right)^2\right]^{\frac{1-n}{2n}} \left(\frac{r}{2}\frac{\partial p}{\partial z} + \frac{C_z}{r}\right) \mathrm{d}r \\ v_\theta(r) &= \frac{r}{K^{1/n}}\int_R^r \left[\left(\frac{r}{2}\frac{\partial p}{\partial z} + \frac{C_z}{r}\right)^2 + \left(\frac{1}{2}\frac{\partial p}{\partial \theta} + \frac{C_\theta}{r^2}\right)^2\right]^{\frac{1-n}{2n}} \left(\frac{1}{2}\frac{\partial p}{\partial \theta} + \frac{C_\theta}{r}\right) \frac{\mathrm{d}r}{r} \end{aligned} \tag{14.6 a, b}$$

Again, as in Section 11.4, we define fluxes:

$$q_z = \frac{1}{R_{ch}}\int_{R_i}^{R_b} r\, v_z\, \mathrm{d}r \tag{14.7a}$$

$$q_\theta = \int_{R_i}^{R_b} v_\theta\, \mathrm{d}r \tag{14.7b}$$

In the flow analysis network (FAN) method, the fluxes are balanced over the various $\Delta z \, \Delta\theta$ elements.

The machine is considered to consist of two separate regions:

(I) between the screw/rotor and the machine barrel

(II) between the rotors.

Flux balances must be established for both regions. The flux balances are similar to those used by the same authors [9] (see Chapter 11) to analyze flow in tangential counter-rotating twin screw extruders. The detailed dimensions and clearances considered in the calculations are given in Table 14.1. Flux patterns on the machine rotors are shown in Figure 14.3.

Table 14.1: Detailed Geometry of the Continuous Mixer Used in Calculations

Barrel diameter	101.6 mm
Center-to-center distance	107.95 mm
Total rotor length diameter ratio	5
Screw section	
Length	152.4 mm
Root diameter	63.5 mm
Pitch	57.15 mm
Helix angle	10.70°
Rotor section	
Total length	355.6 mm
Clearance	
Maximum	20.32 mm
Minimum	2.54 mm
Forward pumping	
Length	101.6 mm
Helix angle	30°
Backward pumping	
Length	152.4 mm
Helix angle	–30°
Neutral	
Length	101.6 mm
Helix angle	0°

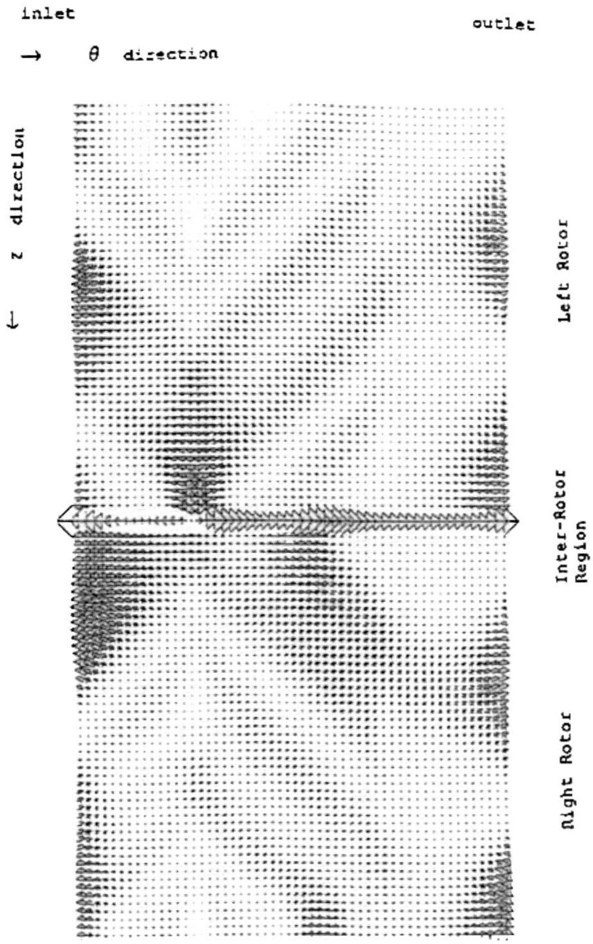

Figure 14.3: Computed flux patterns on rotors by Bang and White [8]

Screw characteristic curves for the various sections of the machine rotor and screws were computed. These are shown in Figure 14.4 a, b, c, d. The curves indicate the forward pumping character of the screws and the first part of the rotor shown in Figure 14.4 a, b. The final two sections of the rotor tend to pump backward as indicated in Figure 14.4 c, d.

For a starved machine, which is the usual case, the machine is fully filled at the exit gate, but not throughout its length. The FAN analysis described above is only valid for fully filled regions.

Bang and White [8] analyzed flow in both fully filled and starved machines. They used computed screw characteristics curves for the fully filled sections of the continuous mixer. They were able to compute pressure profiles along the machine axis and the position where the pressure falls to atmospheric as shown in Figure 14.5.

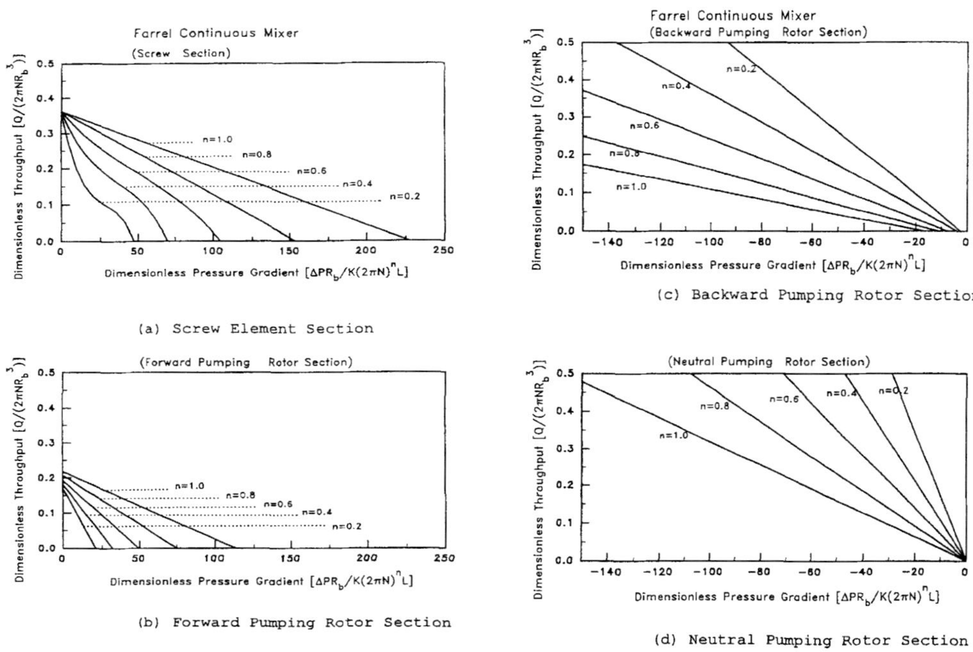

Figure 14.4: Calculated screw characteristic curves for sections of the continuous mixer [8]; a) screw element section; b) forward pumping rotor section; c) backward pumping rotor section; d) neutral rotor section

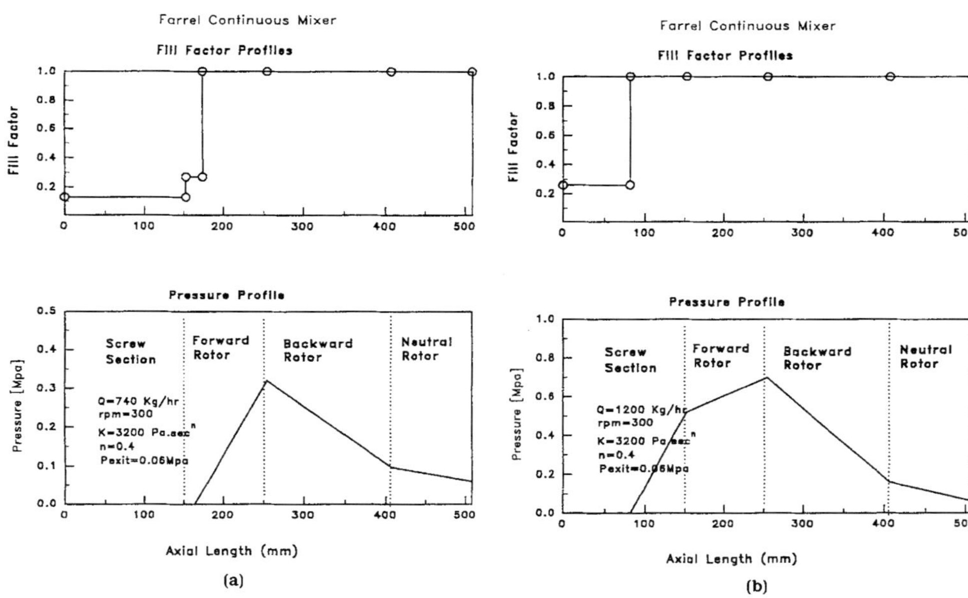

Figure 14.5: Pressure profiles calculated for a starved flow Farrel Continuous Mixer® by Bang and White [8] for conditions described in figures

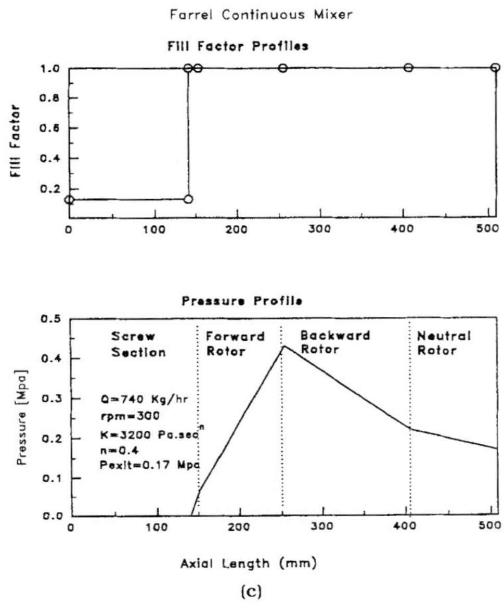

Figure 14.5: (continued)

References

1. L. N. Valsamis and E. L. Canedo, *Int. Polym. Process.*, **4**, 247 (1989).
2. E. L. Canedo and L. N. Valsamis, *SPE ANTEC Tech. Papers*, **35**, 116 (1989).
3. E. L. Canedo and L. N. Valsamis, *SPE ANTEC Tech. Papers*, **37**, 141 (1991).
4. E. L. Canedo and L. N. Valsamis, in Mixing and Compounding Theory and Practice, edited by I. Manas-Zloczower and Z. Tadmor, Hanser, Munich (1994).
5. E. L. Canedo and L. N. Valsamis, in Plastics Compounding, edited by D. B. Todd, Hanser, Munich (1998).
6. Z. Tadmor and C. G. Gogos, Principles of Polymer Processing, 2nd ed., Wiley-Interscience, N.Y. (2006).
7. M. H. Kim and J. L. White, *Int. Polym. Process.*, **7**, 15 (1992).
8. D. S. Bang and J. L. White, *Polym. Eng. Sci.*, **37**, 1210 (1997).
9. D. S. Bang and J. L. White, *Int. Polym. Process.*, **12**, 278 (1997).

15 Experimental Studies of Continuous Mixers

15.1 Introduction

There have been very few experimental investigations of the characteristics of continuous mixers. None of these involves the early machines of Ellermann [1–3], including the DSM [4] (see Sections 13.4 to 13.7). All of the studies concern the Farrel Continuous Mixer® [5] and Farrel's former licensee Kobe Steel's Nex-T® mixer (Section 13.7).

The investigations described involve two research groups:

- Hold, Canedo and Valsamis of Farrel Corporation [6–9], and
- White, Galle, Liu and Bumm at the University of Akron [10–12].

In Section 15.2, we will describe studies of the distribution of a polymer and its state within machines. We turn in Section 15.3 to the consideration of residence time distributions. Dispersive mixing in continuous mixers is considered in Section 15.4 and blending polymer melts in Section 15.5

15.2 Distribution of Polymer in Continuous Mixers

The distribution of polymer in a Farrel Continuous Mixer® was reported by Hold [6] and by Valsamis and Canedo [7] in 1984/1989. They report starved flow of pellets in the pair of non-intermeshing counter-rotating screws below the hopper. The machine is fully filled with polymer in front of the orifice gate and on the rotors, with melting primarily occurring on the forward pumping section of the rotors (sees Figure 15.1).

The distribution of polymer in a Kobe Steel (Kobelco) Nex-T® continuous mixer (Figure 15.2 a) was investigated by Galle and White [10, 11] about a decade later. The results are similar to those found for the Farrel Continuous Mixer®. Starved flow of pellets occurs in the slightly intermeshing screws near the hopper. The region in front of the orifice gate rotor section is largely filled with polymer melt, with melting occurring in the initial forward pumping rotor section (see Figure 15.2 b).

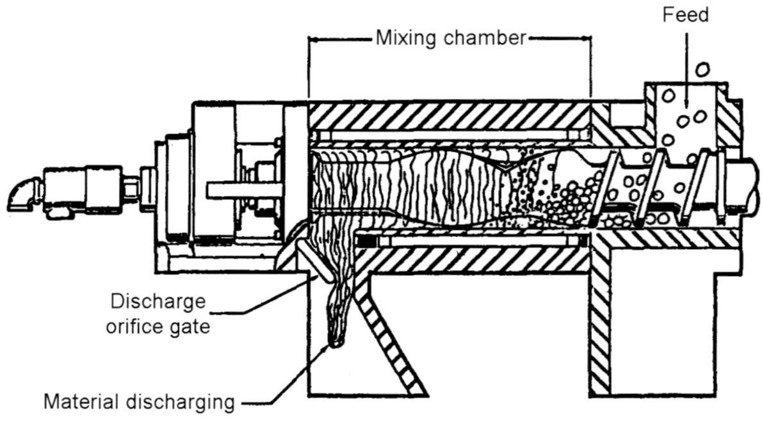

Figure 15.1: Operation of a Farrel Continuous Mixer® [6] (reprinted with permission of John Wiley and Sons)

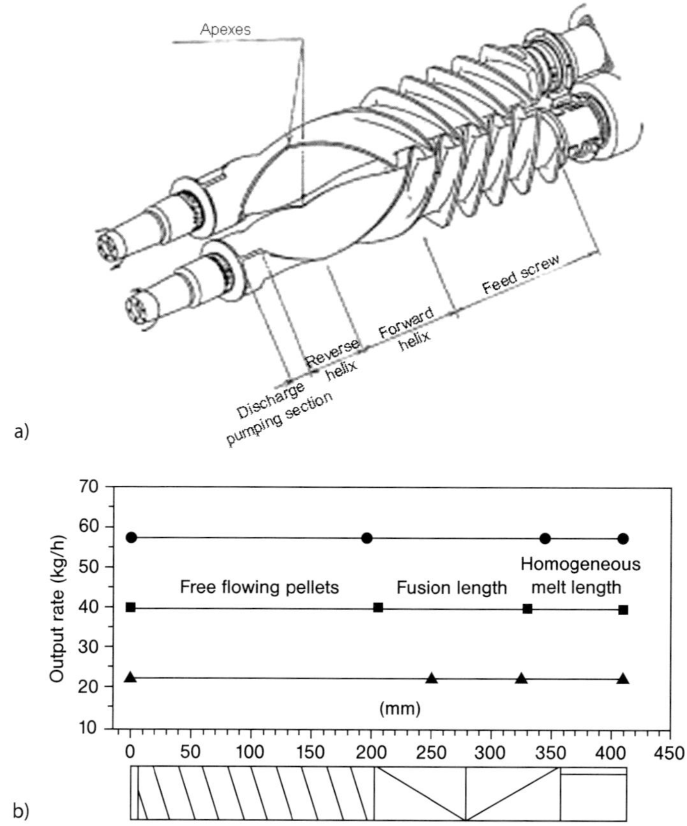

Figure 15.2: Solids conveying and melting in a Kobelco Nex-T® [11]; a) rotor design; b) melting process

15.3 Residence Time Distribution

Valsamis and Canedo [8] reported residence time distributions for a Farrel Continuous Mixer®. They made comparisons to published results for a Werner and Pfleiderer modular co-rotating twin screw extruder and found the residence time distributions to be quite similar.

Subsequently, Galle and White [10, 11] reported on residence time distributions in the Nex-T® continuous mixer. They described the influence of feed rate and rotor RPM on the differential residence time distributions. This is shown in Figures 15.3 and 15.4. Increasing the throughput through the machine moves the residence times to shorter times and narrows the distribution making the behavior approach plug flow. Increasing the rotor speed has a similar but less striking effect.

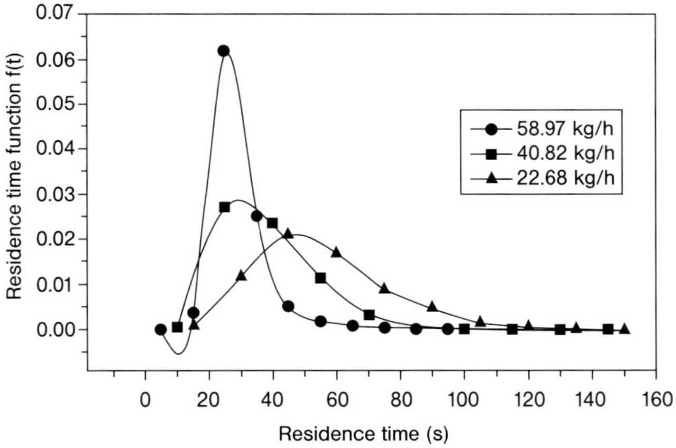

Figure 15.3: Differential residence time distribution of Kobelco Nex-T® continuous mixer at different throughput rates [11]

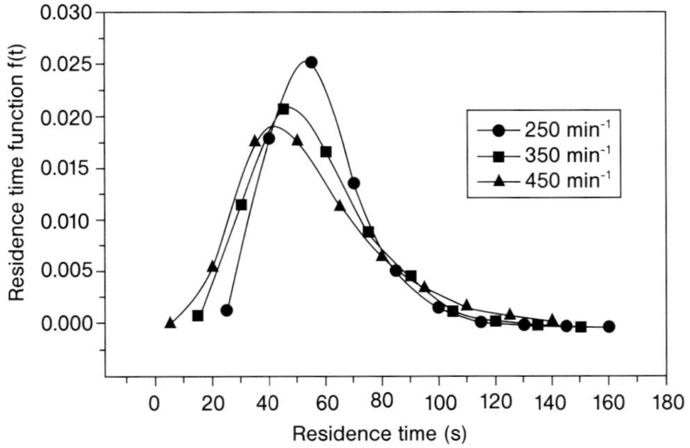

Figure 15.4: Differential residence time distribution of Kobelco Nex-T® continuous mixer at different rotor speeds [11]

15.4 Dispersive Mixing in Continuous Mixers

The rate of dispersive mixing in a Nex-T® continuous mixer, as opposed to a Brabender Plasti-Corder® fitted with tangential four flighted rotors [13] to resemble an industrial internal mixer, was studied by White et al. [12]. Comparisons were made for carbon black, silica, and calcium carbonate mixed into various elastomers, including ethylene propylene terpolymer, acrylonitrile-butadiene copolymer, and butadiene-styrene copolymer. Ten and twenty volume percent compounds were prepared. Agglomerate sizes were determined by scanning electron microscopy. The final agglomerate sizes achieved for the two different types of mixers were similar. When comparing mixing times (i.e., batch mixer mixing time vs. continuous mixer residence time) to achieve particular agglomerate sizes, they found that shorter times were achieved in a continuous mixer.

The results described in the above paragraph were interpreted in terms of the semi-empirical equation (compare Section 6.5)

$$\frac{d\bar{d}}{dt} = -k\,\bar{d} + k'\left(\frac{1}{\bar{d}^2}\right) \tag{15.1}$$

where $\bar{d}$ is the mean agglomerate size, k is a breakdown coefficient and k' a reagglomeration coefficient. The value of k for the continuous mixers was three to four times larger than the one for the batch internal mixer.

This behavior was associated with the ability of polymers to escape the nip between the moving rotor and the mixing chamber wall in a starved batch mixer. In a continuous mixer, all the polymer and filler must pass through the fully filled region in front of the orifice exit and mixing rotor.

15.5 Blending Different Polymers

The blending of isotactic polypropylene and polystyrene in a 75/25 weight ratio in a Nex-T® continuous mixer was investigated by Galle and White [11]. They reported that the size distribution of the dispersed polystyrene phase was a function of axial position in the machine of process variables. Experiments were done with different feed rates and rotor speeds. It was found that dispersed phase size reduction occurred at positions along the rotor axis. The dispersed phase dimensions were smaller at low feed rates and at high rotor rotation speeds (see Figures 15.5 and 15.6).

In these experiments, the continuous mixer was fed into a short L/D single screw extruder where the molten compounds were pumped toward a die exit. This extruder seemed to reduce the difference in dispersed phase sizes from the various continuous mixer operating conditions to similar values.

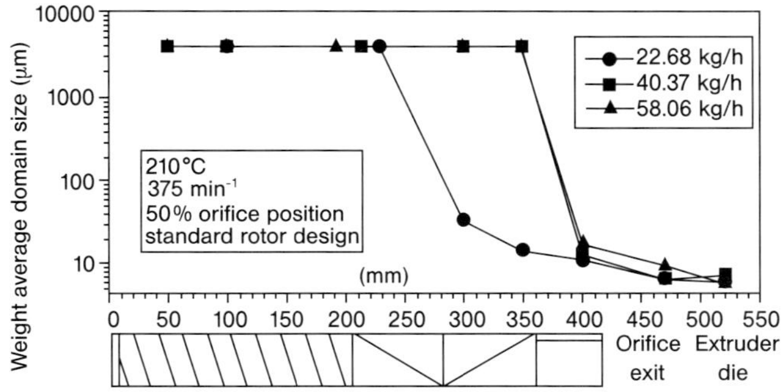

Figure 15.5: Phase size reduction of disperse phase size of polystyrene in isotactic polypropylene-polystyrene 75/25 blends in a Kobelco Nex-T® continuous mixer as a function of throughput rate [11]

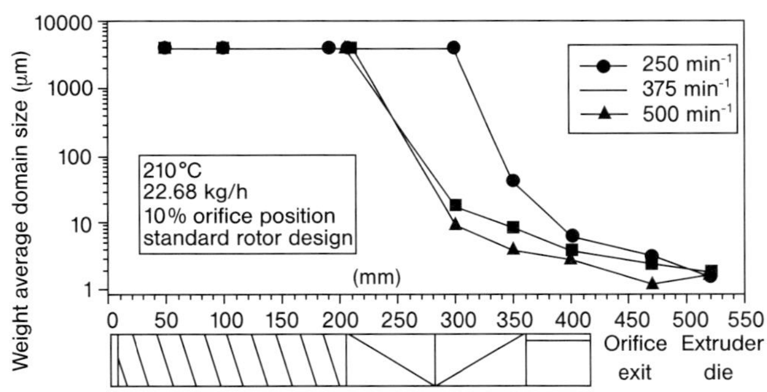

Figure 15.6: Phase size reduction of disperse phase size of polystyrene in isotactic polypropylene-polystyrene 75/25 blends in a Kobelco Nex-T® continuous mixer as a function of rotor speed [11]

References

1. Anonymous (W. Ellermann), German Patent (filed January 31, 1941) 750,509 (1945).
2. W. Ellermann, German Patent (filed July 4, 1951) 879,164 (1953).
3. W. Ellermann, German Patent (filed July 31, 1953) 935,634 (1953).
4. W. Proksch, *Kunststoffe und Gummi*, **3**, 426 (1964).
5. E. H. Ahlefeld, A. J. Baldwin, P. Hold, W. A. Rapetzki and H. R. Scharer, U.S. Patent (filed May 15, 1962) 3,154,908 (1964); U.S. Patent (filed July 24, 1964) 3,239,878 (1966).
6. P. Hold, *Adv. Polym. Technol.*, **4**, 281 (1984).
7. E. L. Canedo and L. N. Valsamis, *SPE ANTEC Tech Papers*, **35**, 116 (1989).
8. L. N. Valsamis and E. L. Canedo, *Int. Polym. Process.*, **4**, 247 91989).
9. E. L. Canedo and L. N. Valsamis in Mixing and Compounding of Polymers, edited by I. Manas-Zloczower and Z. Tadmor, Progress in Polymer Processing Series No.4, Hanser, Munich (1995).
10. F. M. Galle and J. L. White, *SPE ANTEC Tech Papers*, **43**, 271 (1997).
11. F. M. Galle and J. L. White, *Int. Polym. Process.*, **14**, 241 (1999).
12. J. L. White, D. Liu and S. H. Bumm, *Appl. Polym. Sci.*, **102**, 3940 (2006).
13. J. W. Cho, P. S. Kim, L. Pomini and J. L. White, *Kautschuk Gummi Kunstst*, **50**, 496 (1997).

16 Kokneter

16.1 Introduction

The reciprocating single screw compounder, more commonly known as the Kokneter, Ko-Kneader or Kneader, is a widely used continuous compounding machine that competes with the twin screw extruder in many applications. This is essentially the case for mixing and compounding in the polymer and food industries. It seems useful to include a chapter in this book, describing this machine and its characteristics.

In Section 16.2, we will describe the development of the Kokneter. The basic design of the machine was due to Heinz List [1–5] in the mid 1940s. In Section 16.3, we describe the modern Kokneter. The basic flow mechanics and flow simulations of the modern Kokneter are considered in Section 16.4. The modeling of a composite modular Kokneter is treated in Section 16.5. We describe experimental studies of this machine in Section 16.6.

16.2 Technological Development

The original "Kokneader" brought together several threads in the engineering literature:

- screw extruder with pins in their barrels
- screw extruders with reciprocating screws and
- screw extruders with self-wiping action.

The idea of placing pins into the barrel of a screw extruder is an old idea dating at least to the turn of the century. A 1901 German patent by Casimir Wurster [6] described a pin barrel "Fleischschneidemaschine". This machine was intended for a butcher shop, where the pins were used to knead meat moving through a screw channel towards a die, which would produce 'Hackfleisch' (or ground beef). A 1930 U.S. patent application by F. B. Anderson [7] of the V. D. Anderson Company of Cleveland, Ohio for a sausage stuffing machine had a section containing cap screws entering from the barrel. In both the Wurster [6] and the Anderson [7] patents, the screw flights contain slices through which the pins pass.

Screw processing machines with reciprocating screws were first introduced for injection molding machines in the early 1940s [8, 9]. The primary inventions seems to be by H. Beck of the I.G. Farbenindustrie in Ludwigshafen (before 1925 and later BASF).

The 1940s was a period during which there was great concern regarding the development of self-wiping machinery in the polymer processing industry. This concern was first described in patents of Wunsche [10] and Easton [11, 12] earlier in the century, in intermeshing counter-

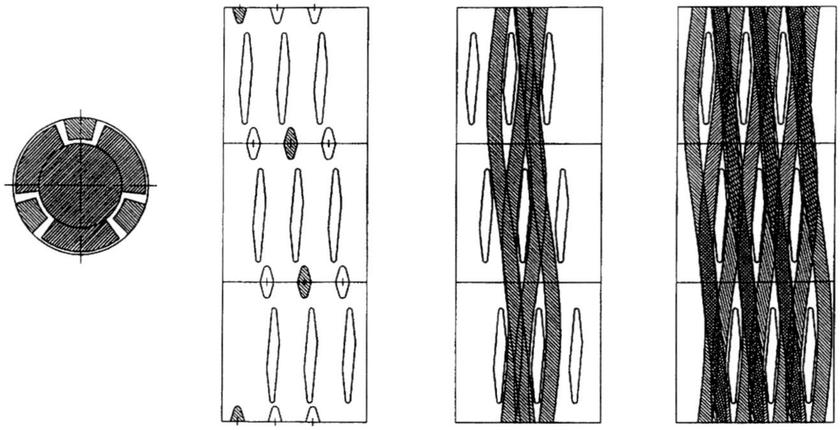

Figure 16.1: Motions of pins through flights in Kokneter

rotating twin screw machines of the 1930s [13, 14], and the intermeshing co-rotating twin screw machines of the late 1930s and early 1940s [15, 16]. Heinz List, who had a degree in mechanical engineering from the University of Munich, was fully aware and agreed with these directions. He was working in the I.G. Farbenindustrie's Leverkusen (now Bayer AG) facilities. He was frustrated by the management's crude attitudes towards mixing technology. As the war came to a close, he resigned from the I.G. Farbenindustrie and migrated down the Rhine to Switzerland. Here he drew up a patent application, describing his ideas about new self-wiping continuous mixing machines and filed it in Switzerland on August 20, 1945 [1]. He later filed patent applications around the world including the United States [2] and Germany [3].

List's patent application describes various designs of self-wiping mixing machines, one of which is the Kokneter. It contained a single screw with slices in its flights that rotated and oscillated. The barrel contained large, specially shaped pins. The machine operated in such a way that as the screw rotated and oscillated, the pins would pass through the pin slices in a self-wiping manner. This is shown in Figure 16.1. List licensed his Kokneter to Buss AG of Pratteln, Switzerland, except for North America, where he licensed Baker Perkins. List himself was employed by Buss AG.

Improvements in the design of the Kokneter were devised by Buss engineers in succeeding years. In a 1961 patent application, Gubler [17] described a modified Kokneter that is divided into a plasticizing section and a kneading/mixing section. The former section had deep pins in the barrel and the latter had set back pins to reduce the level of mastication (see Figure 16.2).

In a 1962 German and a 1963 U.S. patent application [18], Sutter of Buss AG described a modular Kokneter design in which screw-like elements are fitted together to form a modular screw. The screw was combined with a barrel containing pins. This is shown in Figure 16.3.

In a 1965 Swiss and a 1966 U.S. patent applications [19], Gresch of Buss AG describes a machine of this type with barriers present in the barrel and vent ports. The barriers induce vent pressure drops and the development of starved regions under vent ports, where an applied vacuum induces devolatlization (Figure 16.4).

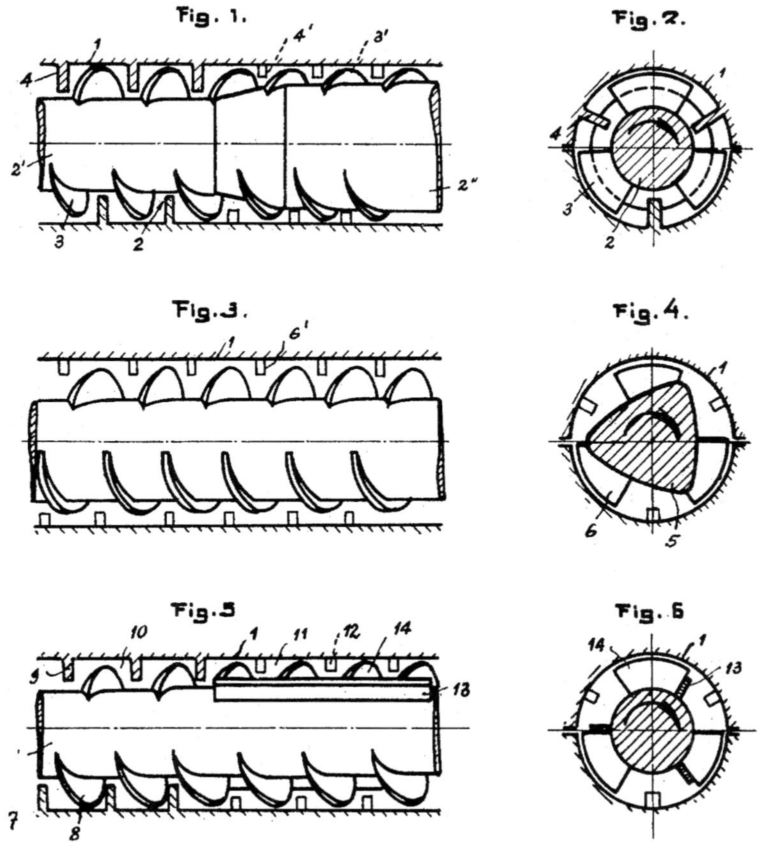

Figure 16.2: Gubler's 1961 patent drawing showing a Kokneter with mixing and kneading sections with pins and plasticizing section with setback pins [17]

A basic problem with the Kokneter is that it provides an oscillating output. Buss and Baker Perkins engineers gave considerable attention to overcoming this problem. This is seen in a 1959 patent application [20] of Geier and Irving of Baker Perkins, which describes an elevated section following the section containing pins and partial flights, which is in turn followed by a right-handed screw and a left-handed screw. The intention was to damp out the effects of oscillation. A 1962 Netherlands and a 1963 U.S. patent application by Schuur [21] of Shell Oil, describe a machine with an oscillating screw with pins and two entry ports that has right-handed screw elements under one hopper and left-handed associated with the other. This problem also led to 1963 and 1965 patent applications of List [22] and List and Ronner [23] of Buss AG, who describe a machine in which the screw rotates and the barrel is divided into two sections; the section closest to the hopper has pins and oscillates producing self-wiping character. The material to be mixed flows first through the oscillating barrel, where mixing occurs and then through a barrel section into a stationary barrel, where flow oscillations are damped out.

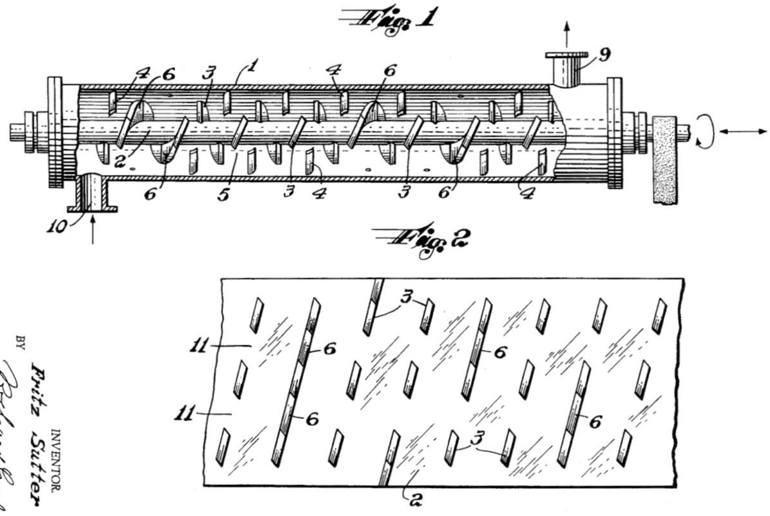

Figure 16.3: Sutter's 1963 modular Kokneter patent drawing [18]

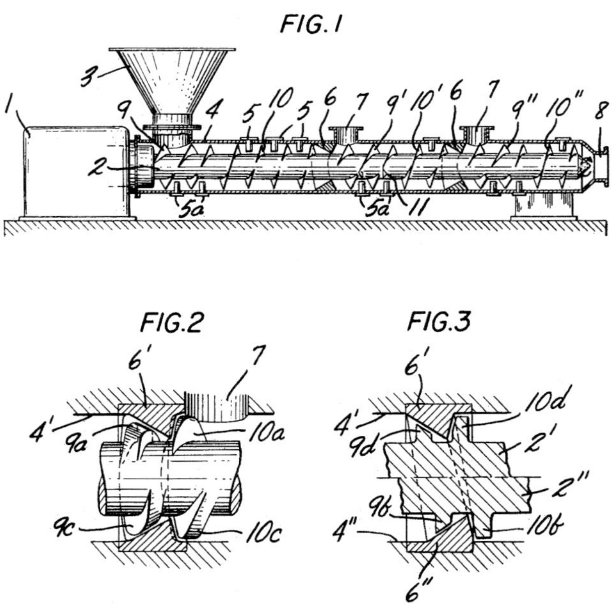

Figure 16.4: Gresch 1966 patent drawing showing a modular Kokneter with barriers and vent ports in the barrel [19]

In the late 1960s, List left Buss AG and formed a new company, List AG, which designs processes and plants. It is also located in Pratteln, Switzerland. Here, List developed many new mixing machines for the chemical industry.

In the mid 1990s, Buss AG was combined with Werner and Pfleiderer in the new Coperion organization. It was later decided that the two mixing machinery companies were not compatible and Buss AG became independent.

16.3 Modern Kokneter

The modern Buss Kokneter is a continuous mixing machine with a modular screw construction in a barrel, containing specially shaped pins and entry and devolatilization ports. It is available with screw/diameter sizes of 46 mm up to 200 mm; in the past oil-heated machines have been available, but today they are electrically heated. The Buss Kneader is usually combined with a crosshead screw extruder or a gear pump to eliminate the fluctuating flow associated with the oscillation of the screw.

The Kokneter has three distinct types of screw modules (see Figure 16.5). The EZ element is a screw element with one screw and one row of pins. This is intended as a transporting element. The KE element is a screw element with three parallel slices and three parallel rows of pins. The ST element is a combination of a KE element and a barrier fitted into the barrel. These elements may, in large part, be placed in any desired order along the axis of the screw shaft.

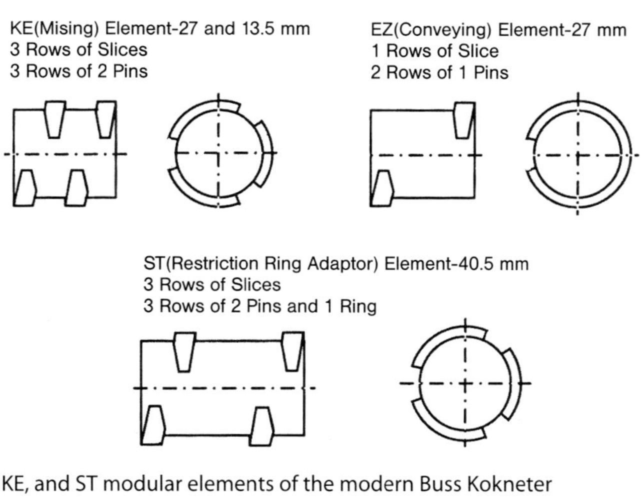

Figure 16.5: EZ, KE, and ST modular elements of the modern Buss Kokneter

16.4 Flow Modeling

The earliest flow simulations of fluid models in a Buss Kokneter took place between 1987 and 1991 with papers of Booy and Kafka [24], Elemans and Meijer [25], and Brzoskowski et al. [26]. Booy and Kafka [24] and Brzoskowski et al. [26] modeled detailed flow patterns and pressure fields around the pins. Elemans and Meijer [25] attempt a broader view of machine behavior. This was essentially continued in subsequent papers of Lyu and White [27–32] between 1995 and 1997.

First, it is useful to show that oscillating outputs result from an oscillating screw. The equations of motion in the screw channel are

$$0=-\frac{\partial p}{\partial x_1}+\frac{\partial \sigma_{12}}{\partial x_2}$$
$$0=-\frac{\partial p}{\partial x_3}+\frac{\partial \sigma_{32}}{\partial x_2} \qquad (16.1\text{ a, b})$$

The boundary conditions are:

$$v_1(H)=U_1(t)=\pi\, D\, N\,\cos\phi+\frac{\mathrm{d}S(t)}{\mathrm{d}t}\sin\phi$$
$$v_3(H)=U_3(t)=-\pi\, D\, N\,\sin\phi+\frac{\mathrm{d}S(t)}{\mathrm{d}t}\cos\phi \qquad (16.2\text{ a, b})$$

where N represents the screw rotation speed, $S(t)$ the stroke of the oscillating screw, ϕ is the helix angle, D the screw diameter, and H the screw channel depth. The solution will clearly have the form

$$\underset{\sim}{v}=v_1(x_2,t)\,\underset{\sim}{e}_1+0\,\underset{\sim}{e}_2+v_3(x_2,t)\,\underset{\sim}{e}_3 \qquad (16.3)$$

If we presume a Newtonian model, i.e., $\sigma_{ij}=\eta\,\partial v_i/\partial x_j$, where η is a constant viscosity. The velocity field is readily shown to be (compare Section 3.3)

$$v_1(x_2,t)=U_1(t)\frac{x_2}{H}-\frac{H^2}{2\,\eta}\left(-\frac{\partial p}{\partial x_1}\right)\left[\frac{x_2}{H}-\left(\frac{x_2}{H}\right)^2\right]$$
$$v_3(x_2,t)=U_3(t)\frac{x_2}{H}-\frac{H^2}{2\,\eta}\left(-\frac{\partial p}{\partial x_3}\right)\left[\frac{x_2}{H}-\left(\frac{x_2}{H}\right)^2\right] \qquad (16.4\text{ a, b})$$

The throughput is:

$$Q=W\int_0^H v_1\,\mathrm{d}x_2=\frac{H\,W\,U_1(t)}{2}-\frac{H^3\,W}{12\,\eta}\frac{\Delta p}{L} \qquad (16.5)$$

If a die is located at the end of the machine, it would have pressure drop characteristics

$$Q = \frac{k}{\eta} \Delta p \tag{16.6}$$

The pressure drop Δp can be eliminated between Eqs. (16.5) and (16.6). This yields

$$Q = \frac{1}{2} \frac{W\, H\, U_1(t)}{1 + \dfrac{H^3\, W}{12\, k\, L}} \tag{16.7}$$

which shows, together with Eq. (16.2 a), that the flow rate leaving the die oscillates. Thus, a gear pump or a starved crosshead extruder is needed to make the flow rate uniform.

Lyu and White [27–32] have sought to model flow in different types of Buss Kokneter flow elements. Different approaches were used. An appropriate analytical method [27] was used as well as the Flow Analysis Network (FAN) method of Section 5.3 [28]. They presume in the FAN analysis a power law fluid with the shear viscosity

$$\eta = K \left[\left(\frac{\partial v_1}{\partial x_2} \right)^2 + \left(\frac{\partial v_3}{\partial x_2} \right)^2 \right]^{\frac{n-1}{2}} \tag{16.8}$$

They considered the oscillations of the screws, but neglected the influence of the pins on the pressure field. They proceeded by balancing the flattened Cartesian coordinate fluxes q_1 and q_3

$$\begin{aligned} q_1 &= \int_0^H v_1\, \mathrm{d}x_2 = \int_0^H \!\! \int \left(x_2' \frac{\partial p}{\partial x_1} + C_1 \right) \frac{1}{\eta}\, \mathrm{d}x_2'\, \mathrm{d}x_2 \\ q_3 &= \int_0^H v_3\, \mathrm{d}x_2 = \int_0^H \!\! \int \left(x_2' \frac{\partial p}{\partial x_1} + C_3 \right) \frac{1}{\eta}\, \mathrm{d}x_2'\, \mathrm{d}x_2 \end{aligned} \tag{16.9 a, b}$$

The fluxes are balanced over the finite difference mesh as

$$\begin{aligned} & q_1(i-1, j)\, \Delta x_3 + q_3(i, j+1)\, \Delta x_1 \\ = \; & q_1(i+1, j)\, \Delta x_3 + q_3(i, j-1)\, \Delta x_1 \end{aligned} \tag{16.10}$$

The computed flux maps for EZ and KE elements are shown in Figure 16.6 and Figure 16.7. These maps indicate the extensive backward leakage of the KE mixing element as compared to the EZ element. Backward leakage is higher for more highly non-Newtonian fluids with lower values of the power law index.

Screw characteristic curves relating throughput, pressure development, and screw speed have been determined for the EZ and KE elements for Newtonian and power law fluids. Theses are

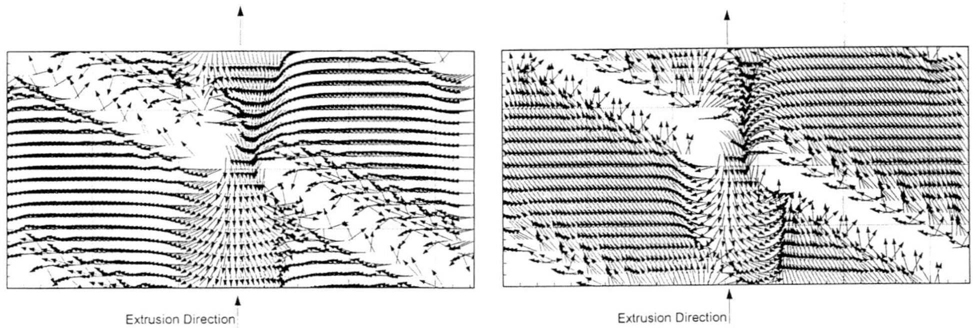

a) Forward Motion
$\Delta p = 2.4$ MPa, $Q = 75{,}276$ cm^3/h
Power Law Index $n = 0.4$

b) Backward Motion
$\Delta p = 0.9$ MPa, $Q = 74{,}506$ cm^3/h
Power Law Index $n = 0.4$

Figure 16.6: Calculated flux distribution in an EZ element; a) forward motion; b) backward motion [28]

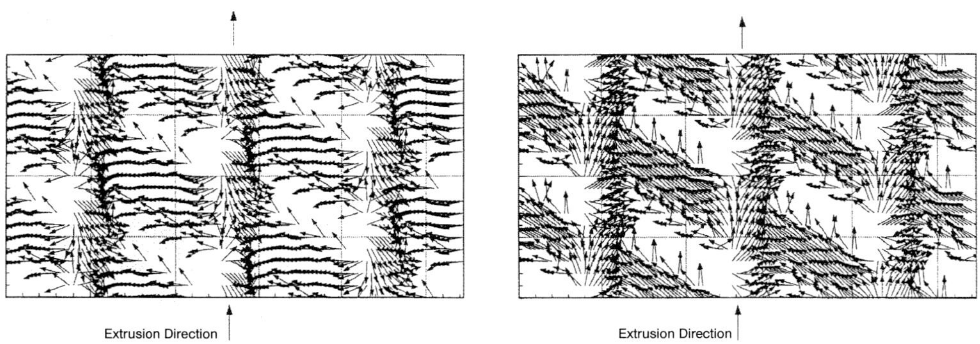

a) Forward Motion
$\Delta p = 0.89$ MPa, $Q = 72{,}482.4$ cm^3/h
Power Law Index $n = 0.4$

b) Backward Motion
$\Delta p = 0.84$ MPa, $Q = 72{,}147.6$ cm^3/h
Power Law Index $n = 0.4$

Figure 16.7: Calculated flux distribution in a KE element; a) forward motion; b) backward motion [28]

shown in Figures 16.8 and 16.9 for both an analytical model of Lyu and White [27] and their numerical simulation [28] based on Eqs. (16.8) to (16.10). It can be seen that the pumping characteristics for Newtonian fluids are superior to those for non-Newtonian power law fluids. The pumping for the forward motion is superior to that for the backward motion.

The influence of viscoelasticity on flow in a Kokneter was also modeled by Lyu and White [30] using linear viscoelastic models. Non-isothermal flow in a Kokneter was also modeled by Lyu and White [29, 32] This was accomplished by extending the modeling of Section 5.6 to an oscillating flow. Beginning with the energy equation in the form

$$\rho\, c\left(\frac{\partial T}{\partial t}+v_1\frac{\partial T}{\partial x_1}+v_3\frac{\partial T}{\partial x_3}\right)=k\frac{\partial^2 T}{\partial x_2^2}+\left(\sigma_{12}\frac{\partial v_1}{\partial x_2}+\sigma_{32}\frac{\partial v_3}{\partial x_2}\right) \tag{16.11}$$

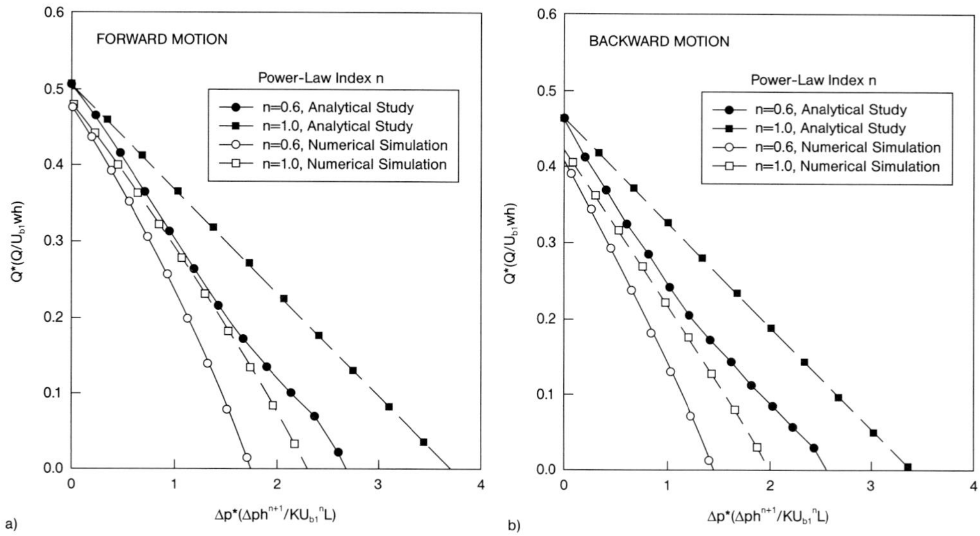

Figure 16.8: Calculated and measured screw characteristic curves for an EZ element; a) forward motion; b) backward motion [28]

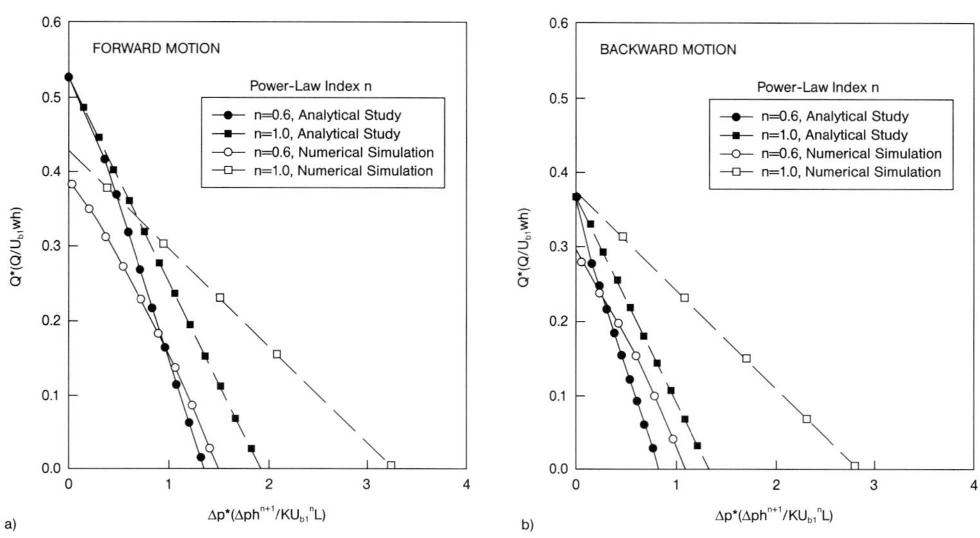

Figure 16.9: Calculated and measured screw characteristic curves for a KE element; a) forward motion; b) backward motion [28]

They derive by integration with respect to x_2

$$\rho c\left(\frac{\partial \tilde{T}}{\partial t}+q_1\frac{\partial \bar{T}}{\partial x_1}\right)=h_s\,(T_s-\bar{T})+h_b\,(T_b-\bar{T})+\sigma_{12}\big|_b\,U_1+\sigma_{32}\big|_b\,U_3-q\frac{dp}{dx_1} \tag{16.12}$$

where

$$\tilde{T}=\int_0^H T(x_1,x_2,x_3,t)\,dx_2 \tag{16.13}$$

$$\bar{T}=\frac{1}{q}\int_0^H v_1(x_2,t)\,T(x_1,x_2,x_3,t)\,dx_2 \tag{16.14}$$

h_s and h_b are screw and barrel heat transfer coefficients. Equation (16.12) may be further integrated with respect to the transverse direction x_3 to give

$$\rho c\left(W H\frac{\partial \tilde{\tilde{T}}}{\partial t}+Q\frac{\partial \bar{\bar{T}}}{\partial x_1}\right)=W\,[h_s\,(T_s-\bar{\bar{T}})+h_b\,(T_b-\bar{\bar{T}})]+\sigma_{12}\big|_b\,U_1+\sigma_{32}\big|_b\,U_3-Q\frac{dp}{dx} \tag{16.15}$$

$\tilde{\tilde{T}}$ and $\bar{\bar{T}}$ being $\tilde{T}$ and $\bar{T}$ averaged over the transverse direction.

16.5 Theory for Composite Modular Machines

As discussed in Section 16.3, modern Kokneters have composite modular structures. As the throughput of compounding machines is known, it is possible to calculate the fluid mechanics of these machines backward from the machine exit towards the hopper, as was done for various types of twin screw extruders as discussed in Sections 5.8, 8.7, and 11.9 using screw characteristic curves such as those shown in Figures 16.9. Computed axial fill factor and axial pressure profiles are shown in Figure 16.10 a and b.

The calculations for the Kokneter are more complex and difficult than those for twin screw extruders because of the oscillation of the screw and the attachment of devices, such as crosshead screws at the exit of the Kokneter. In the case of the starved crosshead extruder, we need to calculate backward from the extruder die backward to the Kokneter and then through the Kokneter.

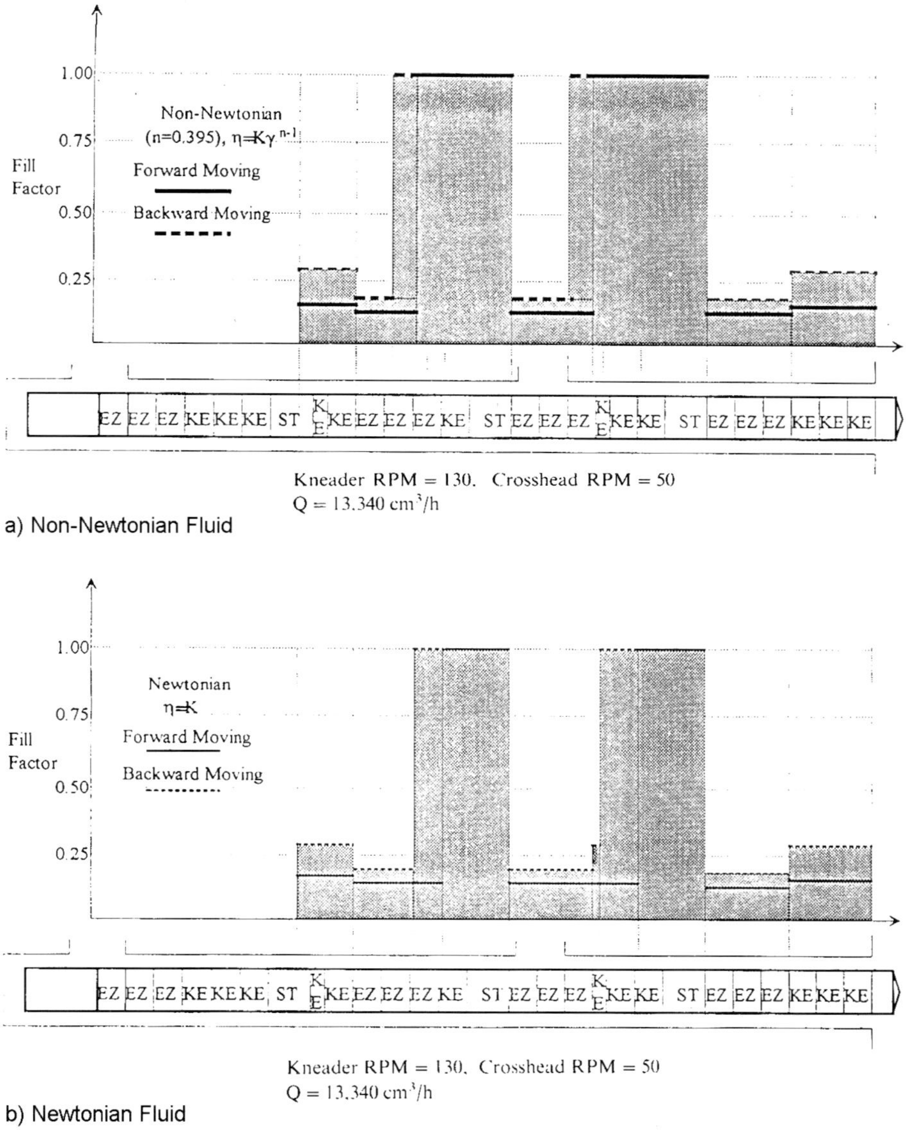

Figure 16.10: Axial fill factor and pressure profile for modular Kokneter; a) fill factor profile; b) pressure profile [28]

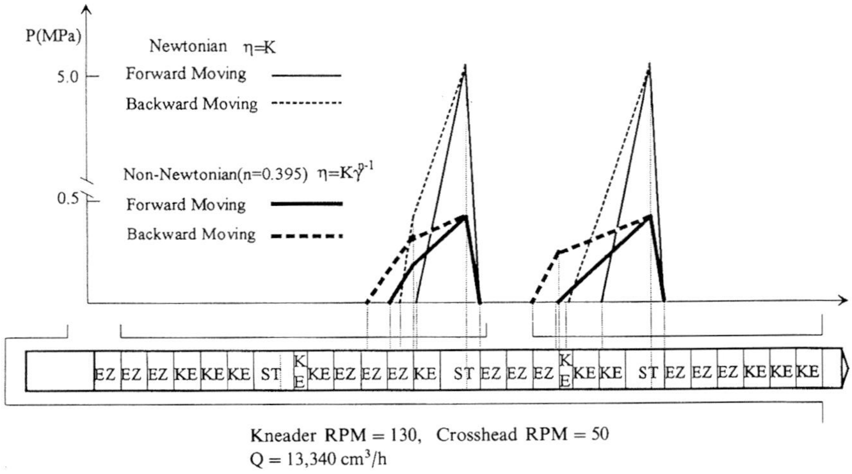

Figure 16.10: (continued)

16.6 Experimental Studies

There have been relatively few published experimental studies of the Kokneter. Following the 1950 paper of List [4], which largely described the machine, the earliest papers are by Schneider [33] in 1951, Timm et al. [34] in 1965, and Todd and Hunt [35] in 1973. There have been various experimental investigations since these studies [25, 27–29, 32, 36–46]. Most early papers were technological in character.

The first basic experimental study was published in 1990 by Elemans and Meijer [25]. These authors describe experiments carried out on a 46 mm Kokneter screw with the screw fitted into a polymethyl methacrylate (Plexiglas®) annular cylinder with small glass tubes that each act as an open manometer attached to it. They describe flow experiments using a Newtonian silicone oil (η = 1.0 Pa-s) and a parafinic oil (0.2 Pa-s) under isothermal conditions at room temperature. They measured throughput versus screw speed, throughput versus pressure generation as well as local fill lengths. With these measurements, they were able to construct screw characteristic curves, i.e., plots of throughput Q versus pressure drop Δp at specific screw speeds for the two Newtonian oils in different modular Kokneter elements. Filled lengths were determined in different screw elements with different processing conditions.

In 1995-8, Lyu and White [27–29, 32] described experimental studies with molten polypropylene in a 46 mm Kokneter connected to a crosshead screw extruder. They measured fill factor profiles, temperature profiles, conditions of melting and residence time distributions as a function of screw configuration and processing conditions.

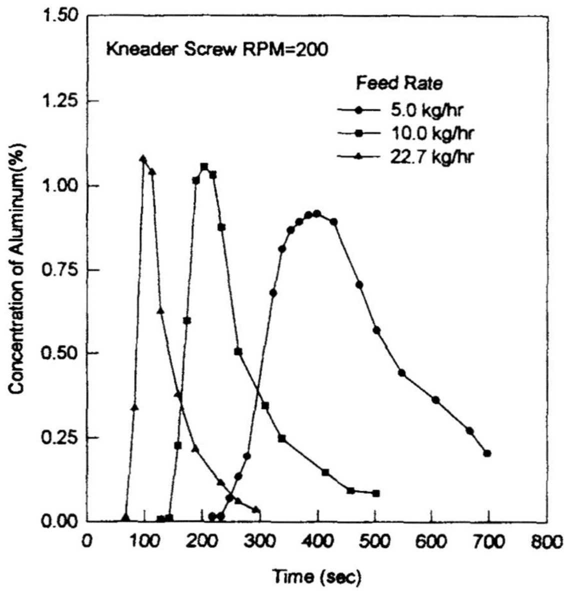

Figure 16.11: Residence time distribution for a Kokneter showing the effect of feed rate [45]

Residence time distributions for the Kokneter were measured by Lyu and White [31] and Shon et al. [45]. In particular, they looked at the effect of throughput and screw speed for an isotactic polypropylene melt. The primary effect was seen with the throughput. Increasing the feed rate both decreased the residence time and narrowed the residence time distribution as shown in Figure 16.11. Increasing screw speed seems to have a similar effect but the variations were smaller as shown in Figure 16.12.

Lyu and White [31] sought to determine melting and flow mechanism in a Kokneter attached to a crosshead extruder during its operation. They made the first study of melting in a Kokneter noting that it occurred between the hopper and the first ST ring. The ST elements with their restriction rings, which usually exist throughout the machine, were found to produce fully filled melt regions immediately behind them on the hopper side. This seems associated with their poor pumping characteristics.

Studies of fiber breakage in compounding have been published by Franzen et al. [43] and Shon and White [45] using the Kokneter and other compounding machines. The extent of breakage depends not only on the type of machine, but on the modular configurations of the screws and operating conditions such as screw speed. In the later and more comprehensive study of Shon and White [46] it was found that the Kokneter generally induced less fiber breakage than the other compounding machines.

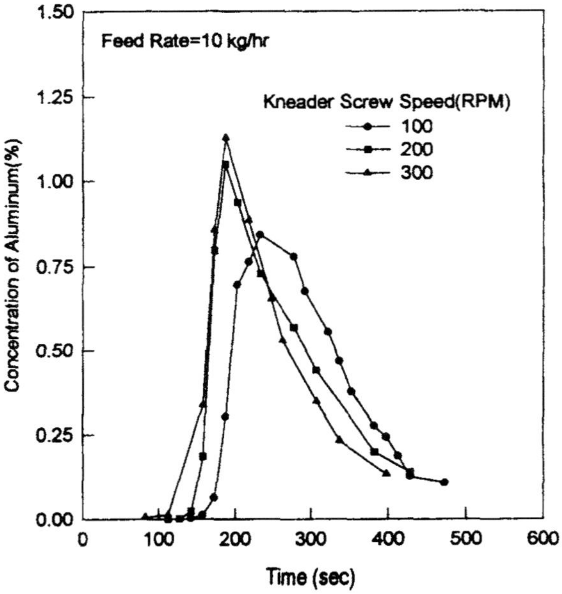

Figure 16.12: Residence time distribution for a Kokneter showing the effect of screw speed [45]

References

1. H. List, Swiss Patent (filed August 20, 1945) 247,704 (1947).
2. H. List, U.S. Patent (filed August 19, 1946) 2,505,125 (1950).
3. H. List, German Patent (filed January 5, 1949) 944,727 (1956).
4. H. List, *Kunststoffe*, **40**, 185 (1950).
5. H. List, *Private Communication* (1988).
6. C. Wurster, German Patent (filed April 30, 1901) 137,813 (1901).
7. F. B. Anderson, U.S. Patent (filed November 14, 1930) 1,848,236 (1932).
8. H. P. M. Quillery, French Patent (filed February 8, 1939) 855,885 (1940).
9. H. Beck, German Patent (filed December 16, 1943) 858,310 (1952).
10. A. Wunsche, German Patent (filed September 12, 1901) 131,392 (1902).
11. R. W. Easton, British Patent (filed September 25, 1916) 109,663 (1917).
12. R. W. Easton, British Patent (filed June 2, 1920) 1,468,379 (1923).
13. S. Kiesskalt, H. Tampke, K. Winnacker and E. Weingaertner, German Patent (filed July 26, 1935) 652,990 (1937).
14. P. Leistritz and F. Burghauser, German Patent (filed December 1, 1935) 682,787 (1939).
15. R. Colombo, Italian Patent (filed February 6, 1939) 370,578 (1939).

16. W. Meskat and R. Erdmenger, German Patent (filed July 7, 1944) 868,668 (1953).
17. E. Gubler, U.S. Patent (filed February 6, 1961) 3,189,324 (1965).
18. F. Sutter, U.S. Patent (filed September 19, 1963) 3,219, 320 (1965).
19. W. Gresch, U.S. Patent (filed January 26, 1966) 3,367,635 (1968).
20. H. F. Geier and H. F. Irving, U.S. Patent (filed March 9, 1959) 3,023,455 (1962).
21. G. Schuur, U.S. Patent (filed September 5, 1963) 3,224,739 (1965).
22. H. List, U.S. Patent (filed October 10, 1963) 3,317,959 (1964).
23. H. List and F. Ronner, U.S. Patent (filed January 21, 1965) 3,347,528 (1967).
24. M. L. Booy and F. Kafka, *SPE ANTEC Tech. Papers*, **33**, 140 (1987).
25. P. H. M. Elemans and H. E. H. Meijer, *Polym. Eng. Sci.*, **30**, 893 (1990).
26. R. Brzoskowski, T. Kumazawa and J. L. White, *Int. Polym. Process.*, **6**, 136 (1991).
27. M. Y. Lyu and J. L. White, *SPE ANTEC Tech. Papers*, **41**, 208 (1995); *Int. Polym. Process.*, **10**, 305 (1995).
28. M. Y. Lyu and J. L. White, *Int. Polym. Process.*, **11**, 208 (1996).
29. M. Y. Lyu and J. L. White, *SPE ANTEC Tech Papers*, **42**, 160 (1996); *J. Reinf. Plastics Comp.*, **16**, 1445 (1997); *Int. Polym. Process.*, **12**, 104 (1997).
30. M. Y. Lyu and J. L. White, *Polym. Eng. Sci.*, **37**, 623 (1997).
31. M. Y. Lyu and J. L. White, *Polym. Eng. Sci.*, **38**, 1366 (1998).
32. J. L. White and M. Y. Lyu, *Polym. Plast. Technol. Eng.*, **37**, 385 (1998).
33. E. Schneider and D. H. M. Brooks, *Brit. Plastics*, p. 481 (1957)
34. Th. Timm, D. Stolzenberg and H. Fettback, *Kautschuk und Gummi*, **18**, 206 (1965).
35. D. B. Todd and J. W. Hunt, *SPE ANTEC Tech. Papers*, **19**, 577 (1973).
36. K. Stade, *Polym. Eng. Sci.*, **17**, 50 (1977).
37. K. Stade, *Polym. Eng. Sci.*, **18**, 107 (1978).
38. S. Jakopin and P. Franz, *Adv. Polym. Technol.*, **3**, 365 (1983).
39. D. M. Kalyon and M. Bouazza, *SPE ANTEC Tech. Papers*, **29**, 778 (1983).
40. D. M. Kalyon and M. Hallouch, *SPE ANTEC Tech. Papers*, **31**, 1206 (1985).
41. P. Schnottale, *Kautsch Gummi Kunstst.*, **38**, 116 (1985).
42. D. B. Todd, *SPE ANTEC Tech. Papers*, **33**, 128 (1987).
43. B. Franzen, C. Klason, J. Kubat and T. Kitano, *Composites*, **20**, 65 (1989).
44. H. Thommen, *Plastverarbeiter*, **44** (1) 12 (1993)
45. K. Shon, D. H. Chang and J. L. White, *Int. Polym. Process.*, **14**, 44 (1999).
46. K. Shon and J. L. White, *Polym. Eng. Sci.*, **39**, 1757 (1999).

17 Reactive Extrusion

17.1 Introduction

Twin screw extruders have been widely used as a continuous chemical reactor. It has been widely used for both post-reactor polymer modification and polymer functionalization as well as polymerization. In this chapter we will describe efforts in this direction. A very good earlier review of this literature was given by Brown [1] in 1992.

We begin this chapter in Section 17.2 by discussing the earliest efforts in this area (1943–1955). We then describe efforts beginning in the late 1950s to modify existing polymers by thermal degradation, grafting, halogenation, transesterification, and cross-linking processes. These are the contents of Sections 17.3 to 17.7. We then turn to polymerization in twin screw extruders, describing efforts on anionic polymerization of lactams, converting trioxane to polyoxymethylene, polymerization of polyurethanes, anionic polymerization of lactones, polymerization of polyetherimide, and various block copolymers (Section 17.8 to 17.14).

17.2 Origins

Efforts on reactive extrusion date to the early 1940s. An early German patent application [2], filed in July 1943, describes three polymerization processes being carried out in a pair of intermeshing co-rotating twin screw extruders (Figure 17.1). These include the polymerization of butadiene, and the reaction of diisocyanates and diols. The extruders seems to be a specially modified machines manufactured by Colombo and LMP [3, 4] in Turin, Italy which had been commercially available since 1939 (see Section 4.3). No commercial processes were developed.

In the US, Stober and Amos [5] of Dow Chemical described the polymerization of styrene by free radical mechanisms in a single screw extruder in a 1950 patent. The screw rotation speed was one RPM and a 120–200 °C temperature profile existed along the extruder. The residence time was 18 hours.

As we described in Sections 4.5 and 4.6, Farbenfabriken Bayer actively developed intermeshing self-wiping co-rotating twin screw extruders from the late 1940s. In a 1955 paper, Riess [6] described a Bayer process for esterifying cellulose that involved using a series of modular co-rotating twin screw extruder system (Figure 17.2).

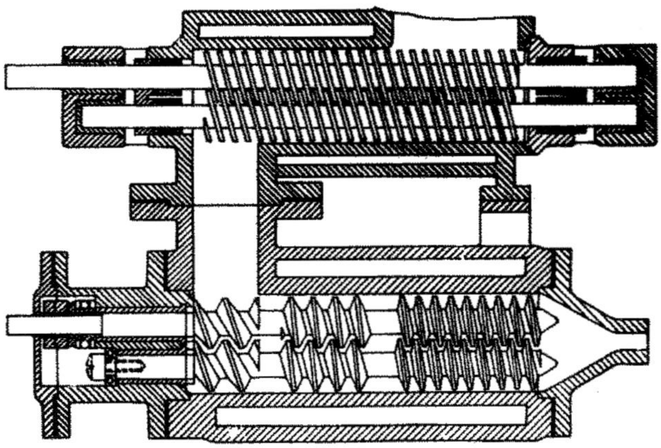

Figure 17.1:
1943 co-rotating twin screw extrusion system for reactive extrusion introduced by I.G. Farbenindustrie [2]

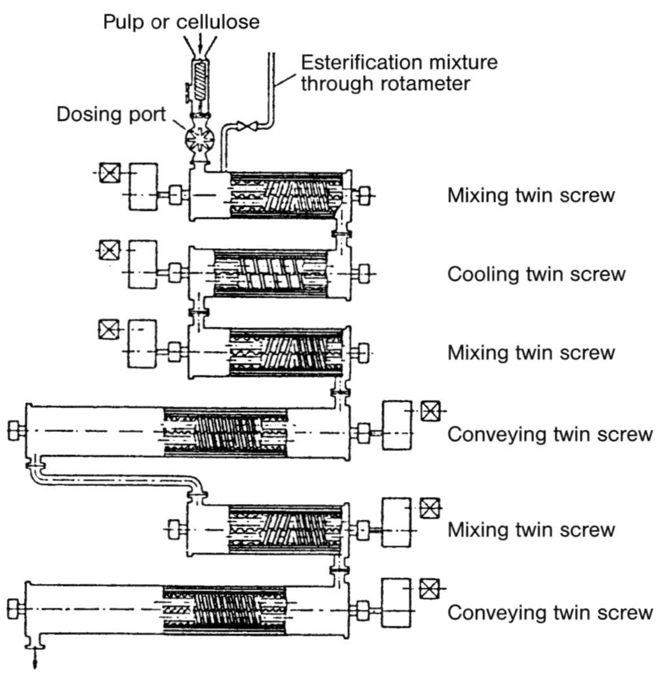

Figure 17.2:
1955 modular co-rotating twin screw process system for esterifying cellulose [6] by Bayer (reprinted with permission of Chemie Ingenieur Technik)

17.3 Polyolefin Controlled Degradation

Soon after the introduction of isotactic polyolefins by Natta and his coworkers at Montecatini of in the 1950s, it was found that these polymers, notably isotactic polypropylene, could be degraded by thermal/oxidative means and its molecular weight distribution be controlled [7]. Subsequent patent applications by Green and Pieski [8] of Dupont and Trieschmann et al. [9] of BASF and Roberts et al. [10] of ICI (Imperial Chemical Industries) in 1961/1962 described degradation processes in single screw extruders. In succeeding years, various researchers [11–14] developed variant single screw extrusion processes for degrading and controlling (narrowing) the molecular weight distribution of polypropylene. This involved heat, oxygen, peroxides, and various sulfides. The reaction involved for thermal degradation is basically

$$\mathrm{PM}_{n+m}\mathrm{P}' \rightarrow \mathrm{PM}_n^{\bullet} + \mathrm{P}'\mathrm{M}_m^{\bullet} \qquad \text{(Structure 17.1)}$$

for the addition of peroxides

$$\begin{aligned} &\mathrm{ROOR}' \rightarrow \mathrm{RO}^{\bullet} + \mathrm{R}'\mathrm{O}^{\bullet} \\ &\mathrm{RO}^{\bullet} + \mathrm{PM}_{n+m}\mathrm{P}' \rightarrow \mathrm{PM}_{n+m}^{\bullet}\mathrm{P}' \qquad \text{(Structure 17.2)} \\ &\mathrm{PM}_{n+m}^{\bullet}\mathrm{P}' \rightarrow \mathrm{PM}_n^{\bullet} + \mathrm{P}'\mathrm{M}_m \end{aligned}$$

There have been many studies of this technology by various industrial and academic research groups [15–18].

17.4 Grafting Reactive Monomers onto Polyolefins

In 1960, Nowak and Jones [19, 20] of Dow Chemical filed patent applications on grafting various monomers onto polyethylene and polypropylene melts in various screw extruders, including single screw machines and Welding Engineers tangential counter-rotating machines. Peroxides were used to generate free radicals for the initiation of the reaction. The monomers included acrylic acid, methacrylic acid, styrene, vinyl toluene, dichlorostyrene, and acrylonitrile. If we accept formula (17.2) we may surmise where M_2^* is grafted onto polymer $(M_1)_n$.

$$\mathrm{P(M_1)}_n^* + \mathrm{M}_2^* \rightarrow \mathrm{P(M_1)}_n - \mathrm{M}_2^* \qquad \text{(Structure 17.3)}$$

In a subsequent patent Nowak [21] of Dow Chemical pre-irradiated polyethylene powder by a beam of high-speed electrons from Van der Graaff accelerator and did not use peroxides to graft acrylic acid and methacrylic acid onto polyethylene during reactive extrusion using a Welding Engineer tangential counter-rotating twin-screw extruder.

In succeeding years, investigators with many different companies including Asahi Kasei [22], Eastman Kodak [23], Exxon [24–28], Chemplex [29], Rohm and Haas [30], and Uniroyal

Table 17.1: Early Investigations of Monomers Grafted onto Polymers in Screw Extrusion Machines.

Monomers	Polymers	Extruders	Patent applications	Inventors
Acrylic acid Methacrylic acid	Polyethylene	Single Screw Welding Engineering tangential twin screw	1960	Nowak and Jones [15] (Dow Chemical)
Styrene Vinyl toluene Dichloro styrene Styrene/acrylonitrile (Maleic anhydride mentioned)	Polyethylene	Single screw	1960	Jones and Nowak [16] (Dow Chemical)
Lauryl methacrylate Methyl methacrylate	Polypropylene Ethylene/propylene Block copolymer	Single screw	1968	Asahi Kasei [18]
Maleic anhydride	Polyethylene	–	1967/1970	McConnell [19] (Eastman Kodak)
Acrylic acid Maleic anhydride	Polypropylene Other polyolefins	Single screw	1972	Steinkamp and Grail [24] (Exxon)

Chemical [27, 28] prepared graft copolymers. The patents cite the use of single screw extruders [18, 24–28], tangential counter-rotating [31], and modular co-rotating [29, 30, 32, 33] twin screw extruders. The monomers and polymers most extensively studied are summarized in Table 17.1. Kowalski [34] has reviewed Exxon's experiences with grafting by reactive extrusion.

In more recent years, there have been many fundamental studies of grafting in extrusion, particularly in modular co-rotating twin screw extruders [35–43]. Chang and White [43] compared the peroxide induced grafting of maleic anhydride onto isotactic polypropylene in various continuous extrusion machines including

- a modular self-wiping co-rotating twin screw extruder,
- a tangential counter-rotating twin screw extruder,
- an intermeshing counter-rotating twin screw extruder, and
- a Kobelco continuous mixer.

17.5 Halogenating Aliphatic Hydrocarbon Polymers

There is an extensive patent literature related to halogenating butyl rubber in twin screw extruders by Exxon researchers [45–47], notably Newman and Kowalski. These authors claim they may similarly halogenate various polyolefins and hydrocarbon elastomers.

Kowalski [34, 48] reviewed Exxon's experiences in halogenation processes in reactive extrusion, including more fundamental studies, notably by L. Erwin. It was concluded that the Welding Engineers' modular tangential counter-rotating machine was most suitable for carrying out the process because of its superior mixing capability, particularly in comparison to single screw extruders. There are metallurgical problems in this process because of the halogenation of iron. The extruders used must be constructed out of special alloys.

17.6 Transesterification of Ethylene Vinyl Acetate Copolymer

Polyvinyl acetate, polyvinyl alcohol and ethyl acetate-vinyl alcohol (EVAL) copolymers are synthesized by first synthesizing polyvinyl acetate by free radical polymerization of vinyl acetate:

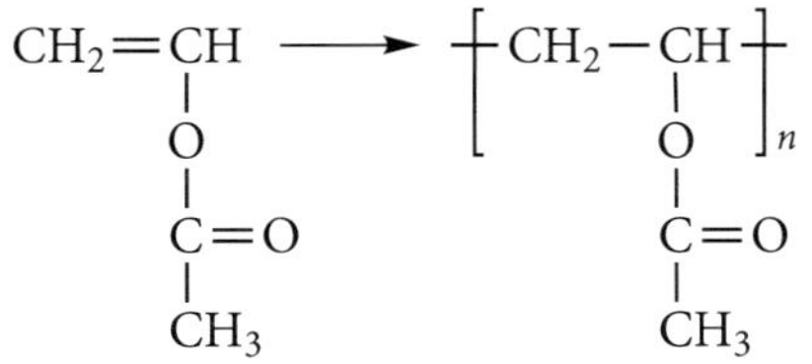

(Structure 17.4)

Polyvinyl alcohol and EVAL are then produced using transesterification with an alcohol

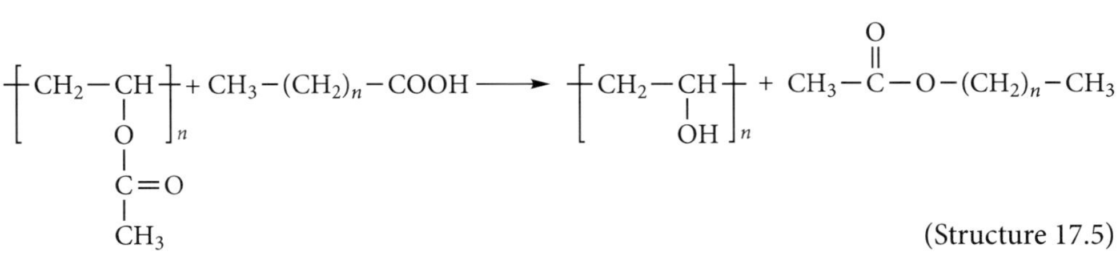

(Structure 17.5)

Lambla and his coworkers [49, 50] and later P. J. Kim and White [51] carried out this reaction in a modular co-rotating twin screw extruder. The latter authors used a twin screw extruder with a transparent barrel. They used four different screw configurations and compared the residence times and levels of conversion.

17.7 Cross-Linking Thermoplastic Elastomers

The 1970s and 1980s saw the development of polypropylene-ethylene propylene terpolymer (EPDM) thermoplastic elastomers in which the EPDM was crosslinked during mixing, a process called "dynamic vulcanization". This was developed notably by Uniroyal Chemical and especially Monsanto in the USA. Early experimental studies were done on laboratory batch mixers and commercial Farrel Banbury mixers. Subsequently, the process was carried out in modular co-rotating twin screw extruders and it was found that superior and more uniform products were formed. In a 1986 patent, Abdou-Sabet and Shen [52] of Monsanto described the blending of polypropylene and EPDM with dynamic vulcanization by a methylol phenolic curing agent and a stannous chloride cure accelerator in a modular co-rotating twin screw extruder.

17.8 Anionic Polymerization of Lactams

We now turn to polymerization in twin screw extruders. Anionic polymerization of caprolactam with sodium metal was studied by Joyce and Ritter [53] of DuPont as early as 1939. This reaction was, however, slow. In the late 1950s, Mottus et al. [54] with Monsanto and Wichterle, Sebenda, and Kralicek [55, 56] with the Institute of Macromolecular Chemistry in Prague discovered the accelerating effect of N-acyl activators such as N-acetylactam in the polymerization caprolactam.

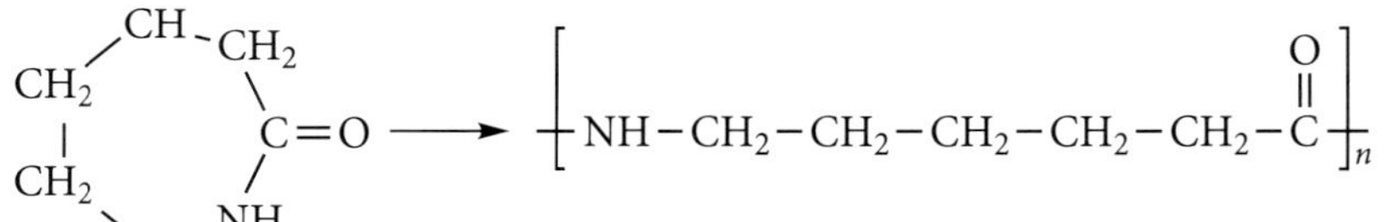

(Structure 17.6)

The first discussion of anionic polymerization of lactams by reactive extrusion is in the 1959 patent application by Wichterle et al. [56] in a single screw extruder. Five years later (in 1964), Illing and Zahradnik [57] associated with Werner and Pfleiderer, filed a patent for the anionic polymerization of caprolactam in a modular co-rotating twin screw extruder. This was also described in the subsequent papers by Illing [58, 59]. In later years Tucker and Nichols [60] described the anionic polymerization of caprolactam in a Welding Engineers modular tangential counter-rotating twin screw machine.

A third generation of investigations began with papers by Menges et al. [61, 62] of the RWTH Aachen in 1987/1989. These involved the first investigation of the influence of process variables. They described both polymerization and subsequent mechanical degradation of polycaprolactam in a co-rotating twin screw extruder. With the retirement of Menges, this research program was continued by his successor, W. Michaeli [63, 64]. More recent studies have been reported by White et al. [65–68]. It has been found possible to also polymerize lauryl lactam [67, 68].

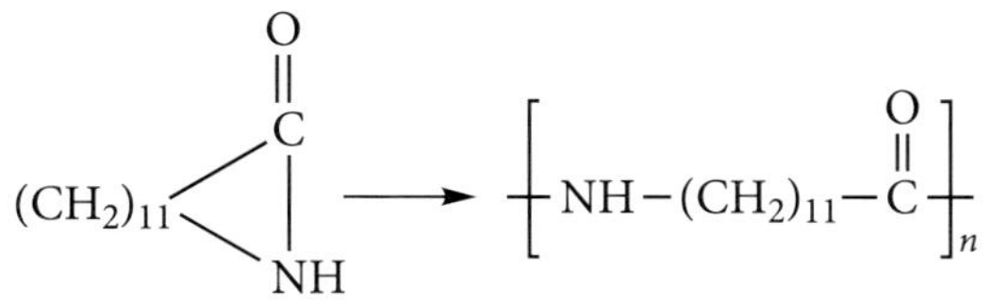

(Structure 17.7)

and to copolymerize caprolactam and lauryl lactam [67]. The experiments of the Menges-Michaeli team and those in White et al.'s laboratories used modular co-rotating twin screw extruders.

It has been possible to shape the newly polymerized polyamide extrudates as they emerge from twin screw extruders and make filaments, sheet, and moldings in a continuous manner beginning with monomer. The anionic polymerization of lactams has not been commercialized.

17.9 Polymerization of Polyurethanes

The 1943 I.G. Farbenindustrie patent application [2] mentioned earlier discussed the polymerization of polyurethanes in a co-rotating twin screw extruder. Bayer AG which had invented polyurethanes (in Leverkusen in the I.G. Farbenindustrie period) and had subsequently developed the modular co-rotating twin screw extruder was also deeply involved in the development of reactive extrusion polymerization polyurethanes.

In the 1960s, there were various patent applications [69, 70] for polymerization of polyurethanes in twin screw extrusion processes. These were filed by Mobay (Monsanto-Bayer USA joint venture) and Upjohn. In the next decade (1974), Ullrich, Meisert and Eitel [71] of Bayer AG in Leverkusen filed a patent application on the formation of elastomeric polyurethanes in modular co-rotating twin screw extruders. Subsequent Bayer AG patents on polymerization and modification of polyurethanes using reactive extrusion were obtained by Quiring et al. [72] and Goyert et al. [73–76].

17.10 Ring-Opening Polymerization of Trioxane

Ring-opening polymerization of trioxane in twin screw extruder has a long history. In the 1960s and 1970s, there were various studies for polymerization in modular co-rotating twin screw extruders [77–80], notably by Seddon et al. [77, 79], Fisher et al. [78], and Semanchik and Braunstein [80] of Celanese, which described the polymerization of trioxane to produce polyoxymethylene.

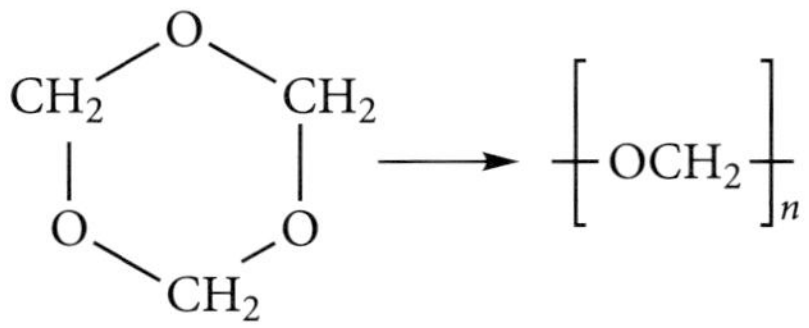

(Structure 17.8)

as well as copolymers with ethylene oxide

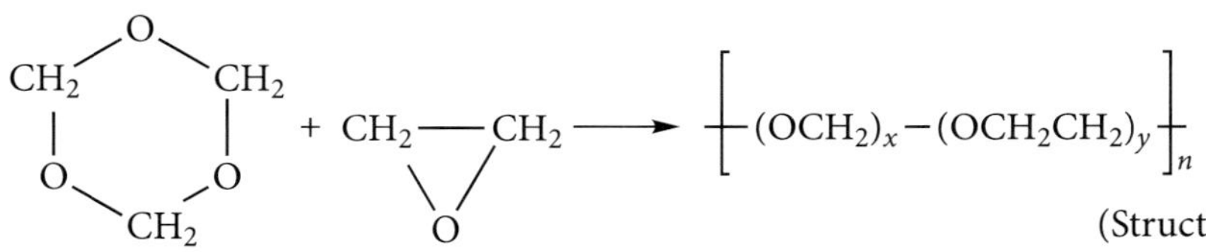

(Structure 17.9)

A variant technology was described by Sugio et al. [81] of Mitsubishi Gas Chemical. A subsequent investigation was published by Todd [82] of Baker Perkins.

17.11 Anionic Polymerization of Lactones

Ring opening polymerizations of caprolactone have been investigated by various researchers [83–86].

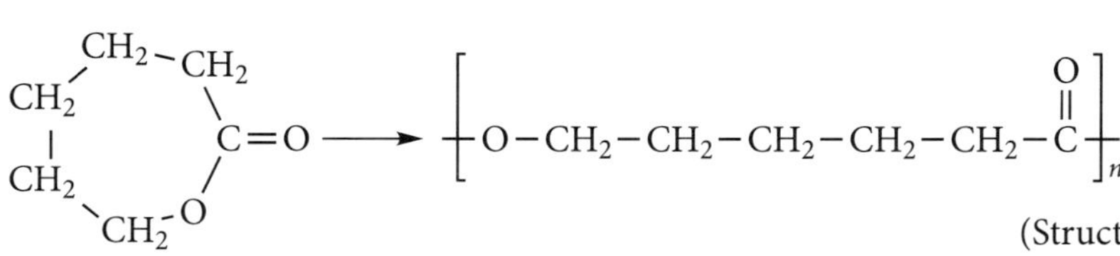

(Structure 17.10)

They used anionic polymerization initiators such as titanium n-propoxide and aluminum triisopropoxide in a modular co-rotating twin screw extruder. B. J. Kim and White [86] discuss apparent mechanical degradation of the polymer along the length of the twin screw extruder.

17.12 Polymerization of Polyetherimide

In the 1970s and early 1980s, Takekoshi and Kochanowski [87, 88], Banucci and Mellinger [89], and Schmidt and Lovgren [90, 91] of General Electric described in various patents the use of a modular co-rotating twin screw extruder to react bisphenol A dianhydride with different aromatic amines to produce polyetherimides.

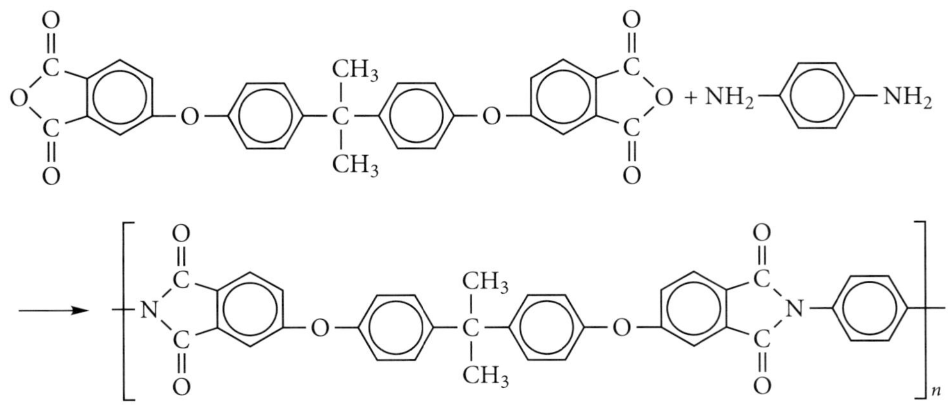

(Structure 17.11)

17.13 Butadiene-Styrene Block Copolymers

A 1973 patent by Sutter et al. [92] of Bayer AG synthesized styrene-butadiene and styrene-isoprene copolymers in a modular co-rotating twin screw extruder using a butyl lithium initiator.

In 1993 papers, Michaeli and his coworkers [93–95] at the RWTH Aachen described the anionic polymerization of styrene and, more strikingly, the block copolymerization of styrene-isoprene block copolymers in modular co-rotating twin screw extruders. They used a butyl lithium anionic polymerization initiator. Various types of copolymers were produced by using both simultaneous introduction of the two monomers and introduction of styrene into the hopper with downstream addition of isoprene.

Subsequently Michaeli et al. [96] synthesized polyisoprene in a modular co-rotating twin screw extruder to which they subsequently (anionically) grafted styrene monomer also in a twin screw extrusion process.

17.14 Lactam-Lactone Block Copolymers

B. J. Kim and White [97] and I. Kim and White [98, 99] have described the synthesis of lactam-lactone block copolymers in a modular co-rotating twin screw extruder. Caprolactam or lauryl lactam was first polymerized using sodium hydride/acetyl caprolactam and subsequently caprolactone was added downstream. Poly lauryl lactam-b-caprolactam-b-caprolactone was also produced in this way.

17.15 Final Remarks

There are few useful critiques on reactive extrusion in twin screw extruders. However, certain issues are clear: Advantages include high levels of local mixing which can be produced by suitable modular screw designs and choices of elements especially mixing elements, as well as possible special designs of feed ports.

There are, however, disadvantages. Reaction times need to be small because of the low residence times of twin screw extruders. Heat transfer is poor, especially as screw diameters increase. The choice of reactions used in twin screw extruders suggest this as they include polymerization of lactams, lactones, and urethanes, where the number and types of bonds does not change or grafting where the number of reactions are small. In these cases the heats of reaction will not overwhelm the twin screw extruder's heat transfer capabilities.

The critiques of Biesenberger [100] and Todd [101] make useful reading and are recommended to the reader.

References

1. S. B. Brown in "Reactive Extrusion: Priciples and Practice", edited by M. Xanthos, Hanser, Munich (1992).
2. Anonymous (BASF), German Patent (filed July 24, 1943) 895,058 (1953).
3. R. Colombo, Italian Patent (filed February 6, 1939) 370,578 (1939).
4. S. A. Liguna, Swiss Patent (filed January 3, 1941) 220,550 (1942).
5. K. E. Stober and J. L. Amos, U.S. Patent 2,530,409 (1950).
6. K. Riess, *Chem. Ing. Tech.*, **27**, 457 (1955).
7. D. Maragliano and E. D. Giulio, U.S. Patent (filed December 6, 1956) 3,013,003 (1961).
8. R. E. Greene and E. T. Pieski, U.S. Patent (filed January 4, 1961) 3,144,436 (1964).
9. H. G. Trieschmann, H. Moeller, G. Schmidtthomee, C. Alt and R. Herbeck, British Patent (filed August 8, 1963) 1,042,178 (1966): First applied in Germany (August 31, 1962).

10. J. F. L. Roberts, E. Walker and G. Scott, U.S. Patent (filed March 15, 1962) 3,143,584 (1964).
11. R. C. Kowalski, U.S. Patent (filed September 15, 1966 and October 15, 1969) 3,563,972 (1971).
12. J. C. Staton, J. P. Keller, R. C. Kowalski and J. W. Harrison, U.S. Patent (filed December 19, 1966) 3,551,943 (1971).
13. R. C. Kowalski, U.S. Patent (filed December 19, 1966; August 26, 1969) 3,608,001 (1971).
14. A. T. Watson, H. L. Wilder, K. W. Bartz and R. A. Steinkamp, U.S. Patent (filed November 21, 1973) 3,898,209 (1975).
15. C. Tzoganakis, J. Vlachopoulos and A. E. Hamielec, *Polym. Eng. Sci.*, **28**, 170 (1988).
16. D. Swanda, R. Len and S. T. Balke, *J. Appl. Polym. Sci.*, **35**, 1033 (1988).
17. A. Pabedinskas, W. R. Cluett and S. T. Balke, *Polym. Eng. Sci.*, **29**, 993 (1989).
18. S. H. Ryu, C. G. Gogos and M. Xanthos, *Adv. Polym. Technol.*, **11**, 121 (1992).
19. R. M. Nowak and G. D. Jones, U.S. Patent (filed October 10, 1960) 3,177,269 (1965).
20. G. D. Jones and R. M. Nowak, U.S. Patent (filed October 10, 1960) 3,177,270 (1965).
21. R. M. Nowak, U.S. Patent (filed April 22, 1963) 3,270,090 (1966).
22. Asahi Kasei, British Patent (filed January 8, 1968) 1,217,231 (1970).
23. R. L. McConnell, U.S. Patent (filed December 27, 1967 and November 23, 1970) 3,658,948 (1972).
24. R. A. Steinkamp and T. J. Grail, U.S. Patent (filed April 3, 1972) 3,862,265 (1975).
25. K. W. Bartz, J. J. Higgins, A. J. Berejka and A. J. DiCresce, U.S. Patent (filed April 3, 1972) 3,868,433 (1975).
26. R. A. Steinkamp and T. J. Grail, U.S. Patent (filed July 10, 1974) 3,953,655 (1976).
27. R. A. Steinkamp and T. J. Grail, U.S. Patent (filed July 10, 1974) 4,001,172 (1977).
28. K. W. Bartz, J. J. Higgins, A. J. Berejka and A. J. DiCresce, U.S. Patent (filed September 27, 1974) 3,987,122 (1976).
29. W. C. L. Wu, L. J. Krebaum and J. Machonis, U.S. Patent (filed December 18, 1972) 3,873,643 (1975).
30. S. W. Caywood, U.S. Patent (filed April 10, 1973) 3,884,882 (1975).
31. W. H. Staas, European Patent (filed January 21, 1981) 0 033 220 (1981).
32. R. Biwsack, P. Rempel, H. Korber and D. Neuray, U.S. Patent 4,260,690 (1981).
33. P. G. Andersen, U.S. Patent (filed November 12, 1982) 4,476,283 (1984).
34. R. C. Kowalski in "Reactive Extrusion: Principles and Practice", edited by M. Xantos, Hanser, Munich (1992).
35. K. J. Ganzueld and L. P. B. M. Janssen, *Polym. Eng. Sci.*, **32**, 467 (1992).
36. Y.-J. Sun, G. H. Hu and M. Lambla, *J. Appl. Polym. Sci.*, **57**, 1043 (1995).
37. G. Samay, T. Nagy and J. L. White, *J. Appl. Polym. Sci.*, **56**, 1423 (1995).
38. B. J. Kim and J. L. White, *Int. Polym. Process.*, **10**, 213 (1995).
39. B. J. Kim and J. L. White, *Polym. Eng. Sci.*, **37**, 576 (1997).
40. J. Cha and J. L. White, *Polym. Eng. Sci.*, **41**, 1227 (2001).
41. J. Cha and J. L. White, *Polym. Eng. Sci.*, **41**, 1448 (2001).
42. J. Cha and J. L. White, *Polym. Eng. Sci.*, **43**, 1830 (2003).
43. D. H. Chang and J. L. White, *J. Appl. Polym., Sci.*, **90**, 1755 (2003).
44. N. F. Newman and R. C. Kowalski, U.S. Patent (filed September 30, 1981) 4,384,072 (1983).

45. N. F. Newman and R. C. Kowalski, U.S. Patent (filed April 1, 1983) 4,486,575 (1984) and N. F. Newman and R. C. Kowalski, U.S. Patent (filed April 1, 1983) 4,501,859 (1985).
46. R. C. Kowalski and N. F. Newman, European Patent Application (filed April 6, 1983) 0,122,340 (1984).
47. R. C. Kowalski, W. M. Davis, N. F. Newman, Z. A. Foroulis and F. P. Baldwin, European Patent Application (filed April 3, 1984) 0,124,278 (1984).
48. R. C. Kowalski, *Chem. Eng. Prog.*, (May) p. 67 (1989).
49. A. Bouilloux, J. Druz and M. Lambla, *Polym. Proc. Eng.*, **4**, 235 (1986).
50. M. Lambla, J. Druz and A. Bouilloux, *Polym. Eng. Sci.*, **27**, 1221 (1987).
51. P. J. Kim and J. L. White, *J. Appl. Polym. Sci.*, **54**, 33 (1994).
52. S. Abdou-Sabet and K. Shen, U.S. Patent 4,594,390 (1985).
53. R. M. Joyce and D. M. Ritter, U.S. Patent (filed February 7, 1939) 2,251,519 (1941).
54. E. H. Mottus, R. M. Hedrick and J. M. Butler, U.S. Patent (filed December 13, 1956) 3,017,391 (1960).
55. J. Sebenda and J. Kralicek, *Coll. Czech. Chem. Comm.*, **23**, 766 (1958).
56. O. Wichterle, J. Sebenda and J. Kralicek, British Patent (filed December 4, 1959) 904, 229 (1962) and U.S. Patent (filed December 15, 1959) 3,200,095 (1965).
57. G. Illing and F. Zahradnik, U.S. Patent (filed July 21, 1964) 3,371,055 (1968).
58. G. Illing, *Kunststofftechnik*, **7**, 351 (1968).
59. G. Illing, *Mod. Plastics* (August) 70 (1969).
60. C. S. Tucker and R. J. Nichols, *Plastics Engineering* (May) p. 27 (1987).
61. G. Menges and T. Bartilla, *Polym. Eng. Sci.*, **27**, 1216 (1987).
62. G. Menges, U. Berghaus, M. Kalwa and G. Speuser, *Kunststoffe*, **79**, 1344 (1989).
63. U. Berghaus and W. Michaeli, *SPE ANTEC Tech. Papers*, **36**, 1929 (1990).
64. W. Michaeli and A. Grefenstein, *Polym. Eng. Sci.*, **35**, 1485 (1995).
65. H. Kye and J. L. White, *J. Appl. Polym. Sci.*, **52**, 1249 (1994).
66. H. Kye and J. L. White, *Int. Polym. Process.*, **11**, 310 (1996).
67. S. K. Ha and J. L. White, *Int. Polym. Process.*, **13**, 136 (1998).
68. I. Kim and J. L. White, *J. Appl. Polym. Sci.*, **97**, 105 (2005).
69. B. F. Frye, K. A. Pigott and J. H. Saunders, U.S. Patents 3,233,025 (1966).
70. K. W. Rausch and T. R. McClellan, U.S. Patent 3,642,964 (1966)
71. M. Ullrich, E. Meisert and A. Eitel, U.S. Patent (filed July 17, 1974) 3,963,679 (1976).
72. B. Quiring, G. Niederdellmann, W. Goyert and H. Wagner, U.S. Patent 4,245,081 (1981).
73. W. Goyert, E. Meisert, W. Grimm, A. Eitel, H. Wagner, G. Niederdellmann and B. Quiring, U.S. Patent (filed December 13, 1979) 4,261,946 (1981).
74. W. Goyert, A. Awater, W. Grimm, K. Ott, W. Oberkirch and H. Wagner, U.S. Patent 4,317,890 (1982).
75. W. Goyert, W. Grimm, A. Awater, H. Wagner and B. Krüger, U.S. Patent 4,500,671 (1985).
76. W. Goyert, J. Winkler, H. Perrey and H. Heidingsfeld, U.S. Patent 4,762,884 (1988).
77. R. M. Seddon, W. H. Russell and K. B. Rollins, U.S. Patent 3,253,818 (1966).
78. G. J. Fisher, F. Brown and W. E. Heinz, U.S. Patent 3,254,053 (1966).
79. R. M. Seddon and L. D. Scarbrough, U.S. Patent 3,442,866 (1969).

80. M. Semanchik and D. M. Braunstein, U.S. patent 4,105,637 (1978).
81. A. Sugio, T. Furusawa, K. Tanaka, T. Umemura and H. Urabe, U.S. Patent 4,115,369 (1978).
82. D. B. Todd, *Polym. Plast. Eng.*, **6**, 15 (1988).
83. J. Gimenez, M. Boudris, P. Cassagnau and A. Michel, *Int. Polym. Process.*, **15**, 20 (2000).
84. A. Poulesquen, B. Vergnes, P. Cassagnau and J. Gimenez, *Int. Polym. Process.*, **16**, 31 (2001).
85. B. J. Kim and J. L. White, *Int. Polym. Process.*, **17**, 33 (2002).
86. B. J. Kim and J. L. White, *J. Appl. Polym. Sci.*, **94**, 1007 (2004).
87. T. Takekoshi and J. E. Kochanowski, U.S. Patent 3,833,546 (1974).
88. T. Takekoshi and J. E. Kochanowski, U.S. Patent 4,011,198 (1977).
89. E. G. Banucci and G. A. Mellinger, U.S. Patent 4,073,773 (1978).
90. L. R. Schmidt and E. M. Lovgren, U.S. Patent 4,421, 907 (1983).
91. L. R. Schmidt and E. M. Lovgren, U.S. Patent 4,443,591 (1984).
92. H. Sutter and F. Haas, U.S. Patent 3,780,139 (1973).
93. W. Michaeli, H. Höcker, U. Berghaus and W. Frings, *J. Appl. Polym. Sci.*, **48**, 871 (1993).
94. W. Michaeli, A. Grefenstein and W. Frings, *Adv. Polym. Technol.*, **12**, 25 (1993).
95. W. Michaeli, W. Frings, H. Höcker and U. Berghaus, *Int. Polym. Process.*, **8**, 308 (1993).
96. W. Michaeli, W. Höcker, W. Frings and A. Oatman, *SPE ANTEC Tech. Papers*, **40**, 62 (1994).
97. B. J. Kim and J. L. White, *J. Appl. Polym. Sci.*, **88**, 1429 (2003).
98. I. Kim and J. L. White, *J. Appl. Polym. Sci.*, **96**, 1875 (2005).
99. I. Kim and J. L. White, *J. Appl. Polym. Sci.*, **90**, 3797 (2003).
100. J. A. Biesenberger in "Reactive Extrusion: Principles and Practice", edited by M. Xanthos, Hanser, Munich (1992).
101. D. B. Todd in "Reactive Extrusion: Principles and Practice", edited by M. Xanthos, Hanser, Munich (1992).

18 Devolatilization and Dewatering

18.1 Introduction

Considerations of devolatilization in extrusion operations dates back to the 19th century and it has been an important area of concern from the middle of the 20th century as the rise of the chemical industry has led to increased numbers of applications. Devolatilization is most effectively carried out using twin screw extrusion.

The beginnings of extrusion devolatilization involved single screw extrusion. By the middle of the 20th century, German and American engineers realized the advantages of twin screw extruders for devolatilization problems in the chemical industry. In Section 18.2, we will describe the earliest efforts of devolatilization with single screw extrusion. Section 18.3 considers the development of multi-screw devolatilization. The modeling of devolatilization is considered in Section 18.4 and experimental studies are summarized in Section 18.5.

18.2 Early Extrusion Devolatilization

The earliest commercial extrusion of filaments, profiles, and wire coating involved ram extrusion [1]. When single screw extruders were introduced to the thermoplastics (gutta percha) industry in the 1870s, it was considered that its major advantage was the elimination of air bubble defects in wire coating [2]. These were associated with air pockets in extrudates produced by the ram extruder.

As the years passed and the requirements for eliminating defects increased, new technologies needed to be introduced. These involved introducing barriers and channel depth increases into screw extruders to produce starved flow and then applying vacuums. This is seen in a 1919 patent of Price shown in Figure 18.1. The barriers and channel depth increase caused pressure drops and negative pressure gradients leading to the pressure falling to atmospheric

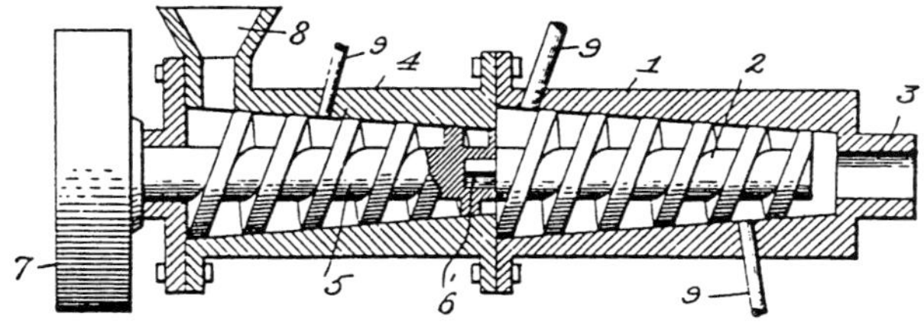

Figure 18.1: Devolatilization in a single screw extruder after Price [3]

and starvation developing. A vacuum could then be applied to the melt surface which would release dissolved gases and other volatile additives.

18.3 Beginnings of Twin Screw Extrusion Dewatering and Devolatilization

The pioneer of twin screw continuous dewatering and devolatilization research in the chemical industry was Walter Meskat of the I.G. Farbenindustrie, whose activities date to the mid 1930s [4]. In the early 1940s, I.G. Farbenindustrie engineers sought to carry out dewatering in multiple-screw extruders and found them to be more effective. Meskat [5] and Meskat and Erdmenger [6] described the use of two and three intermeshing co-rotating twin screw machines for dewatering in patent applications in October 1943 and July 1944 (see Figure 18.2). These patents were claimed by Farbenfabriken Bayer after the war.

There was also activity in the US during this period. This was most strikingly seen in a May 1945 patent application by L. B. Fuller [7] of Welding Engineers, which describes devolatilization in a modular tangential counter-rotating twin screw extruder. This design was shown in Figure 10.2.

Bayer post-war efforts regarding devolatilization in intermeshing co-rotating twin screw extruders continued the I.G. Farbenindustrie activities in the 1940s, as witnessed by a December 1951 German patent application of Winkelmuller et al. [8]. The design of the invention of this patent can be seen in Figure 18.3.

The modular self-wiping co-rotating twin screw extruder described in the December 1950 German patent application of Meskat and Pawlowski [9] (see Figure 4.10) is capable of devolatilization, although this is not explicitly discussed in the patent text. The same is true of the later modular machine of Erdmenger [10]. A 1962 paper by Erdmenger [11] reviews

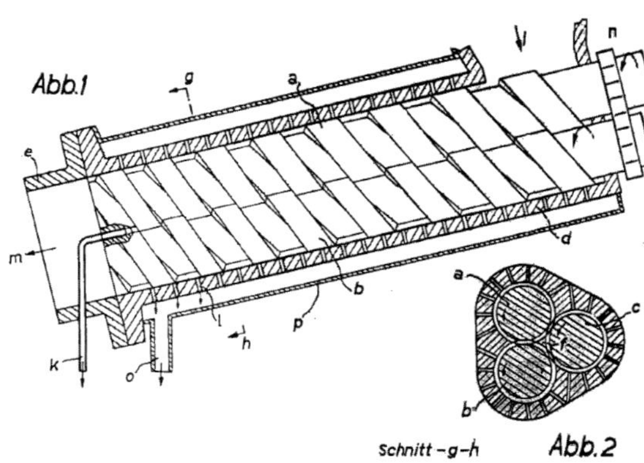

Figure 18.2: Meskat and Erdmenger [6] July 1944 patent application for a three intermeshing co-rotating screw dewatering extruder

various devolatilization technologies including twin screw machines, the Welding Engineers technology of Fuller [7], and later Bayer activities.

Bayer and Werner and Pfleiderer developed a four-screw machine (VDS-V: Vierwellige Dichtprofilschnecke zum Verdampfen) that is described in a patent of Erdmenger and Oetke [12] and papers by Erdmenger and Ullrich [11, 13, 14] (Figure 18.4). This machine was described again in a book by Herrmann [15] of Werner and Pfleiderer. The machine, however, never became commercially successful, presumably because of its cost (four screws) and not because of its performance, which was undoubtedly very good because of the four screws, as shown under a vacuum and the large interfacial area it exhibited.

Today the co-rotating twin screw machines are most widely used for this application.

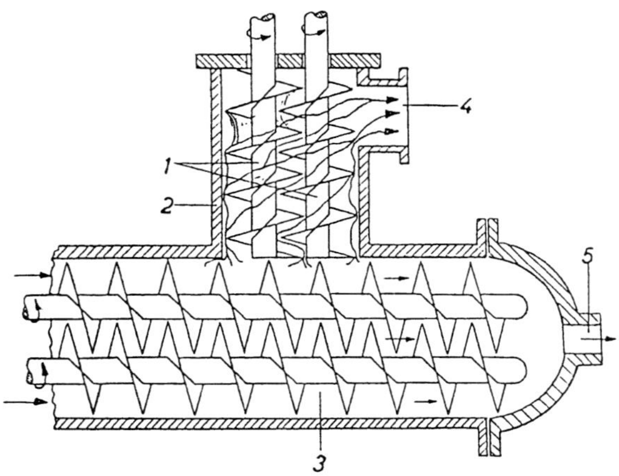

Figure 18.3: Winkelmuller et al. [8] devolatilization twin screw extruder

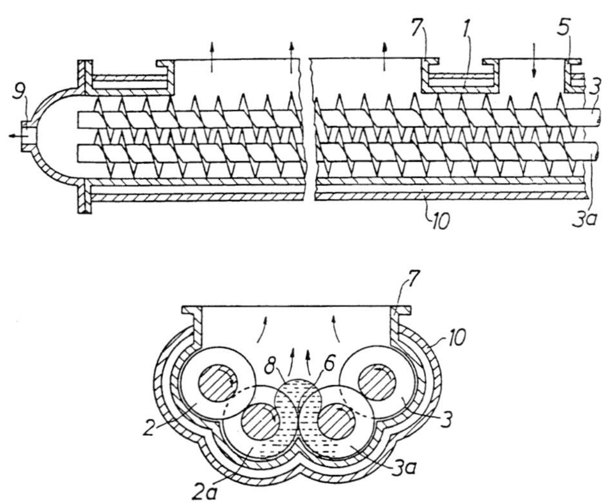

Figure 18.4: Erdmenger and Oetke [12] four-screw devolatilization extruder

18.4 Modeling of Devolatilization

Efforts at modeling devolatilization of polymer melts in screw extruders begin in the 1960s with papers by Latinen [16] of Monsanto and Coughlin and Canevari [17] of Esso Research and Engineering (Exxon). These models are based on describing the mass transfer mechanism. Latinen [16] presumes that diffusion resistances in the polymer solution side of the interface is dominant and the interfacial and gas resistances are negligible. Further studies of models for single screw extruders were given in succeeding years by Roberts [18] and Biesenberger and Kessidis [19, 20].

From the early 1980s, models specifically related to co-rotating twin screw extruders, were given by Werner [21], Collins et al. [22–24], Biesenberger and his coworkers [25, 26] and Sakai et al. [27, 28]. Foster and Lindt [29] described the modeling of devolatilization in a tangential counter-rotating twin screw extruder.

The basic analyses of all authors derive from the original paper of Latinen [16], though here we follow Keum and White [30]. Consider a coordinate z along the screw channel. The molar flux (per unit area) devolatilization in length dz is d$N_T(z)$, which may be expressed

$$\mathrm{d}N_\mathrm{T}(z) = k(z)\,\mathrm{d}a(z)\,[\bar{c}(z) - c_\mathrm{i}] \tag{18.1}$$

where d$a(z)$ is the differential interfacial area, $\bar{c}(z)$ is the molar concentration of volatiles in the polymer melt and c_i the concentration at the vapor-liquid interface. A mass balance on this component over the same length dz leads to

$$\mathrm{d}N_\mathrm{T}(z) = -\,Q\,\mathrm{d}\bar{c}(z) \tag{18.2}$$

Combining this with Eq. (18.1) we obtain

$$\frac{\mathrm{d}\bar{c}}{\mathrm{d}z} = -\frac{k}{Q}\frac{\mathrm{d}a}{\mathrm{d}z}[\bar{c}(z) - c_\mathrm{i}] \tag{18.3a}$$

$$= -\frac{k}{Q}\,\Pi(z)\,[\bar{c}(z) - c_\mathrm{i}] \tag{18.3b}$$

where Π is a perimeter of the polymer melt in the starved screw channels. Equation (18.3b) may be solved to give:

$$\bar{c}(z) - c_\mathrm{i} = [\bar{c}(0) - c_\mathrm{i}]\,e^{-\int_0^z \frac{k}{Q}\Pi(z)\,\mathrm{d}z} \tag{18.4}$$

The value of the integral varies, depending on the detailed design of the extruder. One of the major problems in applying Eq. (18.4) is to specify $\Pi(z)$.

A major issue that arose in the 1980s [24, 31–35] was the realization that foaming can occur during devolatilization. This greatly enhances the mass transfer rates through increasing the interfacial area available for devolatilization. This behavior can be handled with the above formulation if one suitably interprets the interfacial area "a" and its gradient "da/dz" as well as the mass transfer coefficient k.

The first effort at specifically modeling foaming in devolatilization was by Amon and Denson [32] in 1989. They describe a 'cell model' in which the foam is divided into spherical microscopic units of equal mass, which consists of a liquid shell and a concentric spherical bubble. Expansion of the bubbles occurs by diffusion of dissolved gas from the supersaturated shell of solution into the bubble. This theory and other theories of foaming in devolatilization are critically discussed by Lee and Biesenberger [36]. A second generation of cell models of foaming was given by Yang et al. [37] in 1996/1998.

18.5 Experimental Studies of Devolatilization

The first published experimental study of devolatilization by Latinen [16] in 1962 involved the removal of styrene monomer from polystyrene in a single screw extruder. The detailed design of the screw is not described. A subsequent paper by Coughlin and Canevari [17] in 1969 describes the removal of xylene and methanol from molten isotactic polypropylene in a single screw extruder. Again, the detailed design is not given but they presumably follow the description of Section 18.2.

The first experimental study of devolatilization in twin screw extruders was by Todd [38] of Baker Perkins in 1974, using a self-wiping co-rotating twin screw extruder. He studied thiophene-polystyrene and cyclohexane-polyethylene systems. Werner [21] of Werner and Pfleiderer described a study of removing ethylene from molten low density polyethylene in a co-rotating twin screw extruder. In 1985 Collins et al. [23] of the University of Delaware in the first academic study investigated the devolatilization of freon from a polybutene melt in a similar co-rotating twin screw extruder. Subsequently, Biesenberger and his coworkers [25, 26] at the Polymer Processing Institute at the Stevens Institute of Technology studied the devolatilization of poly(methyl methacrylate) in a co-rotating twin screw extruder and compared it to an investigation using a single screw extruder. These studies are summarized in Table 18.1.

In a 1989 paper, Foster and Lindt [34] described an investigation of devolatilization, emphasizing the occurrence and transition to foaming in a tangential counter-rotating twin screw extruder. They cited experiments with polystyrene/ethylbenzene and polymethyl methacrylate/methyl methacrylate, and described the bubble free and foaming regimes. They subsequently [29] described a study of determining mass transfer coefficients in the bubble free region of the polystyrene/ethylbenzene system. In a later paper [39], they gave a more detailed investigation of the transition region between the two regimes.

Basic studies of devolatilization and foaming phenomena in laboratory apparatus and a single screw extruder were made by Tadmor and his coworkers [35, 40]. They studied the devolatilization of polystyrene and polyethylene melts and emphasized the phenomenon of foaming.

Table 18.1: Experimental Studies of Devolatilization

System	Type of extruder	Mass transfer coefficient (mm/s)	Researchers
Polystyrene/styrene	Single screw		Latinen [16]
Polypropylene/xylene	Single screw	2.04	Coughlin and Canevari [17]
Polypropylene/methanol	Single screw	4.95	
Polystyerne/styrene	Co-rotating twin screw		Todd [38]
Polyethylene/cyclohexane			
Polystyerne/styrene	Co-rotating twin screw	0.8	Werner [21]
Polystyrene/styrene	Single screw	0.15–0.23	Biesenberger and Kessidis [20]
Polybutene/freon	Co-rotating twin screw		Collins et al.[23]
PMMA/MMA	Single screw Co-rotating twin screw		Notorgiacomo and Biesenberger [25]
Polystyrene/ethylbenzen PMMA/MMA	Tangential counter-rotating twin screw		Foster and Lindt [34]
Polystyrene/ethylbenzen	Tangential counter-rotating twin screw	0.38	Foster and Lindt [39]
PMMA/MMA	Single screw Co-rotating twin screw	0.122–0.391	Biesenberger et al.[26]
Polyethylene/nonane	Co-rotating twin screw	0.60	Keum and White [30]
Polyethylene/xylene		0.76	
Polyethylene/hexanol		1.52	

In a subsequent paper, Maffettone and Denson [41] investigated the phenomenon of gas entrainment in a co-rotating twin screw extruder using a low molecular weight polybutene. A more recent study using a modular co-rotating twin screw extruder by Keum and White [30] considered the devolatilization of molten polyethylene of nonane (C_9H_{20}), xylene, and hexanol.

There has been relatively little consideration in the literature regarding the interaction between the suspended or dissolved liquid and the polymer melt. Most attention has been given by Todd [38] and by Keum and White [30]. Todd cites the importance of the Flory-Huggins interaction parameter [42] in this regard and the latter authors choose to study three volatile liquids with varying compatibility with molten polyethylene (i.e., nonane, xylene, and hexanol) and discovered that it is much easier to remove the hexanol as compared to nonane, with the xylene being intermediate.

References

1. R. A. Brooman, English Patent, 10,582 (1845).
2. M. Gray, British Patent (filed December 10, 1879) 5,056 (1879).
3. R. B. Price, U.S. Patent 1,156,096 (1919).
4. K. Riess and W. Meskat, *Chem. Ing. Tech.*, **23**, 205 (1951).
5. W. Meskat, German Patent (filed October 17, 1943) 852,203 (1952).
6. W. Meskat and R. Erdmenger, German Patent (filed July 28, 1944) 872,732 (1953).
7. L. B. Fuller, U.S. Patent (filed May 15, 1945) 2,625,199 (1952).
8. W. Winkelmuller, R. Erdmenger, S. Neidhardt and B. Fortuna, German Patent 915,689 (1954).
9. W. Meskat and J. Pawlowski, German Patent (filed December 10, 1950) 949,162 (1956).
10. R. Erdmenger, U.S. Patent (filed August 17, 1959) 3,122,356 (1964).
11. R. Erdmenger, *Chem, Ing. Tech.*, **34**, 751 (1962).
12. R. Erdmenger and W. Oetke, German Auslegeschrift (filed March 16, 1960) 1,111,154 (1961).
13. R. Erdmenger, *Chem. Ing. Tech.*, **36**, 175 (1964).
14. R. Erdmenger and M. Ullrich, *Chem. Ing. Tech.*, **42**, 1 (1970).
15. H. Herrmann, "Schneckenmaschinen in der Verfahrenstechnik", Springer, Berlin (1972).
16. G. A. Latinen, *ACS Advances in Chemistry Series*, **34**, 235 (1962).
17. R. W. Coughlin and G. P. Canevari, *AIChE J.*, 15, 560 (1969).
18. G. W. Roberts, *AIChE J.*, **16**, 878 (1970).
19. J. A. Biesenberger, *Polym. Eng. Sci.*, **20**, 1015 (1980).
20. J. A. Biesenberger and G. Kessidis, *Polym. Eng. Sci.*, **22**, 832 (1982).
21. H. Werner, "Devolatilization of Plastics", VDI Verlag, Berlin (1980).
22. G. P. Collins, C. D. Denson and G. Astarita, *Polym. Eng. Sci.*, **23**, 323 (1983).
23. G. P. Collins, C. D. Denson and G. Astarita, *AIChE J.*, **31**, 1288 (1985).
24. C. D. Denson in "Advances in Chemical Engineering", Academic Press, N.Y. (1983).
25. V. Notorgiacomo and J. A. Biesenberger, *SPE ANTEC Tech papers*, **34**, 71 (1988).
26. J. A. Biesenberger, S. K. Dey and J. Brizzolara, *Polym. Eng. Sci.*, **30**, 1493 (1990).
27. T. Sakai, N. Hashimoto and K. Kataoka, *Int. Polym. Process.*, **8**, 218 (1993).
28. N. H. Wang, T. Sakai and N. Hashimoto, *Int. Polym. Process.*, **10**, 296 (1995).
29. R. W. Foster and J. T. Lindt, *Polym. Eng. Sci.*, **30**, 424 (1990).

30. J. M. Keum and J. L. White, *Int. Polym. Process.*, **19**, 101 (2004).
31. J. A. Biesenberger and D. H. Sebastian, "Principles of Polymerization Engineering", Wiley, N.Y. (1983).
32. M. Anon and C. D. Denson, *Polym. Eng. Sci.*, **24**, 1026 (1984).
33. J. A. Biesenberger and S. T. Lee, *Polym. Eng. Sci.*, **27**, 510 (1987).
34. R. W. Foster and J. T. Lindt, *Polym. Eng. Sci.*, **29**, 178 (1989).
35. R. J. Albalak, Z. Tadmor and Y. Talmon, *AIChe J.*, **36**, 1313 (1990).
36. S. T. Lee and J. A. Biesenberger, *Polym. Eng. Sci.*, **29**, 782 (1989).
37. C. T. Yang, T. G. Smith, D. I. Bigio and C. Anolick, *SPE ANTEC Tech. Papers*, **42**, 350 (1996); *AIChE J.*, 43, 1861 (1997); *IEC Res.*, **37**, 1964 (1998).
38. D. B. Todd, *SPE ANTEC Tech. Papers*, **20**, 472 (1974).
39. R. W. Foster and J. T. Lindt, *Polym. Eng. Sci.*, **30**, 621 (1990).
40. A. Tukachinsky, Y. Talmon and Z. Tadmor, *AIChE J.*, **40**, 670 (1994).
41. P. L. Maffettone and C. D. Denson, *J. Polym. Eng.*, **13**, 175 (1994).
42. P. J. Flory, *"Principles of Polymer Chemistry"*, Cornell, Ithaca (1953).

19 Comparisons and Concluding Remarks

19.1 Introduction

In this book we have presented the technology, flow mechanisms, and modeling of each of the individual classes of machines as well as the Buss Kokneter. It seems important that we discuss comparisons of these machines in terms of their flow mechanisms, residence time distributions, machine characteristics, applications, and business success.

In Section 19.2, we will consider the pumping and mixing mechanisms of the various types of twin screw machines. In Section 19.3 we will compare the relative residence time distributions. Specific machine characteristics are discussed in Section 19.4. Applications of the various machines are described in Section 19.5. We discuss the commercial success of the various machine designs in Section 19.6.

19.2 Flow Mechanisms

When we describe flow mechanisms, we must be careful to clearly distinguish between the mechanism that pumps the process fluid through the machine, the mechanism by which distributive mixing occurs, and the mechanism by which dispersive mixing occurs.

We first consider pumping. There are two major mechanisms, by which pumping of viscous liquids occurs: drag flow and positive displacement. The drag flow mechanism is due to the relative motion of a moving surface past a stationary surface. This is the mechanism by which the single screw extruder, Buss Kokneter, tangential counter-rotating twin screw extruder, self-wiping co-rotating twin screw extruder, and continuous mixers operate. Positive displacement involves placing the liquid in one or more chambers which move along the machine and in time push the material out of the pump. This is the mechanism of gear pumps, lobe pumps, and intermeshing counter-rotating twin and multiple screw pumps.

The mechanisms of distributive mixing vary significantly among the various twin screw extruders and continuous mixers. However, it generally involves the non-screw elements. An exception to this is the tangential counter-rotating twin screw extruder where the leakage flows between the screws are the primary mixing mechanism. In the modular co-rotating twin screw extruder there are various poorly or non-pumping elements, such as kneading disc blocks, gear-shaped mixing elements, TME, and ZME elements. The Leistritz AG intermeshing counter-rotating twin screw extruders have a similar range of mixing elements. In continuous mixers, one has rotor sections that provide distributive mixing. The Buss Kokneter with pins, screw slices, and an oscillating motion, all of which provide distributive mixing.

It is important to realize that little distributive mixing occurs in starved regions. The various poorly pumping elements described above and barrier rings in the barrel also lead to fully filled regions of the screw, not only in these elements but in front of them (hopper direction). Here, the drag flow is supplemented by a pressure flow and complex velocity/stress fields, which aid in the distributive mixing process.

Dispersive mixing involves the breaking up of particle agglomerates and the breakup of droplets of dispersed melt phases. This is generally associated with high levels of shear and elongational stresses during the mixing process, which occur in kneading disc blocks in modular co-rotating twin screw extruders and between the screws in intermeshing counter-rotating twin screw extruders. Dispersive mixing seems to be accomplished best in intermeshing counter-rotating twin screw machines, where the material experiences high stresses between screws as well as in mixing elements. The tangential counter-rotating twin screw, while showing excellent distributive mixing, performs poorer than the modular intermeshing counter-rotating and co-rotating twin screw extruders in dispersive mixing.

Some words should be said about how well the different machines perform these mixing operations. All of the modular twin screw extruders and modular Buss Kokneter have advocates among their manufacturers and the engineers and chemists in firms that use such machinery. All agree that these continuous mixers are much better distributive mixers than batch internal mixers. There are supporters of superiority of modular self-wiping co-rotating twin screw extruders, while others believe the Buss Kokneter is superior and still others prefer tangential counter-rotating twin screw machines. There seem to be no quantitative comparative studies. Further modular design of the screws is very important and can affect the results.

The situation is clearer with regard to dispersive mixing. J. L. White performed studies with different machines for some years. First, he have found that at specified residence times continuous mixers mix better than batch internal mixers at the same mixing time [1]. Second, while modular construction is very important, intermeshing counter-rotating machines of Leistritz design seemed to outperform other continuous machines as a dispersive mixer [2–4].

19.3 Residence Time Distributions

It is inherent in the machine design that positive displacement pumps, such as gear pumps, lobe pumps and intermeshing counter-rotating twin screw pumps/extruders, should have narrower residence time distributions than drag flow pumps. Indeed, gear pumps are widely used to feed special additives and steady the output of single screw and co-rotating twin screw machines. It should not be surprising that almost all investigators have found intermeshing counter-rotating twin screw extruders exhibiting narrower residence time distributions than single screw or co-rotating twin screw extruders [5–8]. The Buss Kokneter seems to have the broadest residence time distribution of the machines studied by Shon et al. [8].

Residence time distributions can vary with modular screw design. As mixing elements, such as kneading disc blocks, replace screw pumping elements, residence time distributions broaden [9].

Tangential counter-rotating twin screw extruders have the ability to exhibit very broad residence time distributions. This can be achieved by making one of the screws pump backwards towards the hopper, as described by Biesenberger and Todd [10].

19.4 Machine Characteristics

The various machines discussed in this book have some very different characteristics that should be emphasized. One example of this is the Buss Kokneter that inherently produces only oscillatory flows.

Other unique characteristics exist. In intermeshing counter-rotating twin screw machines, the polymer melt being processed passes between the screws. By Reynolds hydrodynamic lubrication theory, high pressures are developed in the small clearances between the rotating screws. The pressure increases with screw speed. This results in the screws bending outward and contacting the barrel [11]. For this reason, there are severe limitations on screw speeds for these machines. This can be mitigated by increasing the clearance between the screws.

In intermeshing co-rotating machines, the fluid mechanics lead to the effect of the melt passing around the circumference of the two screws and no lubrication pressures develop between the screws. Modern self-wiping co-rotating twin screw extruders can operate at thirty times the screw speeds (RPM) compared to intermeshing counter-rotating machines. However, there are still limitations on the length of two self-wiping rapidly rotating steel screws.

Tangential counter-rotating screws have no hydrodynamic lubrication pressures between their screws and no limitations in screw length. These are clear advantages for their application. Another advantage of tangential counter-rotating twin screw machines is they larger contact area for devolatilization than other twin screw machine designs.

19.5 Applications

The intermeshing counter-rotating twin screw extruders are unique in their primary application for extrusion of pipes and profiles, especially from polyvinyl chloride. They are extremely well suited for this application as positive displacement pumps with narrow residence time distributions and high productivities.

The modular self-wiping co-rotating twin screw extruder is the most widely used twin screw extruder for compounding and reactive extrusion. This is in part obvious from the large number of apparently successful manufacturers around the world. It should of course be realized that in many of these applications other types of twin screw extruders would perform as well. The Buss Kokneter and tangential counter-rotating twin screw extruder have also been widely used for reactive extrusion by various industrial firms. The Farrel Continuous Mixer® and Buss Kokneter have been widely used for compounding.

19.6 Commercial Success of Various Machines

It is very clear from estimating the machine sales from the various continuous mixing machinery manufacturers, or more simply by noting the number of machinery manufacturers of each general type of machine, that the dominant compounding-devolatilization-reactive extrusion machine is the modular self-wiping co-rotating twin screw extruder.

The question arises as to why counter-rotating machines overwhelmingly dominate batch internal mixer technology in the world, while self-wiping co-rotating machines dominate continuous mixers. The batch mixer question is seemingly easier to address. Batch mixing machines were always counter-rotating. In the rubber industry, the first industrial mixer was the counter-rotating two-roll mill. The first commercial batch internal mixer by Pfleiderer [12, 13] of Werner and Pfleiderer was counter-rotating as was Kempter's mixer [14] for rubber. The Banbury Mixer was developed by a Werner and Pfleiderer engineer to solve operational problems by introducing a ram and redesigning the rotors [15), the machine being ultimately manufactured by the Birmingham Iron Foundry (now Farrel Inc.). F. A. Banbury subsequently designed a much superior machine to anything that existed previously [16, 17]. The subsequent intermeshing counter-rotating mixer, the Shaw intermix, was also counter-rotating [18].

This brings us to the continuous co-rotating machines. These machines were commercially developed at the end of the 1930s by Roberto Colombo [19] in Turin, Italy and were in succeeding years used to extrude thermosets and polyvinyl chloride. Some machines were sold to the I.G. Farbenindustrie which later (1943) filed a patent for their application to reactive extrusion [20]. Erdmenger [21], in his memoir, seems to mention these activities in Leverkusen in 1943 in connection with K. Sigwart and polyurethane. It was in this period that Meskat's group, which included Erdmenger and A. Geberg [21, 22], began their activities with co-rotating twin screw extruders in the Wolfen Works for applications to devolatilization [23, 24]. The machines used could have been of LMP manufacture. As the years passed Erdmenger, Geberg, and Meskat increasingly became convinced of the superiority of properly designed self-wiping intermeshing machines for many applications. The LMP machine was not self-wiping. Geberg worked out the geometry to make the machine self-wiping in 1944 [22]. They decided they would have to have these specially manufactured for them.

After the war, Erdmenger and Meskat obtained positions with the revived Bayer organization in Leverkusen and Dormagen respectively, where they both gave considerable attention to the improvements of the co-rotating machine. Erdmenger now devised kneading disc block machines [25] for mastication and mixing and Meskat and Pawlowski [26] devised modular screw machines. Erdmenger [27] then developed modular machines with screw and kneading disc block elements.

It must be emphasized that Erdmenger and Meskat were not alone in these activities. They had solid support from first the I.G. Farbenindustrie and later Bayer AG, especially their engineering departments. Kurt Riess who was their leader both in the Wolfen Works and in Leverkusen was highly supportive [28–30]. There were many other talented engineers who lent their support to Erdmenger and Meskat in the 1940s and 1950s, including A. Geberg and J. Pawlowski mentioned above. The Bayer organization has always been proud of their role in developing this machine with good reason and this comes through in the 1972 book of Herrmann [31] and especially the 2007 volume edited by Kohlgrüber [32].

It was only in the mid 1950s that the modular co-rotating twin screw extruder was licensed to Werner and Pfleiderer GmbH, who now found themselves manufacturing both counter-rotating batch internal mixers and continuous co-rotating twin screw mixers. The latter machine rapidly became a much larger sales product. The license was exclusive and Werner and Pfleiderer could count on the large Bayer legal staff for protection from potential competitors.

Machinery manufacturers are inherently much smaller than chemical companies. They have much smaller financial resources and fewer engineers. In difficult economic times, they suffer much more. Their customers can usually delay purchases of new machinery. The European machinery companies we mentioned in this volume, such as Berstorff, Buss AG and Leistritz AG, despite the merits of their inventions, could not compete against the Bayer/Werner and Pfleiderer combine. Leistritz developed an intermeshing counter-rotating twin screw machine as a competitor for the continuous market. Both Berstorff and Leistritz began to manufacture and market similar self-wiping co-rotating twin screw extruders as soon as the Bayer patents expired.

Werner and Pfleiderer came to the United States with their own and the Bayer developed technology in the 1970s. The American manufacturers of (counter-rotating) continuous mixers, Farrel Inc. and Welding Engineers, found it very difficult to compete. Berstorff and Leistritz now also began marketing co-rotating twin screw extruders, both in Europe and the US. By the 1990s, both Farrel and Welding Engineers had begun to make co-rotating twin screw extruders. Farrel was now also making a co-rotating twin screw extruder and a counter-rotating continuous mixing machine as well as a counter-rotating batch mixers.

In Asia, Werner and Pfleiderer was less successful. In Japan, widely diversified reasonably large machinery companies such as Japan Steel Works, Kobe Steel and Toshiba Machine, entered the market in the 1970s and 1980s producing a range of different continuous mixing machines. Kobe Steel manufactures the same distribution of machinery products as Farrel Inc.

References

1. J. L. White, D. Liu and S. H. Bumm, *J. Appl. Polym. Sci.*, **102**, 3940 (2006).
2. S. Lim and J. L. White, *Int. Polym. Process.*, **9**, 33 (1994).
3. K. Shon, D. Liu and J. L. White, *Int. Polym. Process.*, **20**, 322 (2005).
4. K. Shon, S. H. Bumm and J. L. White, *Polym. Eng. Sci.*, **48**, 757 (2008).
5. T. Sakai, *Gose Jushi*, **29**, 7 (1978).
6. C. Rauwendaal, *SPE ANTEC Tech. Papers*, **27**, 618 (1981).
7. T. Sakai, N. Hashimoto and N. Kobayoshi, *SPE ANTEC Tech. Papers*, **33**, 146 (1987).
8. K. Shon, D. H. Chang and J. L. White, *Int. Polym. Process.*, **14**, 45 (1999).
9. P. J. Kim and J. L. White, *Int. Polym. Process.*, **9**, 108 (1994).
10. J. A. Biesenberger and D. B. Todd, U.S. Patent (filed November 19, 1992) 5,372,418 (1994).
11. J. L. White and A. Adewale, *Int. Polym. Process.*, **10**, 15 (1995).
12. P. Pfleiderer, German Patent (filed April 4, 1879) 10,164 (1880).
13. P. Pfleiderer, German Patent (filed June 10, 1881) 18,797 (1882).
14. F. Kempter, German Patent (filed November 16, 1913) 279,649 (1914).
15. F. H. Banbury, U.S. Patent (filed January 15, 1916) 1,200,070 (1916).
16. F. H. Banbury, U.S. Patent (filed November 18, 1916) 1,227,522 (1917).
17. F. H. Banbury, U.S. Patent (filed April 20, 1917) 1,279,220 (1918).
18. R. T. Cooke, British Patent (filed June 14, 1934) 641,685 (1937).
19. R. Colombo, Italian Patent (filed February 6, 1939) 370,578 (1939).
20. Anonymous, German Patent (filed July 24, 1943) 895,058 (1953).
21. R. Erdmenger, "Schneckenmaschinen für die Hochviskos-Verfahrenstechnik", Bayer AG, Leverkusen (1978).
22. M. Ullrich in "Co-Rotating Twin-Screw Extruders: Fundamentals, Technology and Applications", edited by K. Kohlgrüber, Hanser, Munich (2007).
23. W. Meskat, German Patent (filed October 17, 1943) 852,203 (1952).
24. W. Meskat and R. Erdmenger, German Patent (filed July 28, 1944) 872,732 (1953).
25. R. Erdmenger, German Patent (filed September 29, 1949) 813,154 (1951); German Patent (filed July 28, 1953) 940,109 (1956).
26. W. Meskat and J. Pawlowski, German Patent (filed December 10, 1950) 949,162 (1956).
27. R. Erdmenger, U.S. Patent (filed August 17, 1959) 3,122,356 (1964); *Chem. Ing. Tech.* **34**, 751 (1962); *ibid*, **36**, 175 (1964).
28. K. Riess and W. Meskat, *Chem. Ing. Tech.*, **23**, 205 (1951).
29. K. Riess and R. Erdmenger, *VDI Zeitschr.*, **93**, 633 (1951).
30. K. Riess, *Chem. Ing. Tech.*, **27**, 457 (1955).
31. H. Herrmann, "Schneckenmaschinen in der Verfahrenstechnik", Springer, Berlin (1972).
32. K. Kohlgrüber, Editor, "Co-Rotating Twin Screw Extruders: Fundamentals, Technology and Applications", Hanser, Munich (2007).

Author Index

Subject Index

HANSER

Co-Rotating Twin-Screw Extruder

Klemens Kohlgrüber
Co-Rotating Twin-Screw Extruder
369 pages. 390 fig.
ISBN 978-3-446-41372-6

Co-rotating screws are used in many branches of industry for producing, preparing and processing highly viscous materials. These machines usually have modular configurations and are thus quite flexible for adapting to changing tasks and material properties. Well-founded knowledge of machines, processes and material behavior are required in order to design twin-screw extruder for economically successful operations.

This book provides basic engineering knowledge regarding twin-screw machines; it lists the most important machine-technical requirements and provides examples based on actual practice. Better understanding of the processes is emphasized as this is a prerequisite for optimizing twin-screw designs and operating them efficiently.

More Information on Plastics Books and Magazines:
www.kunststoffe.de or **www.hansergardner.com**

HANSER

Solving Extrusion Problems.

Noriega/Rauwendaal
Troubleshooting the Extrusion Process
2nd edition
208 pages
ISBN 978-3-446-42244-5

Troubleshooting extrusion problems is one of the most challenging tasks for everyone involved in extrusion operations. It requires a good understanding of the extrusion process, knowledge of the material properties, good instrumentation, good analysis tools, and a systematic and logical approach.

This book addresses all issues crucial in extrusion troubleshooting. In addition, industrial case studies, richly illustrated with photographs and photomicrographs, are used to provide exemplary approaches to efficient problem analysis and problem solving.

More Information on Plastics Books and Magazines:
www.kunststoffe.de or **www.hansergardner.com**

HANSER

The Resource for the Plastics Engineer's Success!

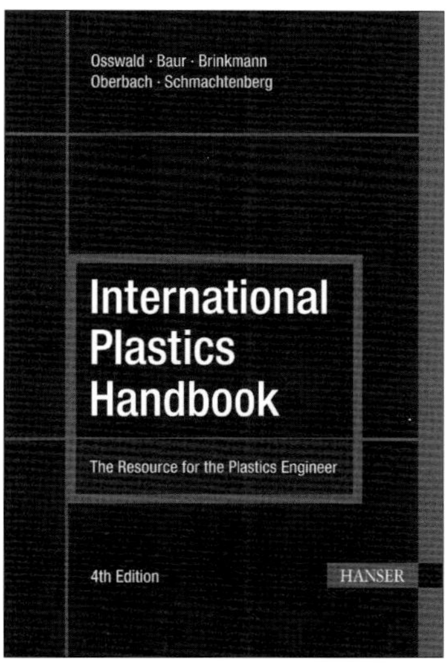

Osswald/Baur/Brinkmann/
Oberbach/Schmachtenberg
International Plastics Handbook
approx. 850 pages
ISBN 3-446-22905-1

From basic materials and theoretical concepts to synthesis, compounding, processing and manufacturing, with detailed descriptions of individual plastics and boundary areas, this handbook contains more than 400 tables of plastics data with ASTM, ISO, and DIN standards, and an International Trade Name Register.

Contents: Material Science of Polymers · Properties and Testing · Plastics Processes · Engineering Design · Materials · Polymer Additives · Material Property Tables · Literature · Polymer Research Institute · Trade Names

More Information on Plastics Books and Magazines:
www.kunststoffe.de or **www.hansergardner.com**